Essentials of Oceanography

TOM GARRISON

ORANGE COAST COLLEGE
UNIVERSITY OF SOUTHERN CALIFORNIA

THOMSON

BROOKS/COLE

Australia • Canada • Mexico • Singapore • Spain • United Kingdom • United States

THOMSON

BROOKS/COLE

Publisher: David Harris
Editor: Keith Dodson
Development Editor: Mary Arbogast
Assistant Editor: Carol Benedict
Editorial Assistant: Melissa Newt
Technology Project Manager: Sam Subity
Marketing Manager: Kelley McAllister
Marketing Assistant: Sandra Perin
Advertising Project Manager: Nathaniel Bergson-Michelson
Project Manager, Editorial Production: Teri Hyde
Print/Media Buyer: Karen Hunt

Permissions Editor: Joohee Lee
Production Service: Joan Keyes, Dovetail Publishing Services
Text Designer: Adriane Bosworth
Photo Researcher: Myrna Engler
Copy Editor: Carol Reitz
Illustrator: Precision Graphics
Cover Designer: Irene Morris
Cover Image: Overview of Crashing Wave, © Guy Motil/CORBIS
Cover Printer: The LeHigh Press
Compositor: G&S Typesetters, Inc.
Printer: Quebecor World

> For more information about our products, contact us at:
> **Thomson Learning Academic Resource Center**
> **1-800-423-0563**
> For permission to use material from this text, contact us by:
> **Phone:** 1-800-730-2214
> **Fax:** 1-800-730-2215
> **Web:** http://www.thomsonrights.com

Library of Congress Control Number: 2003101436

Student Edition: ISBN 0-534-39259-8

Instructor's Edition: ISBN 0-534-49114-6

Brooks/Cole—Thomson Learning
511 Forest Lodge Road
Pacific Grove, CA 93950
USA

Asia
Thomson Learning
5 Shenton Way #01-01
UIC Building
Singapore 068808

Australia/New Zealand
Thomson Learning
102 Dodds Street
Southbank, Victoria 3006
Australia

Canada
Nelson
1120 Birchmount Road
Toronto, Ontario M1K 5G4
Canada

Europe/Middle East/Africa
Thomson Learning
High Holborn House
50/51 Bedford Row
London WC1R 4LR
United Kingdom

Latin America
Thomson Learning
Seneca, 53
Colonia Polanco
11560 Mexico D.F.
Mexico

Spain/Portugal
Paraninfo
Calle/Magallanes, 25
28015 Madrid, Spain

To my family and my students:
My hope for the future.

About the Author

Tom Garrison (Ph.D., University of Southern California) is a professor in the Marine Science Department at Orange Coast College in Costa Mesa, California, one of the largest undergraduate marine science departments in the United States. Dr. Garrison also holds an adjunct professorship at the University of Southern California. He has been named Outstanding Marine Educator by the National Marine Technology Society and was a winner of the prestigious Salgo-Noren Foundation Award for Excellence in College Teaching. Dr. Garrison was also an Emmy Award team participant as writer and science advisor for the PBS syndicated *Oceanus* television series. His widely used textbooks in oceanography and marine science are the college market's best sellers.

Dr. Garrison's interest in the ocean dates from his earliest memories. As he grew up with an admiral as a dad, the subject was hard to avoid! He had the good fortune to meet great teachers who supported and encouraged this interest. Years as a midshipman and commissioned naval officer continued his marine emphasis. Graduate school and 30+ years of teaching have allowed him to pass his oceanic enthusiasm on to more than 65,000 students.

Dr. Garrison is married, has two children (a daughter who teaches 4th grade and a son who is an executive in international trade), a son-in-law, a new granddaughter, and a very patient cadre of teaching assistants. He and his family reside in Newport Beach, California.

John Wilson Cramer IV

Brief Contents

Contents

Preface for Students and Instructors

This book was written to provide an *interesting*, clear, and relatively brief overview of the marine sciences. It is designed for college and university students who are curious about Earth's largest feature, but who may have little or no formal background in science. Indeed, oceanography—the study of the ocean—is an ideal course for students wishing to fulfill general education requirements. Oceanography is broadly interdisciplinary; students are invited to see the connections between astronomy, economics, physics, chemistry, history, meteorology, geology, and ecology—areas of study they once considered separate. It's no surprise that oceanography courses have become increasingly popular in the last decade.

Students bring a natural enthusiasm to their study of this field. Even the most indifferent reader will perk up when presented with stories of encounters with huge waves, tales of exploration under the best and worst of circumstances, evidence that vast chunks of Earth's surface slowly move, micrographs of glistening diatoms, and data showing the economic importance of seafood and marine materials. If pure spectacle is required to generate an interest in the study of science, oceanography wins hands down!

In the end, however, it is subtlety that triumphs. Studying the ocean reminds us of the wonder we felt as children when we first encountered the natural world. The story of the ocean is a story of change and chance; its history is written in the rocks, the water, and the genes of the millions of organisms that have evolved here. My goal is to help the students who use this book to gain an oceanic perspective. *Perspective* means being able to view things in terms of their relative importance or relationship to one another. An oceanic perspective enables you to see this misnamed planet in a new light and helps you plan for its future. You will see that water, continents, seafloors, sunlight, storms, seaweeds, and society are connected in subtle and beautiful ways.

The Plan

The plan of the book is straightforward: Because all matter on Earth—except hydrogen and some helium—was generated in stars, our story of the ocean starts with stars. Have oceans evolved elsewhere? We continue with a brief look at the history of marine science (with some additional historical information sprinkled throughout later chapters). The theories of

Earth structure and plate tectonics are presented next as a base on which to build the explanation of bottom features that follows. A survey of ocean physics and chemistry prepares us for discussions of atmospheric circulation, classical physical oceanography, and coastal processes. Our look at marine biology begins with an overview of the problems and benefits of living in seawater, continues with a discussion of the production and consumption of food, and ends with an ecological survey of marine organisms. The last chapter treats marine resources and environmental concerns.

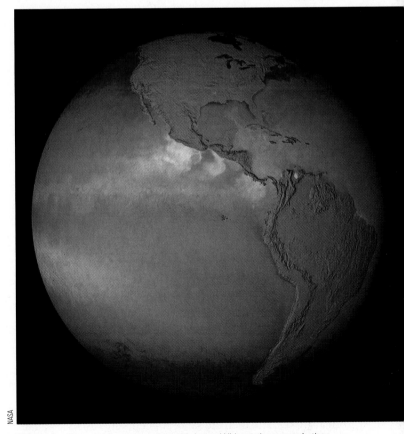

NASA

Figure 1 The world ocean swirls with heat. Without the ocean's thermostatic effects, and its ability to move heat from near the equator to the polar regions, Earth would be a dramatically different (and uninhabitable) place.

NASA/ORBIMAGE

Figure 2 The Gulf Stream, the world's fastest ocean current, jets northeastward into the open Atlantic at Cape Hatteras. Water evaporating from the warm sea surface has condensed into clouds. Warmth transported by the Gulf Stream greatly influences Europe's weather and climate.

Connections between disciplines are emphasized throughout. Marine science draws on several fields of study, integrating the work of specialists into a unified whole. For example, a geologist studying the composition of marine sediments on the deep seabed must be aware of the biology and life histories of the organisms in the water above, the chemistry that affects the shells and skeletons of the creatures as they fall to the ocean floor, the physics of particle settling and water density and ocean currents, and the age and underlying geology of the study area. This book is written to make those connections from the first.

How This Book Is Organized

A broad view of oceanography is presented in 15 chapters, each freestanding (or nearly so) to allow an instructor to assign chapters in any order he or she finds appropriate. Each chapter begins with an attractive **vignette**—a short observation, eyewitness account, or description of marine scientists at work. The chapters are written in an **engaging style** at a level appropriate to students not majoring in science. The number of technical terms is kept to a minimum. Some of the more complex ideas are initially outlined in broad brushstrokes, and then the same concepts are discussed again after you have a clear view of the overall situation. **Measurements** are given in both metric and English systems. At the request of a great many students, the units are written

out (that is, we write "kilometer" rather than "km") to avoid ambiguity and for ease of reading.

The **illustration program** is extensive; the photos, charts, graphs, and paintings have been chosen for their utility, clarity, and beauty. Chapters begin with an **outline** and end with **questions** asked by students (and their **answers**), a **summary,** and a list of important **terms and concepts** that are defined in an extensive **glossary** at the back of the book. As you read, you'll notice **key concepts** in distinctive boxes that look like this:

Key concepts are found in boxes like these.

Study questions are included at the end of each chapter. A brief **annotated bibliography** is available on the book's web site for anyone who wishes to know more about a particular topic.

Appendixes will help you master measurements and conversions, geological time, latitude and longitude, and chart projections. In case you'd like to join us in our life's work, the last appendix discusses jobs in marine science.

The book has been **thoroughly student-tested.** You need not feel intimidated by the concepts presented or the words used to describe them. Students just like you have mastered this material. Read slowly, and go step by step through any parts that give you trouble. Your predecessors have found the ideas presented here to be understandable, useful, inspiring, and applicable to their lives. Best of all, they have found the subject *interesting!*

The Web Site

This book is fully Internet integrated. You'll find its dedicated Web site at:

www.info.brookscole.com/garrison

Internet icons at each head and subhead indicate text-specific information available on the site. They look like this:

Questions asked by students, chapter outlines, sample quizzes, flashcards, links, exercises, animations, and tutorial assistance are also accessible at the site. We will periodically update the Web site to add features and announce important advances in marine science. *The site is open to anyone without cost or subscription.*

Suggestions for Using This Book

1. *Begin with a preview.* Scout the territory ahead; flip through the assigned pages reading only the headings

and subheadings. Look at the figures and read captions that catch your attention. See where we're going.

2. *Keep a pen and paper handy.* Jot down a few questions—any questions—that this quick glance stimulates. *Why* is the deep ocean cold if the inside of Earth is so hot? *What* makes storm conditions during El Niño years? *Where* did sea salt come from? *Will* global warming actually be a problem? *Does* anybody still hunt whales? Writing questions will help you focus when you start studying.

3. *Now read in small but concentrated doses.* Each chapter is written in a sequence and tells a story. The logical progression of ideas is going somewhere. Find and follow the organization of the chapter. Stop occasionally to review what you've learned. Flip back and forth to review and preview.

4. *Think about the key concepts—the ones in the boxes—when you encounter them.* They're pegs to hang your knowledge on.

5. *Strive to be actively engaged!* Write notes in the margins, underline the key ideas (not everything you read!), write more questions, draw on the diagrams, check off subjects as you master them, and make flashcards while you read (if you find them helpful). *Use the book!*

6. *Monitor your understanding.* If you start at the beginning of the chapter, you will have little trouble understanding the concepts as they unfold. But if you find yourself at the bottom of the page having only scanned (rather than understood) the material, stop there and start that part again. Look ahead once again to see where we're going. Remember that other students have been here before, and I have listened to their comments to make the material as clear as I can. *This book was written for you.*

7. *Use the Internet sites and on-line study resources included in the book's website.* We have provided addresses for more than 2,000 Internet sites. If you have access to a computer, fire it up and scan the designated site as you read the associated passage of the book. Take advantage of the practice quizzes to assess your grasp of the chapter's contents and prepare for tests.

8. *Don't miss class.* Listening to lectures and participating in discussions is crucial to effective learning.

9. *Budget your time.* Spending an hour or two each day on this most interesting of subjects is much more effective (and far less agonizing) than frenzied studying the night before an exam.

10. *Enjoy the journey.* Your instructor will be glad to share his or her understanding and appreciation of marine science with you—you have only to ask. Students, instructors, and authors all work together toward a common goal: an appreciation of the beauty and interrelationships a growing understanding of the ocean can provide.

Acknowledgments

Jack Carey at Thomson Learning, the grand master of college textbook publishing, willed the first edition of this book into being. His suggestions have been combined with those of more than 100 peer reviewers and 450 undergraduate students, all of whom gave of their time and knowledge to help me assemble and organize the material. In particular for this edition I'd like to acknowledge the expert advice of Jeffrey Chanton, Florida State University; Brent K. Dugolinsky, State University of New York, Oneonta; Karen Grove, San Francisco State University; David N. Lumsden, University of Memphis; Nancy Mesner, Utah State University; Jay P. Muza, Broward Community College; Ignacio Pujana, University of Texas at Dallas; Karen L. Savage, Moorpark College; Martha R. Scott, Texas A&M University; David W. Townsend, University of Maine. Special thanks should go once again to Donald Lovejoy, Palm Beach Atlantic College, for his detailed assistance. As always, I am greatly indebted to my long-suffering departmental colleagues Dennis Kelly, Jay Yett, and Robert Profeta for putting up with me through this book's gestation. A corps of diligent teaching assistants led by Tim Riddle worked tirelessly on the book's Internet site. Thanks also to Stanley Johnson, our dean, and Gene Farrell, our college president, for supporting and encouraging the faculty to write, engage in community service, and conduct research.

Yet another round of gold medals should go to my family for being patient (well, *relatively* patient) during those years of days and nights when dad was holed up in his dark reference-littered cave listening to really loud Glenn Gould Bach recordings, working late on The Books. Thank you, Marsha, Jeanne, Greg, John, and Grace, for your love and understanding.

Many of the illustrations for this book came from friends and acquaintances who maintained their cooperative cheer through a blizzard of increasingly anxious e-mails and telephone calls. Bruce Hall, teacher and friend, again contributed photos, as did colleagues Norman Cole, Ron Romanosky, Andreas Rechnitzer, and Don Walsh. Herbert Kauainui Kane again donated his beautiful pictures of Hawaiian topics. Deborah Day and Cindy Clark at Scripps Institution, Jutta Voss-Diestelkamp at the Alfred Wegener Institut in Bremerhaven, Robert Headland at Scott Polar Research Institute, and David Taylor at the Centre for Maritime Research in Greenwich dug through their archives one more time. Don Dixon provided paintings, Dan Burton sent photos, and Andrew Goodwillie printed customized charts. Wim van Egmond contributed striking photomicrographs of plankton. Bill Haxby at Lamont-Doherty Earth Observatory provided truly beautiful seabed scans. Michael Latz at Scripps Institution taught me about bioluminescence. Thomas Maher, vice-provost and friend, led my son and me on a personal inspection of the effects of the Gulf Stream and other fluid wonders. The team at Woods Hole Oceanographic Institution was generous, as always, in providing photographs and diagrams for inclusion. The U.S. Coast Guard, U.S. Navy, Army Corps of Engineers, Breitling-SA, Associated Press, The Smithsonian, NASA, JOI, and NOAA were patient and understanding of my needs and deadlines.

Tom Garrison

Figure 3 The largest feature of a poorly named planet, the Pacific Ocean is crossed every day by thousands of air travelers. Keep your shades up—there is much to see!

The Brooks/Cole team performed the customary miracles. The charge was led by Joan Keyes, production editor and champion of every known means of digital communi-cation. The text was again polished by Mary Arbogast, a developmental editor so excellent she is in constant danger of becoming a co-author. Carol Reitz was the heroic copy editor, and Myrna Engler, Joohee Lee, and Samuel Subity helped with photo research, permissions, and Internet matters. Editor Keith Dodson and senior project manager Teri Hyde kept me and my overheated computer and fax machine on the right path. My unending thanks to all.

A Gift

The ocean's greatest gift to humanity is intellectual—the constant analytical challenge its restless mass presents. Let yourself be swept into this book and the class it accompanies. Ask questions of your instructors and teaching assistants, read some of the references, try your hand at the questions at the ends of the chapters. Be optimistic. Take pleasure in the natural world. Please write to me when you find errors or if you have comments. *Above all, enjoy yourself!*

Tom Garrison
Orange Coast College
University of Southern California
tgarriso@mail.occ.cccd.edu

Reviewers

ERNEST E. ANGINO, *University of Kansas*
M. A. ARTHUR, *Pennsylvania State University*
HENRY A. BART, *La Salle University*
STEVEN R. BENHAM, *Pacific Lutheran University*
LATSY BEST, *Palm Beach Community College*
EDWARD BEUTHER, *Franklin and Marshall College*
WILLIAM L. BILODEAU, *California Lutheran University*
JULIE BRIGHAM-GRETTE, *University of Massachusetts at Amherst*
LAURIE BROWN, *University of Massachusetts*
KEITH A. BRUGGER, *University of Minnesota, Morris*
DONALD BUCHANAN, *San Bernadino Valley College*
JEFFREY CHANTON, *Florida State University*
ZANNA CHASE, *Lamont-Doherty Earth Observatory, Columbia University*
KARL M. CHAUFF, *St. Louis University*
D. L. CLARK, *University of Wisconsin*
WILLIAM COCHLAN, *San Francisco State University*
JAMES E. COURT, *City College of San Francisco*
RICHARD DAME, *University of South Carolina, Columbia*
DAVID DARBY, *University of Minnesota at Duluth*
BRENT K. DUGOLINSKY, *State University of New York, Oneonta*
ROBERT J. FELLER, *University of South Carolina, Columbia*
L. KENNETH FINK, JR., *University of Maine*
KATHLEEN FLICKINGER, *Maui Community College*
BRUCE FOUKE, *University of Illinois, Urbana*
DIRK FRANKENBURG, *University of North Carolina, Chapel Hill*
ROBERT R. GIVEN, *Marymount College, Rancho Palos Verdes*
WILLIAM GLEN, *U.S. Geological Survey*
KAREN GROVE, *San Francisco State University*
BARRON HALEY, *West Valley College*
JACK C. HALL, *University of North Carolina at Wilmington*
WILLIAM HAMNER, *University of California, Los Angeles*
WILLIAM B. HARRISON III, *Western Michigan University*
DAVID HASTINGS, *University of British Columbia*
TED HERMAN, *West Valley College*
NANCY HINMAN, *University of Montana*
JOSEPH HOLLIDAY, *El Camino College*
ANDREA HUVARD, *California Lutheran University*
JAMES C. INGLE, JR., *Stanford University*
RONALD E. JOHNSON, *Old Dominion University*
SCOTT D. KING, *Purdue University*
JOHN A. KLASIK, *California Polytechnic University, Pomona*
C. ERNEST KNOWLES, *North Carolina State University*
MICHELLE KOMINZ, *Western Michigan University*
EUGENE KOZLOFF, *Friday Harbor, WA*
LAWRENCE KRISSEK, *Ohio State University*
ALBERT M. KUDO, *University of New Mexico*
FRANK T. KYTE, *University of California, Los Angeles*
LYNTON S. LAND, *University of Texas, Austin*
RICHARD W. LATON, *Western Michigan University*
RUTH LEBOW, *University of California, Los Angeles Extension*
BEN LeFEBVRE, *City College of San Francisco*
DOUGLAS R. LEVIN, *Bryant College*

LARRY LEYMAN, *Fullerton College*
TIMOTHY LINCOLN, *Albion College*
DONALD L. LOVEJOY, *Palm Beach Atlantic College*
DAVID N. LUMSDEN, *University of Memphis*
MICHAEL LYLE, *Tidewater Community College*
JAMES MACKIN, *State University of New York, Stony Brook*
DAVID C. MARTIN, *Centralia College*
ELLEN MARTIN, *University of Florida*
BRIAN McADOO, *Vasser College*
JAMES McWHORTER, *Miami-Dade Community College, Kendall Campus*
GREGORY A. MEAD, *University of Florida*
NANCY MESNER, *Utah State University*
CHRIS METZLER, *Mira Costa College*
RICHARD W. MURRAY, *Boston University*
JAY P. MUZA, *Broward Community College*
JOHN E. MYLROIE, *Mississippi State University*
CONRAD NEWMAN, *University of North Carolina, Chapel Hill*
JAMES G. OGG, *Purdue University*
B. L. OOSTDAM, *Millersville University*
JAN PECHENIK, *Tufts University*
BERNARD PIPKIN, *University of Southern California*
MARK PLUNKETT, *Bellevue Community College*
K. M. POHOPIEN, *Covina, CA*
IGNACIO PUJANA, *University of Texas at Dallas*
RENE S. REVUELTA, *Miami-Dade Community College*
RICHARD G. ROSE, *West Valley College*
JUNE R. P. ROSS, *Western Washington University*
WENDY L. RYAN, *Kutztown University*
ROBERT J. SAGER, *Pierce College, Washington*
KAREN L. SAVAGE, *Moorpark College*
ROBERT F. SCHMALZ, *Pennsylvania State University*
MARTHA R. SCOTT, *Texas A&M University*
DON SEAVY, *Olympic College*
SAM SHABB, *Highline Community College*
WILLIAM G. SIESSER, *Vanderbilt University*
RALPH SMITH, *University of California, Berkeley*
SCOTT W. SNYDER, *East Carolina University*
MORRIS L. SOTONOFF, *Chicago State University*
JAMES F. STRATTON, *Eastern Illinois University*
KENT SYVERSON, *University of Wisconsin*
J. COTTER THARIN, *Hope College*
DAVID W. TOWNSEND, *University of Maine*
STANLEY ULANSKI, *James Madison University*
J. J. VALENCIC, *Saddleback College*
RAYMOND E. WALDNER, *Palm Beach Atlantic College*
JILL M. WHITMAN, *Pacific Lutheran University*
P. KELLY WILLIAMS, *University of Dayton*
BERT WOODLAND, *Centralia College*
JOHN H. WORMUTH, *Texas A&M University*
RICHARD YURETICH, *University of Massachusetts at Amherst*
MEL ZUCKER, *Skyline College*

© Pal Hermansen / The Image Bank /Getty Images

Sunset on an ocean world.

Origins

A MARINE POINT OF VIEW

I believe *Earth* is misnamed.

From space our planet shines a brilliant blue, is white in places covered by clouds and ice, and sometimes swirls with storms. Dominating its surface is a single great ocean of liquid water. This ocean moderates temperature and dramatically influences weather. The ocean borders most of the planet's largest cities. It is a primary shipping and transportation route and provides much of our food. From its floor is pumped about one-third of the world's supply of petroleum and natural gas. The dry land on which nearly all of human history has unfolded is hardly visible from space because nearly three-quarters of the planet is covered by water. *Oceanus* would surely be a better name for our watery home.

Earth, its ocean, and its organisms have changed together over the ages. Life originated in the ocean, developing and flourishing there for more than 3 billion years

1

before venturing onto the unwelcoming continents. All our planet's living things—terrestrial as well as aquatic—carry an ocean within themselves. Their blood, their eggs, and the fluids that bathe their cells are all saline. We ourselves are made largely of water that has about the same relative proportion of salts as the sea. The first nine months of human life are spent in a water world, a warm supportive ocean that cradles from shock and provides a stable and weightless environment for the complex processes of growth and development. After birth, we view the universe through an ocean—the fluid behind the corneas of our eyes is similar to seawater. In a sense we see everything from a marine point of view.

CHAPTER AT A GLANCE

An Ocean World

Ours is not a particularly large planet, and it is not unusual in overall composition. Its sun and its position within the galaxy are unremarkable. What *is* extraordinary is its liquid water ocean.

Traditionally, we have divided the ocean into artificial compartments called *oceans* and *seas* using the boundaries of continents and imaginary lines such as the equator. In fact, there are few dependable natural divisions, only one great mass of water. The Pacific and Atlantic Oceans, the Mediterranean and Baltic Seas, so named for our convenience, are in reality only temporary features of a single **world ocean.**[1] In this book we refer to the world ocean, or simply the **ocean,** *as a single entity*, with subtly different characteristics at different locations but with very few natural partitions. Such a view emphasizes the interdependence of ocean and land, life and water, atmospheric and oceanic circulation, and natural and man-made environments.

The ocean may be defined as the vast body of saline water that occupies the depressions of Earth's surface. More than 97% of the water on or near Earth's surface is contained in the ocean; less than 3% is held in land ice, groundwater, and all the freshwater lakes and rivers (**Figure 1.1**).

On a *human* scale, the ocean is impressively large—it covers 361 million square kilometers (139 million square miles) of Earth's surface.[2] If Earth's contours were leveled to a smooth ball, the ocean would cover it to a depth of 2,686 meters (8,810 feet). The average ocean depth is 4½ times greater than the average land elevation. The Pacific Ocean is Earth's most prominent single feature. **Figure 1.2** summarizes some basic characteristics of the world ocean.

On a *planetary* scale, however, the ocean itself is insignificant. Its average depth is a tiny fraction of Earth's radius—the blue ink representing the ocean on an 8-inch paper globe is proportionally thicker. The ocean accounts for only slightly more than 0.02% of Earth's mass, or 0.13% of its volume. There is much more water chemically trapped within Earth's hot interior than there is in its ocean and atmosphere.

One world ocean covers about 71% of Earth's surface. For our convenience, we separate it into *oceans* and *seas*, but it is really one great intermixing mass of saline water.

---

[1] When an important new term is introduced and defined, it is printed in **boldface type.** These terms are listed at the end of the chapter and defined in the Glossary.

[2] Throughout this book, metric measurements precede American measurements. For a quick review of metric (SI) units and their abbreviations, please turn to Appendix I.

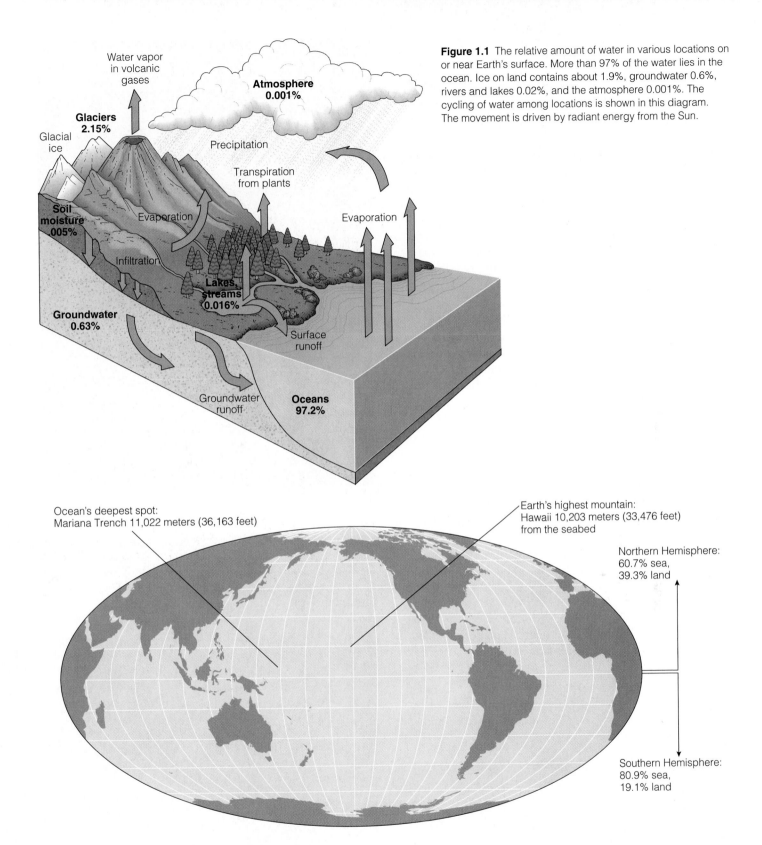

Figure 1.1 The relative amount of water in various locations on or near Earth's surface. More than 97% of the water lies in the ocean. Ice on land contains about 1.9%, groundwater 0.6%, rivers and lakes 0.02%, and the atmosphere 0.001%. The cycling of water among locations is shown in this diagram. The movement is driven by radiant energy from the Sun.

Water vapor in volcanic gases

Atmosphere 0.001%

Glaciers 2.15%

Glacial ice

Precipitation

Transpiration from plants

Soil moisture .005%

Evaporation

Evaporation

Infiltration

Lakes, streams 0.016%

Groundwater 0.63%

Surface runoff

Groundwater runoff

Oceans 97.2%

Ocean's deepest spot: Mariana Trench 11,022 meters (36,163 feet)

Earth's highest mountain: Hawaii 10,203 meters (33,476 feet) from the seabed

Northern Hemisphere: 60.7% sea, 39.3% land

Southern Hemisphere: 80.9% sea, 19.1% land

Area: 361,100,000 square kilometers (139,400,000 square miles)
Volume: 1,370,000,000 cubic kilometers (329,000,000 cubic miles)
Average depth: 3,796 meters (12,451 feet)
Average temperature: 3.9°C (39.0°F)
Average salinity: 34,482 grams per kilogram (0.56 ounce per pound), 3.4%
Average land elevation: 840 meters (2,772 feet)
Age: About 4 billion years
Future: Uncertain

Figure 1.2 The proportion of sea versus land, shown on an equal-area projection of Earth. (An equal-area projection is a map drawn to represent areas in their correct relative proportions.) The average depth of the ocean is 4½ times greater than the average land elevation. Note the extent of the Pacific Ocean, Earth's most prominent single feature.

Marine Science, Oceanography, and the Nature of Science

Marine science (or **oceanography**) is the process of discovering unifying principles in data obtained from the ocean, its associated life-forms, and the bordering lands. It draws on several disciplines, integrating the fields of geology, physics, biology, chemistry, and engineering as they apply to the ocean and its surroundings. Nearly all marine scientists specialize in one area of research, but they also must be familiar with related specialties and appreciate the linkages between them.

- *Marine geologists* focus on questions such as the composition of the inner Earth, the mobility of the crust, the characteristics of seafloor sediments, and the history of Earth's climate. Some of their work touches on areas of intense scientific and public concern, including earthquake prediction and the distribution of valuable resources.

- *Physical oceanographers* study and observe wave dynamics, currents, and ocean–atmosphere interaction. Their predictions of long-term climate trends are becoming increasingly important as pollutants change Earth's atmosphere.

- *Marine biologists* work with the nature and distribution of marine organisms, the impact of oceanic and atmospheric pollutants on the organisms, the isolation of disease-fighting drugs from marine species, and the yields of fisheries.

- *Chemical oceanographers* study the ocean's dissolved solids and gases and their relationships to the geology and biology of the ocean as a whole.

- *Marine engineers* design and build oil platforms, ships, harbors, and other structures that enable us to use the ocean wisely.

Other marine specialists study the techniques of weather forecasting, ways to increase the safety of navigation, methods to generate electricity, and much more. **Figure 1.3** shows marine scientists in action.[3]

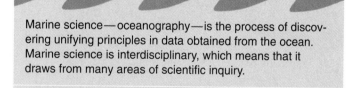

Marine science—oceanography—is the process of discovering unifying principles in data obtained from the ocean. Marine science is interdisciplinary, which means that it draws from many areas of scientific inquiry.

Marine scientists today are asking some critical questions about the origin of the ocean, the age of its basins, and the

---

[3] Would you like to join us? Appendix V discusses careers in oceanography.

Scripps Institution of Oceanography

a

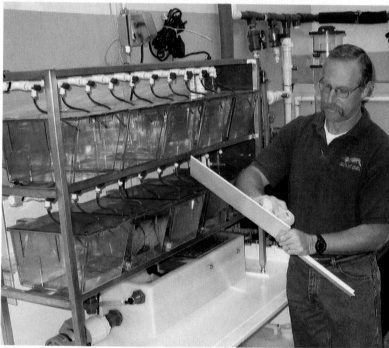

Tom Garrison

b

Figure 1.3 Doing marine science is sometimes anxious, sometimes routine, and always interesting. (**a**) Deploying a data-gathering buoy from the slippery deck of a rolling ship takes courage and timing. (**b**) Routine maintenance and inspections are part of every marine scientist's day.

nature of the life-forms it has nurtured. We are fortunate to live at a time when scientific study may be able to answer some of those questions. **Science** is a systematic *process* of asking questions about the observable world and then testing the answers to those questions. Scientists gather and study information (data), but the information itself is not science. Science interprets raw information by constructing a general explanation with which the information is compatible.

Scientists start with a question—a desire to understand something they have observed or measured. They then form a tentative explanation called a working **hypothesis,** a speculation about the natural world that can be tested and verified or disproved by further observations and controlled experiments. (An **experiment** is a test that simplifies observation in nature or in the laboratory by manipulating or controlling the conditions under which observations are made.) A hypothesis that is consistently supported by observation or experiment is advanced to the status of **theory,** a statement of relationship accepted by most scientists. The largest constructs, known as **laws,** are principles explaining events in nature that have been observed to occur with unvarying uniformity under the same conditions.

Theories and laws in science do not arise fully formed or all at once. Scientific thought progresses as a continuing chain of questioning, testing, and matching theories to observations. A theory is strengthened if new facts support it. If not, the theory is modified or a new explanation is sought. The power of science lies in the ability of the process to operate *in reverse*—that is, in the use of a theory or law to make predictions and anticipate new facts to be observed.

This procedure, often called the **scientific method,** is an orderly process by which theories are verified or rejected. It is based on the assumption that nature "plays fair"—that the rules governing natural phenomena do not change capriciously as our powers of questioning and observing improve. We believe that the answers to our questions about nature are *ultimately knowable.*

There is no one "scientific method." Some researchers observe, describe, and report on some subject and leave it to others to hypothesize. Scientists don't have one single method in common; the general method they employ is a critical attitude about being *shown* rather than being *told,* and taking a logical approach to problem solving. **Figure 1.4** summarizes the main points of the scientific method.

Nothing is ever proven absolutely true by the scientific method. Theories may change as our knowledge and powers of observation change; so all scientific understanding is tentative. Science is neither a democratic process nor a popularity contest. The conclusions about the natural world that we reach by the process of science may not always be comfortable, easily understood, or immediately embraced, but if those conclusions consistently match observations, they may be considered true.

This book shows some of the results of the scientific process as it has been applied to the world ocean. It presents facts, interpretations of facts, examples, stories, and some of the crucial discoveries that have led to our present understanding of the ocean and the world on which it formed. As

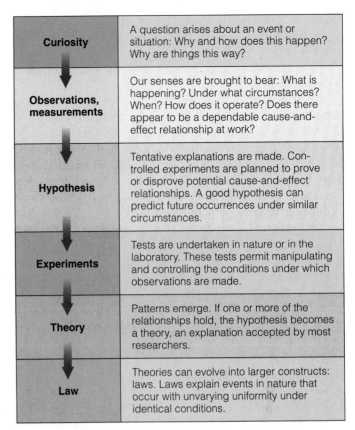

Figure 1.4 An outline of the scientific method, a systematic process of asking questions about the observable world and then testing the answers to those questions. There is not a single "scientific method." Science rests on a critical attitude about being shown rather than being told and then a logical approach to problem solving. Application of the scientific method leads us to truth based on the observations and measurements that have been made—a work in progress, never completed. The external world, not internal conviction, must be the testing ground for scientific beliefs.

the results of science change, so will the ideas and interpretations presented in books like this one.

Science is a *process* of asking questions about the observable world and then testing the answers to those questions. The external world, not internal conviction, is the testing ground for scientific beliefs.

Origins

We have always wondered about our origins—how Earth was formed, how the ocean arose, and how life came to be. In the last 50 years, researchers using the scientific method have determined a tentative age for the ocean, Earth, and the universe. They have developed hypotheses about how matter is assembled, how stars and planets are formed, and even how life may have arisen. Many of the details are still sketchy, of course, but hypotheses have predicted some important recent discoveries in subatomic physics and molecular biology. Perhaps the most dramatic recent discoveries in natural science have been those dealing with the origin and history of the universe.

The universe apparently had a beginning. The **big bang,** as that event is wonderfully named, probably occurred 13.7 billion years ago. All of the mass and energy of the universe are thought to have been concentrated at a geometric point at the beginning of time, the moment when the expansion of the universe began. We don't know what initiated the expansion, but it continues today and will probably continue for billions of years, perhaps forever.

The very early universe was unimaginably hot, but as it expanded it cooled. About a million years after the big bang, temperatures fell enough to permit the formation of atoms from the energy and particles that had predominated up to that time. Most of these atoms were hydrogen, then as now the most abundant form of matter in the universe. About a billion years after the big bang, this matter began to congeal into the first galaxies and stars.

Galaxies and Stars

A **galaxy** is a huge rotating aggregation of stars, dust, gas, and other debris held together by gravity. Our galaxy is named the **Milky Way.** A galaxy very similar to our own is shown in **Figure 1.5.**

The **stars** that make up a galaxy are massive spheres of incandescent gases. They are usually intermingled with diffuse clouds of gas and debris. In spiral galaxies like the Milky Way, the stars are arrayed in spiral arms radiating from the galactic center. Our part of the Milky Way is populated with many stars, but distances within a galaxy are so vast that the star nearest the sun is about 42 trillion kilometers (26 trillion miles) away. Astronomers tell us there are perhaps 50 billion galaxies in the universe and 50 billion stars in each galaxy. Imagine more stars in the Milky Way than grains of sand on a small beach!

Our sun is a typical star. The sun and its family of planets, called the **solar system,** is located about three-fourths of the way out from the galaxy's center. We orbit the galaxy's brilliant core, taking about 230 million years to make one orbit even though we are moving at about 280 kilometers per second (half a million miles an hour). The Earth has made about 20 circuits of the galaxy since the ocean formed in its basins.

Our planet and our sun probably had a common origin. The members of the solar system are thought to have condensed from a thin cloud—the **solar nebula**—that had been enriched with heavy elements created and released by the explosive deaths of nearby stars. By about 5 billion years ago, the solar nebula was a rotating disk-shaped mass of about 75% hydrogen, 23% helium, and 2% other material, including heavier elements, gases, dust, and ice (**Figure 1.6**). Like a spinning skater bringing in her arms, the nebula spun faster as it condensed. Material concentrated near its center became the **protosun.** Much of the outer material eventually became **planets,** the smaller bodies that orbit a star and do not shine by their own light.

The new planets formed in the disk of dust and debris surrounding the young sun through a process known as **accretion**—the clumping of small particles into large masses (**Figure 1.7**). Bigger clumps with stronger gravity pulled in

Figure 1.5 This spiral galaxy in the constellation Coma Berenices is very similar in size and structure to our own Milky Way galaxy. The galaxy, photographed in 1999 by the Hubble Space Telescope, is 62 million light years away and about 56,000 light years in diameter. The stars in the foreground are in our galaxy. Note the clouds of dust and gas that obscure our view of some parts of the spiral arms. Planets and oceans are made of such material.

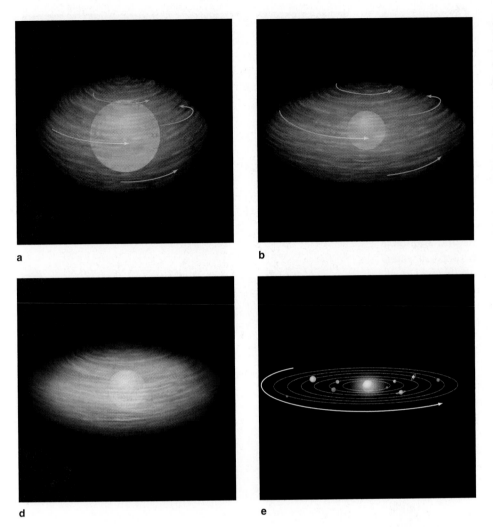

a

b

c

d

e

Figure 1.6 The origin of a star in the spiral arm of a galaxy. As a cloud of gas and dust contracts under the influence of gravity (**a**), the spinning mass forms a disk with a warm, thick center—a protostar (**b**). The center contracts under its own gravity and continues to heat up while the gas and dust in the surrounding disk continue to fall in. As the protostar radiates more heat, it ejects some matter outward from its poles (**c**). When fusion begins in the star's core (**d**), the shock waves disperse much of the remaining matter that has not coalesced into planets, moons, or comets around the periphery. Our solar system (**e**) was formed in this way about 5 billion years ago. (Source: *The Universe Explained*, C. A. Ronan, ed., New York: H. Holt, © 1994.)

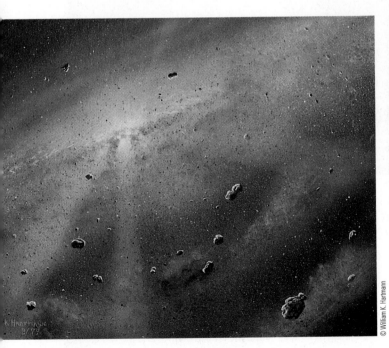

a

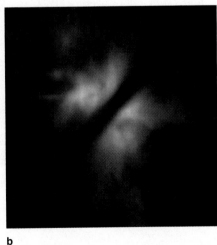

b

Figure 1.7 (**a**) In this artist's conception, our sun begins to shine at the center of a spinning disk of dust, ice, gases, and rock. The accretion phase is in progress: The bits will condense into planets and other bodies. (**b**) A solar nebula about the same size as the one that formed our solar system is seen in this 1998 image from the Hubble Space Telescope. The new star is hidden from direct view by a thick, dark disk crossing the center of the image. (The same sort of disk is represented in Figure 1.6d.) The disk—seen here edge-on—is about 130 billion kilometers (80 billion miles) in diameter, or about 15 times the diameter of Neptune's orbit around our sun. Dark clouds and bright wisps above and below the disk suggest it is still growing from infalling dust and gas.

a

most of the condensing matter. Near the protosun, where temperatures were highest, the first materials to solidify were substances with high boiling points, mainly metals and certain rocky minerals. The planet Mercury, closest to the sun, is mostly iron because iron is a solid at high temperatures. Somewhat farther out, in the cooler regions, magnesium, silicon, water, and oxygen condensed. Methane and ammonia accumulated in the frigid outer zones. Earth's mix of water, silicon–oxygen compounds, and metals results from its position within that accreting cloud. The planets of the outer solar system—Jupiter, Saturn, Uranus, and Neptune—are composed mostly of methane and ammonia ices because those gases can congeal only at cold temperatures.

The period of accretion lasted perhaps 50 to 70 million years. The protosun became a star when hydrogen-fusion temperature was reached. The violence of its new nuclear reactions sent radiation sweeping past the inner planets, clearing the area of excess particles and ending the period of rapid accretion. Gases like those we now see on the giant outer planets may once have surrounded the inner planets, but this rush of solar energy and particles stripped them away.

Earth and Ocean

1.6

The young Earth, formed by the accretion of cold particles, was probably homogeneous throughout. Then, in the midst of the accretion phase, Earth's surface was heated by the impact of asteroids, comets, and other falling debris. This heat, combined with gravitational compression and heat from the decay of radioactive elements accumulating within the newly assembled planet, caused Earth to partially melt. Gravity pulled most of the iron inward to form the planet's core. The sinking iron released huge amounts of gravitational energy, which, through friction, heated Earth even more. At the same time, a slush of lighter minerals—silicon, magnesium, aluminum, and oxygen-bonded compounds—rose toward the surface, forming Earth's crust (**Figure 1.8**). This important process, called **density stratification,** lasted perhaps 100 million years.

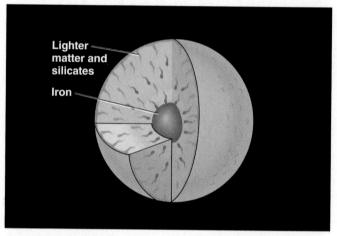

Lighter matter and silicates

Iron

b

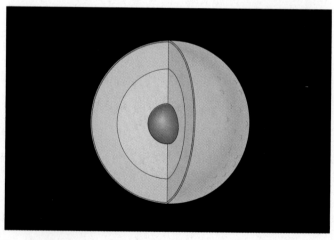

c

Figure 1.8 A representation of the formation of Earth. (**a**) The planet grew by the aggregation of particles. Meteors and asteroids bombarded the surface, heating the new planet and adding to its growing mass. At the time, Earth was composed of a homogeneous mixture of materials. (**b**) Earth lost volume because of gravitational compression. High temperatures in the interior turned the inner Earth into a semisolid mass; dense iron (red drops) fell toward the center to form the core, while less dense silicates move outward. Friction generated by this movement heated Earth even more. (**c**) The result of *density stratification* is evident in the formation of an inner and outer core, a mantle, and the crust.

Then Earth began to cool. Its first surface is thought to have formed about 4.6 billion years ago. That surface did not remain undisturbed for long. A planetary body somewhat larger than Mars smashed into the young Earth. The rocky mantle of the impactor was ejected to form a ring of debris around the Earth, and its metallic core fell into Earth's core and joined with it. Could a similar cataclysm happen today? The issue is addressed in Chapter 12.

Radiation from the energetic young sun then stripped away our planet's outermost layer of gases, its first atmosphere, but soon gases that had been trapped inside the forming planet burped to the surface to form a second atmosphere. This volcanic venting of volatile substances—including water vapor—is called **outgassing** (**Figure 1.9a**). As the hot vapors rose, they condensed into clouds in the cool upper atmosphere. Though most of Earth's water was present in the solar nebula during the accretion phase, recent research suggests that a barrage of icy comets colliding with Earth may also have contributed a portion of the accumulating mass of water, this ocean-to-be (**Figure 1.9b**).

Figure 1.9 Sources of the ocean. (**a**) Outgassing. Volcanic gases emitted by fissures add water vapor, carbon dioxide, nitrogen, and other gases to the atmosphere. Volcanism was a major factor in altering Earth's original atmosphere; the action of photosynthetic plants was another. (**b**) Comets may have delivered some of Earth's surface water. Intense bombardment of the early Earth by large bodies—comets and asteroids—probably lasted until about 3.8 billion years ago.

a

b

Earth's surface was so hot that no water could settle there and no sunlight could penetrate the thick clouds. After millions of years, the upper clouds cooled enough for some of the outgassed water to form droplets. Hot rains fell toward Earth, only to boil back into the clouds again. As the surface became cooler, water collected in basins and began to dissolve minerals from the rocks. Some of the water evaporated, cooled, and fell again. The world ocean was gradually accumulating.

These heavy rains may have lasted for about 25 million years. Large amounts of water vapor and other gases continued to escape through volcanic vents during that time and for millions of years thereafter. The ocean grew deeper. Recent studies suggest that water may have covered Earth's entire surface for some 200 million years before the continents emerged. Although most of the ocean was in place about 4 billion years ago, ocean formation continues very slowly even today: About 0.1 cubic kilometer (0.025 cubic mile) of new water is added to the ocean each year, mostly as steam flowing from volcanic vents and in the form of small cometary fragments.

Earth first had a solid surface about 4.6 billion years ago. The ocean formed later when Earth's surface became cool enough to allow clouds of steam and water vapor to condense and rest on the surface.

The composition of that early atmosphere was much different from today's. Geochemists believe it may have been rich in carbon dioxide and water vapor. Beginning about 3.5 billion years ago, this mixture began a gradual alteration to its present composition, mostly nitrogen and oxygen. This change was brought about by carbon dioxide first dissolving in seawater to form carbonic acid and then combining with crustal rocks. About 1.5 billion years later, the ancestors of today's green plants greatly accelerated the rate of oxygen production by the action of photosynthesis.

The Origin of Life

Life, at least as we know it, would be inconceivable without large quantities of water. Water has the ability to retain heat, moderate temperature, dissolve many chemicals, and suspend nutrients and wastes. These characteristics make it a mobile stage for the intricate biochemical reactions that allowed life to begin and prosper on Earth.

Life on Earth is formed of aggregations of a few basic kinds of carbon compounds. Where did the carbon compounds come from? There is growing consensus that most of the organic (that is, carbon-containing) materials in these compounds were transported to Earth by the comets, asteroids, meteors, and interplanetary dust particles that crashed into our planet during its birth. The young ocean was a thin broth of organic and inorganic compounds in solution.

In laboratory experiments, mixtures of dissolved compounds and gases thought to be similar to Earth's early atmosphere have been exposed to ultraviolet light, heat, and electrical sparks. These energized mixtures produce simple sugars and most of the biologically important amino acids. They even produce small proteins and nucleotides (components of the molecules that transmit genetic information between generations). The main chemical requirement seems to be the absence (or near absence) of free oxygen, a compound that can disrupt any unprotected large molecule.

Did *life* form in these experiments? No, the compounds that formed are only building blocks of life. But the experiments do tell us something about the commonality and unity of life on Earth. It is probably not coincidental that these crucial compounds can be synthesized so easily and are present in virtually all living forms. Those compounds are "permitted" by physical laws and by the chemical composition of this planet. The experiments also underscore the special role of water in life processes. The fact that all life, from a jellyfish to a dusty desert weed, depends on saline water within its cells to dissolve and transport chemicals is certainly significant. It strongly suggests that simple, self-replicating—living—molecules arose somewhere in the early ocean. It also suggests that all life on Earth has a common origin and ancestry.

The early steps in the evolution of living organisms from simple organic building blocks, a process known as **biosynthesis,** are still speculative. Planetary scientists suggest that the sun was faint in its youth. It put out so little heat that the ocean may have been frozen to a depth of around 300 meters (1,000 feet). The ice would have formed a blanket that kept most of the ocean fluid and relatively warm. Periodic fiery impacts by asteroids, comets, and meteor swarms could have thawed the ice, but between batterings it would have reformed. In 2002, chemists Jeffrey Bada and Antonio Lazcano suggested that organic material may have formed and then been trapped beneath the ice—protected from the atmosphere, which contained chemical compounds capable of shattering the complex molecules. The first self-sustaining—living—molecules might have arisen deep below the layers of surface ice, on clays or pyrite crystals at cool mineral-rich seeps on the ocean floor (**Figure 1.10**).

Life probably arose on Earth shortly after its formation, about 4 billion years ago. Life may have arisen in the deep ocean.

A similar biosynthesis cannot occur today. Living things have changed the conditions in the ocean and atmosphere, and those changes are not consistent with any new origin of life. For one thing, green plants have filled the atmosphere with oxygen. For another, some of this oxygen (as ozone) now blocks much of the ultraviolet radiation from reaching the surface of the ocean. And finally, the many tiny organ-

Figure 1.10 An environment for biosynthesis? Weak sunlight and violent conditions on Earth's surface may have favored the origin of life on mineral surfaces near deep-ocean hydrothermal vents similar to the one shown here.

isms present today would gladly scavenge any large organic molecules as food.

How long ago might life have begun? The oldest fossils yet found, from northwestern Australia, are between 3.4 and 3.5 billion years old (**Figure 1.11**). They are remnants of

Figure 1.11 Fossil of a bacteria-like organism (with an artist's reconstruction) that photosynthesized and released oxygen into the atmosphere. Among the oldest fossils ever discovered, this microscopic filament from northwestern Australia is about 3.5 billion years old. Life on Earth began earlier, perhaps between 4 and 4.2 billion years ago.

fairly complex bacteria-like organisms, indicating that life must have originated even earlier, probably only a few hundred million years after a stable ocean formed. Evidence of an even more ancient beginning has been found in the form of carbonaceous residues in some of the oldest rocks on Earth, from Akilia Island near Greenland. These 3.85 billion year old specks of carbon bear a chemical fingerprint that researchers think could have come only from a living organism. Life and Earth have grown old together; each has greatly influenced the other.

The Distant Future of Earth

Our descendants may enjoy another 5 billion years of Earth as we know it today. But then our sun, like any other star, will begin to die. Its red giant phase will engulf the inner planets. Its fiery atmosphere will expand to a radius greater than the orbit of Earth. The ocean and atmosphere, all evidence of life-forms, the crust, and perhaps the whole planet will be recycled into component atoms and hurled by shock waves into space. Our descendants, if any, will have perished or fled to safer worlds. Its fuel exhausted and its energies spent, the sun will cool to a glowing ember and ultimately to a dark cinder. Perhaps a new system of star and planets will form from the debris of our remains.

The history of past and future Earth is shown as a graph in **Figure 1.12.**

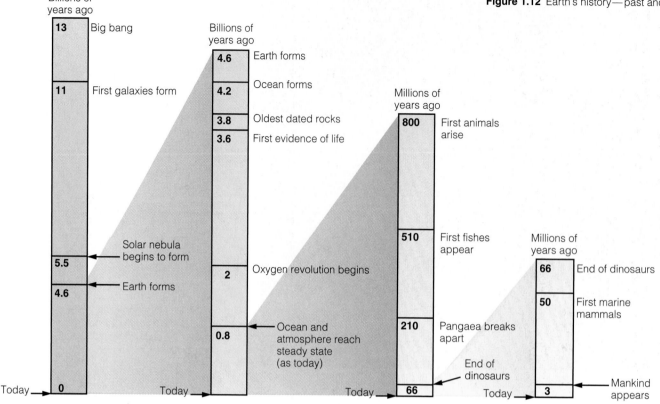

Figure 1.12 Earth's history—past and future.

Other Ocean Worlds

Water planets are probably uncommon in the universe, but water itself is not scarce. In the solar system, for example, Jupiter has hundreds of times more water than Earth does. Astronomers have even located water molecules drifting in free space. *Liquid* water, however, is unexpected.

Consider the conditions necessary for a large permanent ocean of liquid water to form on a planet. An ocean world must move in a nearly circular orbit around a stable star and the distance of the planet from the star must be just right, to provide a temperature environment in which water is liquid. Unlike most stars, a water planet's sun must not be a double or multiple star or else the orbital year would have irregular periods of intense heat and cold. The materials that accreted to form the planet must have included both water and substances capable of forming a solid crust. The planet must be large enough so that its gravity keeps the atmosphere and ocean from drifting off into space.

Are these conditions met anywhere else? Is our pleasant blue world the only water planet in the galaxy? Seventy-seven planets had been discovered outside our solar system by the summer of 2002, and new planets were being discovered at the rate of about two per month! Because the new planets are so far away, we are unable to see details on their surfaces. But two likely nearby candidates have been studied: Jupiter's moon Europa and the planet Mars.

The spacecraft *Galileo* passed close to Europa, a body about the same size as our moon, in early 1997. Photos sent to Earth revealed a cracked, icy crust covering what appears to be a slushy mix of ice and water (**Figure 1.13a**). The jigsaw-puzzle pattern of ice pieces appears to have formed when the ice crust cracked apart, moved slightly, and then froze together again. The underlying ocean is probably kept slushy or liquid by heat escaping from Europa's interior and by gravitational friction of tidal forces generated by Jupiter itself. Though the surface of the ice is as cold as the surface of Jupiter, the liquid interior of the ocean, cradled deep in rocky basins, may be warm enough to sustain life. More missions to Europa are being planned.

Europa may have an icy ocean now, but Mars, a much nearer neighbor, probably had an ocean in the distant past. An ocean may have occupied the low places on the surface of Mars between 3.2 and 1.2 billion years ago when conditions were warmer, and water may have flowed across the surface of Mars in the recent past (**Figure 1.13b**). Where is the water now? Mars has become much colder in the past billion years, perhaps because of the loss of greenhouse gases in the atmosphere. If a large quantity of water is present, it probably lies at or just beneath the surface in the form of permafrost (**Figure 1.13c**).

Other planets may have—or may have had—oceans. At least one of the moons of Jupiter appears to contain an ocean beneath an icy crust.

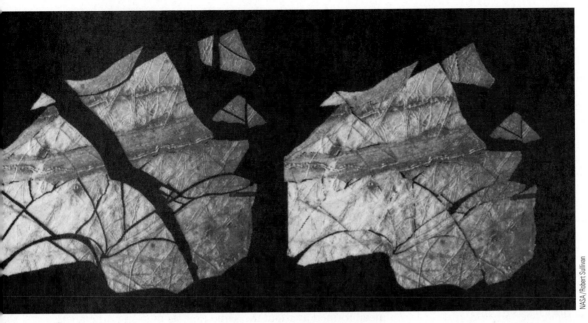

a

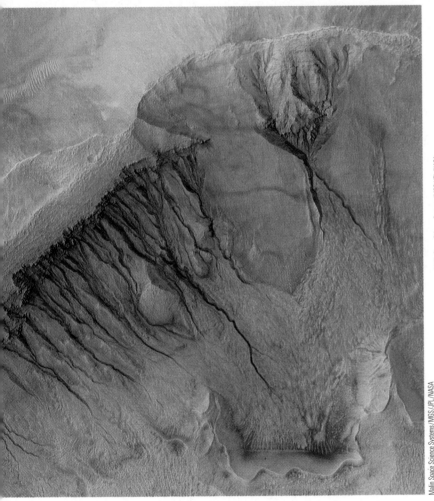

b

c

Figure 1.13 Two ocean worlds? (**a**) A photograph of the icy surface of Europa, a moon of Jupiter, taken by the *Galileo* spacecraft on 20 February 1997. Jupiter's gravitational pull twists Europa, cracking the ice crust and warming the interior. In some areas the ice has broken into large pieces that have shifted away from one another but fit together like pieces of a jigsaw puzzle (as seen here). This process suggests the ice crust is lubricated by slush or liquid water. (**b**) Young gullies on Mars. These valleys, possibly caused by flowing water, are about 3 kilometers (2 miles) wide and line a south-facing wall of a large crater. Researchers suggest that these furrows formed between 10,000 and 1 million years ago. Their recent formation is indicated by the lack of surface craters. (**c**) The edge of the north polar cap of Mars, photographed in April 2001, shows layers of water ice over darker layers of sand. The image is about 5 kilometers (3 miles) across.

And life? Special conditions were necessary for the formation of life on our planet. Earth's gravity is strong enough to retain an ocean but not strong enough to crush the life-forms that came from it. Our planet has a magnetic field provided by an iron core to deflect radiation that would otherwise harm the genetic instructions of living things. A single moon provides gentle tides to encourage life-forms to leave the ocean and reside on land. The atmosphere is relatively clear—so that sunlight penetrates to the surface—but moist enough to form rains and winds that drive air and ocean currents. Furthermore, the upper air contains ozone, which protects large molecules from the most harmful ultraviolet rays. This combination may be rare in the galaxy. We have no clear evidence of life elsewhere. We should enjoy and protect the poorly named planet we call home.

QUESTIONS FROM STUDENTS[4]

1. How did this wet planet get the name "Earth"?

We are terrestrial organisms, so the name is understandable. The modern English word *Earth* is derived from the Anglo-Saxon word *eorthe*, which itself comes from the German word for Earth, *Erde*. The earliest German tribes were land-locked people who had no knowledge of the true extent of the world ocean. We seem to be stuck with "Earth."

2. OK, then where did the word ocean *come from?*

Ocean derives from the Greek word *okeanos* (oceanus), a word that means "outer sea" (in contrast to the Greeks' "inner sea," the Mediterranean). Okeanos was also a mythical Greek titan who was god of the sea before Poseidon and father of the oceanides, or ocean nymphs. The later Latin name for the ocean was *oceanus*. The Latin word evolved into the Middle English term *occean*, in which the double *c* was pronounced like a *k*. This later became the English word *ocean* (which, until this century, was pronounced with a soft *c*, as in the word *celery*). Perhaps the best name for Earth would be Oceanus, but so far I've not had much luck in changing the name!

3. You wrote that "Nothing is ever proven absolutely true by the scientific method." What good is it then? Can't we depend on the process of science?

One philosopher of science has described truth as a liquid: It flows around ideas and is hard to grasp. The progressive improvement in our understanding of nature is subject to the limitations inherent in our observations. As our observations become more accurate, so do our conclusions about the natural world. But because observation (and interpretation of observation) is never perfect, truth can never be absolute. In the 1920s, for example, astronomers assumed that the universe was limited to our own Milky Way galaxy. Observations made with a large new telescope on Mt. Wilson in California by Harlow Shapley and Edwin Hubble allowed them to measure more distant objects. Galaxies were discovered in profusion, "like grains of sand on a beach," in Shapley's words. Thus truth changed its shape.

We learn as we go. We depend on the underlying assumption that nature "plays fair"—that is, is consistent and does not capriciously change the rules as our powers of observation grow. What we have learned so far is of great practical and aesthetic value, and we have only scratched the surface.

4. How do scientists know how old Earth, the solar system, and the galaxy are? How can they calculate the age of the universe?

The age estimates presented in this chapter are derived from interlocking data obtained by many researchers from different sources. One source is meteorites, chunks of rock and metal formed at about the same time as the sun and planets and out of the same cloud. Many have fallen to Earth in recent times. We know from signs of radiation within these objects how long ago they were formed. That information, combined with the rate of radioactive decay of unstable atoms in meteorites, moon rocks, and the oldest rocks on Earth, allows astronomers to make reasonably accurate estimates of how long ago these objects formed. By the way, there is essentially no evidence to support the contention that Earth is between 6,000 and 10,000 years old.

As for the age of the universe itself, in February 2003 astronomers had obtained very accurate measurements of its background radiation—a sort of cosmic echo of the big bang. Their calculations interlock with other pieces of evidence to suggest the universe began 13.7 billion years ago.

CHAPTER SUMMARY

Earth is a water planet, possibly one of few in the galaxy. An ocean covering 71% of its surface has greatly influenced its rocky crust and atmosphere. The ocean dominates Earth, and the average depth of the ocean is about 4½ times the average height of the continents above sea level. Life on Earth almost certainly evolved in the ocean; the cells of all life forms are still bathed in salty fluids.

We have learned much about our planet using the scientific method, a systematic *process* of asking and answering questions about the natural world. Marine science applies the scientific method to the ocean, the planet of which it is a part, and the living organisms dependent on it.

Most of the atoms that make up Earth and its inhabitants were formed within stars. Stars form in the dusty spiral arms of galaxies and spend their lives changing hydrogen and helium to heavier elements. As they die, some stars eject these elements into space by cataclysmic explosions. The sun and the planets, including Earth, probably condensed from a cloud of dust and gas enriched by the

---

[4] Each chapter ends with a few questions students have asked me after a lecture or reading assignment. These questions and their answers may be interesting to you, too.

recycled remnants of exploded stars. Earth formed by the accretion of cold particles about 4.6 billion years ago. Heat from infalling debris and radioactive decay partially melted Earth, and density stratification occurred as heavy materials sank to its center and lighter materials migrated toward the surface. Our moon was formed by debris ejected when a planetary body somewhat larger than Mars smashed into Earth.

The ocean formed later, as water vapor trapped in Earth's outer layers escaped to the surface through volcanic activity during the planet's youth. Comets may also have brought some water to Earth. Life originated in the ocean soon after its formation—life and Earth have grown old together. We know of no other planet with a similar ocean, but water is abundant in interstellar clouds and other water planets are not impossible to imagine.

ON-LINE STUDY RESOURCES

The Web site for this book contains a wealth of helpful study aids. Log on to:

http://info.brookscole.com/garrison

and select *Essentials of Oceanography*, 3rd edition. Select Chapter 1 from the drop-down menu or click on one of these resource areas:

- For study and review, **Chapter at a Glance** gives you an outline of the chapter. **Chapter Summary** allows you to review the chapter's main ideas, and **Glossary** lists concepts and terms for the chapter along with their definitions.

- To text your mastery of the Terms and Concepts to Remember for this chapter, you can use the electronic **Flash Cards,** play the **Concentration** game, or work the **Crossword Puzzle.**

- For practice quizzing, try the multiple-choice **Tutorial Quiz,** the ten-question **True/False Quiz,** or the **Image Analysis Quiz,** which poses questions based on art and photos from the chapter.

TERMS AND CONCEPTS TO REMEMBER

accretion	oceanography
big bang	outgassing
biosynthesis	planets
density stratification	protosun
experiment	science
galaxy	scientific method
hypothesis	solar nebula
laws	solar system
marine science	stars
Milky Way galaxy	theory
ocean	world ocean

STUDY QUESTIONS

1. Why do we say there is one world ocean? What about the Atlantic and Pacific Oceans, the Baltic and Mediterranean Seas?

2. Which is greater: the average depth of the ocean or the average height of the continents above sea level?

3. Can the scientific method be applied to speculations about the natural world that are not subject to test or observation?

4. What are the major specialties within marine science?

5. What is biosynthesis? Where and when do researchers think it might have occurred on our planet? Could it happen again this afternoon?

6. Would you expect ocean worlds to be relatively abundant in the galaxy? Why or why not?

7. Earth has had three distinct atmospheres. Where did each one come from, and what were the major constituents and causes of each?

8. How old is Earth? When did life arise? On what are those estimates based?

9. Where did Earth's heavy elements come from?

10. What is density stratification? What does it have to do with the present structure of Earth?

 Earth Systems Today CD Question

Take a few minutes to install and become familiar with **Earth Systems Today,** the interactive CD-ROM that accompanies this text. At the end of most of the chapters in this book are questions based on this CD-ROM. Answering these questions using the CD-ROM will help you understand basic oceanographic principles.

1. Although no questions specific to Chapter 1 are on the CD-ROM, go to "Geologic Time" in the CD-ROM's main menu, and click on "Changing Earth." Begin the animation. Notice that the span of time included in this animation is only 750 million years (0.75 billion years), but Earth is around 4,600 million (4.6 billion) years old. The slow movement of continents—about as fast as your fingernails grow—combined with the immensity of geological time, gives an appreciation of Earth's great age.

RESOURCES FOR FURTHER READING AND RESEARCH

The Web site for this book contains many ideas for further reading and research. Log on to:

http://info.brookscole.com/garrison

and select *Essentials of Oceanography*, 3rd edition. Select Chapter 1 from the drop-down menu or click on one of these resource areas:

- **Additional Readings** lists the major books and articles consulted in writing this chapter, along with comments from the author about their content and reading level.

- **Hypercontents** takes you to an extensive list of current links to Internet sites with news, research, and images related to individual subjects in the chapter. Just click on the icon that corresponds to a numbered section to see the links for that subject.

- **Internet Exercises** are critical thinking questions that involve research on the Internet with starter URLs provided.

- **InfoTrac College Edition Exercises** leads you to Critical Thinking Projects that use InfoTrac College Edition® as a research tool.

- **Regional InfoTrac College Edition** articles are organized into East Coast, West Coast, and Gulf Coast regions, allowing you to study oceanography on a more local level.

 For more readings, go to InfoTrac College Edition, your on-line research library, at:

http://infotrac.thomsonlearning.com

Students investigate the history of oceanography first hand at the Scripps Institution archives in La Jolla, California. E. W. Scripps looks on.

Tom Garrison

History

MAKING MARINE HISTORY

History is made a day at a time. Sometimes those who make history realize the significance of a single day's effort, but more often the days blend into weeks and months of hard work punctuated only rarely by scientific or artistic insights. Still, individual days are important, and none is more important to a discoverer than the day that began his or her involvement in a field that will become the subject of lifelong study. No matter what follows, that day, perhaps spent with a good teacher and friends, will remain unique in memory.

Student oceanographers get cold, wet, and seasick when they make their first cruises to study the ocean they've been learning about in books and classes. They take samples with probes and buckets and thermometers lowered on lines. They may work from outboard-powered boats or well-equipped research vessels. They struggle with buckets of sediment, steady their microscopes, memorize the interior of a shark, and are the last ones out of the library at night. They write, they talk, and they party. A few of them experience *a flash of commitment* and decide to continue. That is the moment they will remember most vividly—the day they start to make their own history.

With few exceptions, we have no knowledge of how the travelers and scientists described in this chapter became personally interested in advancing marine science, but we do know that, as before, some of today's student oceanographers will contribute to the oceanography texts of tomorrow. Progress in marine science depends on them—and perhaps on you.

Voyaging and Discovery

It has taken a long time for humans to appreciate the nature of the world, but we're naturally restless and inquisitive, and despite the ocean's great size, we have populated nearly every inhabitable place. This fact was aptly illustrated when the European explorers set out to "discover" the world only to be met by native peoples at almost every landfall! Clearly the ocean did not prevent the spread of humanity. The early history of marine science is closely associated with the history of voyaging.

The ocean did not prevent humans from occupying nearly every place on Earth that could sustain them.

Voyaging Begins

Ocean transportation offers people the benefits of mobility and greater access to food supplies. Any coastal culture skilled at raft building or small boat navigation would have economic and nutritional advantages over less skilled competitors. From the earliest period of human history, then, understanding and appreciating the ocean and its life-forms benefited those patient enough to learn.

The first direct evidence we have of **voyaging,** traveling on the ocean for a specific purpose, comes from records of trade in the Mediterranean Sea. The Egyptians organized shipborne commerce on the Nile River, but the first regular ocean traders were probably the Cretans or the Phoenicians, who inherited maritime supremacy in the Mediterranean after the Cretan civilizations were destroyed by earthquakes and political instability around 1200 B.C. Skilled sailors, the Phoenicians carried their wares through the Straits of Gibraltar to markets as distant as Britain and the west coast of Africa. Given the simple ships they used, this was quite an achievement.

The Greeks began to explore outside the Mediterranean into the Atlantic Ocean around 900–700 B.C. (**Figure 2.1**). Early Greek seafarers noticed a current running from north to south beyond Gibraltar. Believing only rivers had currents, they decided that this great mass of water, too wide to see across, was part of an immense flowing river. The Greek name for this river was *okeanos*. Our word *ocean* is derived from **oceanus,** a Latin variant of that root. Phoenician sailors were also very much at home in this "river," but like the Greeks, they rarely ventured out of sight of land.

As they went about their business, early mariners began to record information to make their voyages easier and safer—the locations of rocks in a harbor, landmarks and the sailing times between them, the direction of currents. These first **cartographers** (chart makers) were probably Mediterranean traders who made routine journeys from producing areas to markets. Their first charts (drawn about 800 B.C.) were drawn merely to jog their memory for obvious features along the route. Today's **charts** are graphic representations that depict primarily water and water-related information.[1] (*Maps* primarily represent land.)

In this early time other cultures also traveled on the ocean. The Chinese began to engineer an extensive system of inland waterways, some of which connected with the Pacific Ocean to make long-distance transport of goods more convenient. The Polynesian peoples had been moving easily among islands off the coasts of Southeast Asia and Indonesia since 3000 B.C. and were beginning to settle the mid-Pacific islands. Though none of these civilizations had contact with the others, each developed methods of charting and navigation. All these early travelers were skilled at telling direction by the stars and by the position of the rising or setting sun.

Figure 2.1 A Greek ship from about 500 B.C. Such ships were used for trade to explore the Atlantic outside the Mediterranean.

© Taschen Verlag GMBH / De Espona Infografica SL

Curiosity and commerce encouraged adventurous people to undertake ever more ambitious voyages. But these voyages were possible only with the coordination of astronomical direction finding and knowledge of the shape and size of Earth, advanced shipbuilding technology, accurate graphic charts (not just written descriptions), and, perhaps most important, a growing understanding of the ocean itself. Marine science, the organized study of the ocean, had its origin in the technical studies of voyagers.

The origins of marine science lie in voyaging—traveling on the ocean for a purpose. Technical advancements made during voyaging and later marine exploration led to the rise of scientific oceanography.

Science for Voyaging

Progress in applied marine science began at the **Library of Alexandria** in Egypt. Founded in the third century B.C. by Alexander the Great, the library constituted history's greatest accumulation of ancient writings and could be considered the first university in the world. Written knowledge of all kinds—the characteristics of nations, trade, natural wonders, artistic achievements, tourist sights, investment opportunities, and other items of interest to seafarers—was warehoused around its leafy courtyards. Traders quickly realized the competitive benefit of this information, and librarians welcomed their interest in return for even more information. Here perhaps was the first instance of cooperation between a university and the commercial community, a partnership that has paid dividends for science and business ever since.

The second librarian at Alexandria (from 235 to 192 B.C.) was the Greek astronomer, philosopher, and poet **Eratosthenes of Cyrene.** This remarkable man was the first to calculate the circumference of Earth. The Greek Pythagoreans had realized Earth was spherical by the sixth century B.C., but Eratosthenes was the first to estimate its true size.

Eratosthenes had heard from travelers returning from Syene (now Aswan, site of the great Nile dam) that at noon on the longest day of the year, the sun shone directly onto the waters of a deep vertical well. In Alexandria, he noticed that a vertical pole cast a slight shadow on that day. He measured the shadow angle and found it to be a bit more than 7°, about one-fiftieth of a circle. He correctly assumed that the sun is a great distance from Earth, so the sun's rays would approach Syene and Alexandria in parallel lines. If the sun were directly overhead at Syene but not directly overhead at Alexandria, then the surface of Earth must be curved. But what was the *circumference* of Earth?

---

[1] More on charts and maps may be found in Appendix IV.

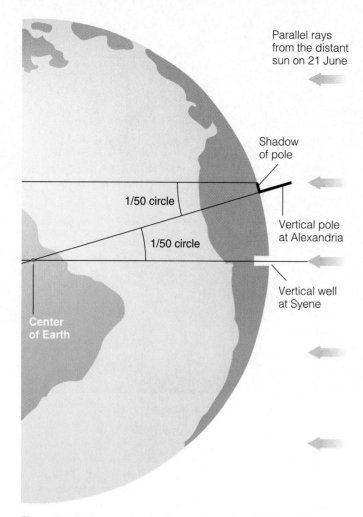

Parallel rays from the distant sun on 21 June

Shadow of pole

1/50 circle

1/50 circle

Vertical pole at Alexandria

Vertical well at Syene

Center of Earth

Figure 2.2 A diagram showing Erastosthenes' method for calculating the circumference of Earth. As described in the text, he used simple geometric reasoning based on the assumptions that Earth is spherical and the sun is very far away. Using this method, he was able to discover the circumference of Earth to within about 8% of its true value. This knowledge was available more than 1,700 years before Columbus began his voyages. (The diagram is not drawn to scale.)

By studying the reports of camel caravan traders, he estimated the distance from Alexandria to Syene at about 785 kilometers (491 miles). Eratosthenes now had the two pieces of information needed to derive the circumference of Earth by geometry. **Figure 2.2** shows his solution. The size of the units of length (stadia) Eratosthenes used is thought to have been 555 meters (607 yards), and historians estimate his calculation, made in about 230 B.C., was accurate to within about 8% of the true value. Within a few hundred years most people in the West who had contact with the library or its scholars knew Earth's approximate size.

Cartography flourished. The first workable charts that represented a spherical surface on a flat sheet were developed by Alexandrian scholars. Latitude and longitude, systems of imaginary lines dividing the surface of Earth, were invented by Eratosthenes. **Latitude** lines were drawn parallel to the equator, and **longitude** lines ran from pole to pole.

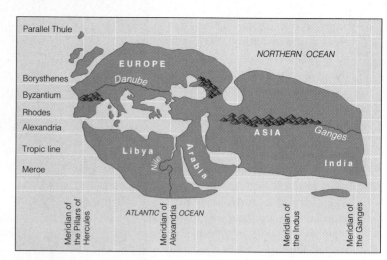

Figure 2.3 The world, according to a chart from the third century B.C. Eratosthenes drew latitude and longitude lines through important places rather than spacing them at regular intervals as we do today. The Alexandrian perception of the world is reflected in the size of the continents and the central position of Alexandria at the mouth of the Nile.

He placed the lines through prominent landmarks and important places to create a convenient, though irregular, grid (**Figure 2.3**). Our present regular grid of latitude and longitude was invented by Hipparchus (c. 165–127 B.C.), a librarian who divided the surface of Earth into 360°.[2] A later Egyptian-Greek, Claudius Ptolemy (A.D. 90–168), "oriented" charts by placing east to the right and north at the top. Ptolemy's division of degrees into minutes and seconds of arc is still used by navigators.

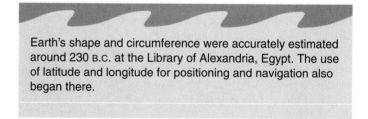

Earth's shape and circumference were accurately estimated around 230 B.C. at the Library of Alexandria, Egypt. The use of latitude and longitude for positioning and navigation also began there.

Ptolemy also attempted to improve on Eratosthenes' surprisingly accurate estimate of Earth's circumference, but he wrongly calculated 1° as about 50 miles instead of the more correct 70 miles. This error, coupled with his mistake of overestimating the size of Asia, greatly reduced the apparent width of the unknown part of the world between the Orient and Europe. Unfortunately for generations of navigators, Eratosthenes' estimate was forgotten while Ptolemy's persisted.

Though it weathered the dissolution of Alexander's empire, the Alexandrian library did not survive the subsequent period of Roman rule. The last librarian was Hypatia, the first notable woman mathematician, philosopher, and scientist. In Alexandria she was a symbol of science and knowledge, concepts the early Christians identified with

---

[2] For more information on latitude and longitude, please see Appendix III.

pagan practices. The mission of the library, as personified by the last librarian, antagonized the governors and citizens of the city of Alexandria. After years of rising tensions, in A.D. 415 a mob brutally murdered Hypatia and burned the library with all its contents. Most of the community of scholars dispersed, and Alexandria ceased to be a center of learning in the ancient world. The academic loss was incalculable, and trade suffered because shipowners no longer had a clearing-house for updating the nautical charts and information upon which they had come to depend. All that remains of the library today is a remnant of an underground storage room. We shall never know the true extent and influence of its collection of more than 700,000 irreplaceable scrolls.

Western intellectual development slackened during the so-called Dark Ages that followed the fall of the Roman Empire in A.D. 476. For almost a thousand years, until the European Renaissance, much of the progress in medicine, astronomy, philosophy, mathematics, and other vital fields of human endeavor was made by Arabs or imported by them from Asia. For example, Arabs used the Chinese-invented compass for navigating caravans over seas of sand. During this time the Vikings raided and explored to the south and west, and the Polynesians continued some of the most extraordinary voyages in history.

Voyages of the Oceanian Peoples

In the history of human migration, no voyaging saga is more inspiring than that of the **Polynesian** colonizations, the peopling of the central and eastern Pacific islands. A profound knowledge of the sea was required for these voyages, and the story of the Polynesians is a high point in our chronology of marine science applied to travel by sea.

The Polynesians are one of four cultures that inhabited some 10,000 islands scattered across nearly 26 million square kilometers (10 million square miles) of open Pacific Ocean (**Figure 2.4**). The Southeast Asian or Indonesian ancestors of the Oceanian peoples, as these cultures are collectively called, spread eastward in the distant past. Although experts differ in their estimates, there is some consensus that by 30,000 years ago New Guinea was populated by these wanderers, and by 20,000 years ago the Philippines were occupied. By around 850 B.C. the so-called cradle of Polynesia—Tonga, Samoa, and the Marquesas and the Society Islands—started to be settled.

For a long and evidently prosperous period, the Polynesians spread from island to island until the easily accessible islands had been colonized. Eventually, however, overpopulation and depletion of resources became a problem. Politics, intertribal tensions, and religious strife shook society. Groups of people scattered in all directions from some of the "cradle" islands during a period of explosive dispersion. Between A.D. 300 and 600, Polynesians successfully colonized nearly every inhabitable island within a vast triangular area (see Figure 2.4). Easter Island was found against prevailing winds and currents, and the remote islands of Hawaii were discovered and occupied. These were among the last places on Earth to be populated.

How did these risky voyages into unexplored territory come about? Religious warfare may have been the strongest stimulus to colonization. If the losers of a religious war were banished from the home islands under penalty of death, their only hope for survival was to reach a distant and hospitable new land. Seafaring had been a long tradition in the home islands, but these trips called for radical new technology. Great dual-hulled sailing ships, some capable of transporting up to 100 people, were designed and built (**Figure 2.5**). Whole populations left their home islands in fleets designed especially for long-distance discovery. In some cases, fire was nurtured on board in case of landfall on an island that lacked volcanic flame. But a new island was only a possibility, a dream. Their gods may have promised the voyagers safe deliverance to new lands, but how many fleets set out from the troubled homelands only to fall victim to storms, thirst, or other dangers?

Yet in that anxious time the Polynesians honed and perfected their seafaring knowledge. To a skilled navigator, a change in the rhythmic waves against the hull could indicate an island out of sight over the horizon. The flight tracks of birds at dusk could suggest the direction of land. The positions of the stars told stories, as did the distant clouds over an unseen island. The sunrise colors, sunset colors, hue of the moon—every nuance had meaning, every detail had been passed in ritual from father to son. The greatest Polynesian minds were navigators, and reaching Hawaii was their greatest achievement.

Of all islands colonized by the Polynesians, Hawaii is farthest away, across an ocean whose guide stars were completely unknown to the southern navigators. The Hawaiian Islands are isolated in the northern Pacific. There are no islands of any significance for more than 2,000 miles to the south. Moreover, Hawaii lies beyond the equatorial doldrums, a hot and often windless stretch across which these pioneers must somehow have paddled. And yet some fortunate and knowledgeable people colonized Hawaii sometime between A.D. 450 and 600. Try to imagine their feelings of relief and justification upon reaching a promised paradise under a new night sky. Think of that first approach to the high islands of Hawaii, the first unlimited drink of fresh water, the first solid Earth after months of uncertainty.

At a time when seafarers of other civilizations sailed beside the comforting bulk of a charted coast, Polynesians looked to the open sea for sustenance, deliverance, and hope. Their great knowledge of the ocean protected them.

The Age of Discovery

Half a world away from their Polynesian counterparts, Renaissance Europeans—having been jolted by internal awakening and external reality—set out to explore the world by sea. They did not undertake exploration for its own sake, however; any voyage had to have a material goal. Trade between East and West had long been dependent on

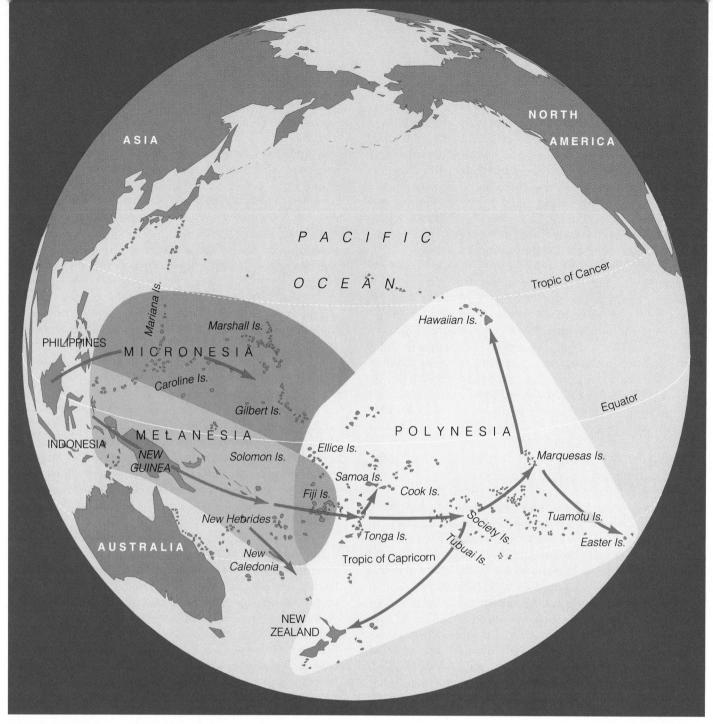

Figure 2.4 The Polynesian triangle. Ancestors of the Polynesians spread from Southeast Asia or Indonesia to New Guinea and the Philippines by about 20,000 years ago. The mid-Pacific islands have been colonized for about 2,500 years, but the explosive dispersion that led to the settlement of Hawaii occurred about A.D. 450–600. Arrows show a possible direction and order of settlement.

arduous and insecure desert caravan routes through the central Asian and Arabian deserts. This commerce was cut off in 1453 when the Turks captured Constantinople, and an alternative ocean route was needed.

A European visionary who thought ocean exploration held the key to great wealth and successful trade was **Prince Henry the Navigator,** third son of the royal family of Portugal (**Figure 2.6**). Prince Henry established a center at Sagres for the study of marine science and navigation ". . . through all the watery roads." Although he personally was not well

traveled (he went to sea only twice in his life), captains under his patronage explored from 1451 to 1470, compiling detailed charts wherever they went. Henry's explorers pushed south into the unknown and opened the west coast of Africa to commerce. He sent out small, maneuverable ships designed for voyages of discovery and manned by well-trained crews. For navigation, his mariners used the **compass**—an instrument (invented in China in the fourth century B.C.) that points to magnetic north. Although Arab traders had brought the compass from China in the twelfth century, navigators still

Figure 2.5 A Polynesian voyaging canoe similar to the ones used for long-range colonization and the discovery of Hawaii. Note the double-hulled design and the small deckhouse for shelter. The crew is seen paddling through the doldrums, a relatively windless area near the equator. Water would have been in short supply, the voyage tense and tiresome.

considered it a magical tool. They concealed the compass in a special box (predecessor of today's binnacle) and consulted it out of public view. Henry's students knew Earth was round, but because of the errors of Claudius Ptolemy they were wrong in their estimation of its size.

A master mariner (and aggressive salesman), **Christopher Columbus** "discovered" the New World quite by acci-

Figure 2.6 Prince Henry of Portugal, the Navigator. In the mid-1400s, Henry established a center at Sagres for the study of marine science and navigation ". . . through all the watery roads."

dent. Native Americans had been living on the continent for about 11,000 years, and the Norwegian Vikings had made about two dozen visits to a functioning colony on the continent 500 years before his noisy arrival, yet Columbus gets the credit. Why? Because his interesting souvenirs, exaggerated stories, inaccurate charts, and promises of vast wealth excited the imagination of royal courts. Columbus made North America a media event without ever sighting it!

Columbus wasn't trying to discover new lands. His intention was to pioneer a sea route to the rich and fabled lands of the East, made famous more than 200 years earlier in the overland travels of Marco Polo. As "Admiral of the Ocean Sea," Columbus was to have a financial interest in the trade routes he blazed. He was familiar with Prince Henry's work and, like all other competent contemporary navigators, knew Earth was spherical. By sailing west, he believed he could come close to his eastern destination, the latitude of which he thought he knew. Because he depended on Ptolemy's data, however, Columbus made the *smallest* estimate of Earth's size by any navigator in modern history; he assumed Earth to be only about half its actual size.

Not surprisingly, Columbus mistook the New World for his goal of India or Japan. He thought that the notable absence of wealthy cities and well-dressed inhabitants resulted from striking the coast too far north or south of his

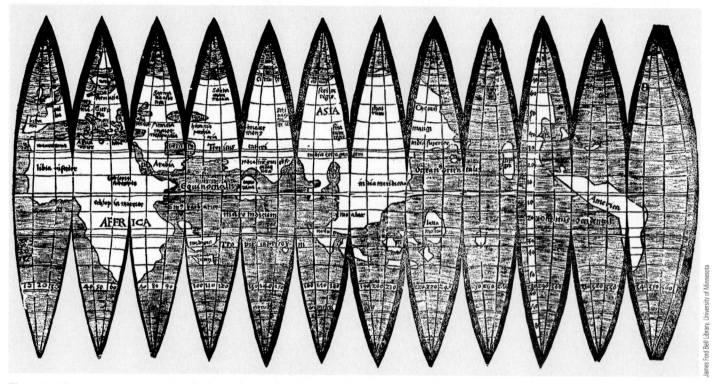

Figure 2.7 The Waldseemüller map, published in 1507—the first map to name America and to show the New World as separate from Asia. The deep gores are designed to form a globe about 10 centimeters (4 inches) in diameter.

desired latitude. He made three more trips to the New World but died still believing that he had found islands off the coast of Asia. He never saw the mainland of North America and never realized the size and configuration of the continents whose future he had so profoundly changed.

Christopher Columbus did not discover North America, and he did not sail around the world.

Other explorers quickly followed, and Columbus's error was soon rectified. Charts drawn as early as 1507 included the New World (**Figure 2.7**). Such charts perhaps inspired **Ferdinand Magellan** (**Figure 2.8**), a Portuguese navigator in the service of Spain, to believe that he could open a westerly trade route to the Orient. Unfortunately, the chart makers estimated that the Americas and the Pacific Ocean were much smaller than they actually are. In the Philippines Magellan was killed, and his men decided to continue sailing west around the world. Only 34 of the original 260 crew survived, returning to Spain three years after they had set out. But they had proved it was possible to circumnavigate the globe.

The Magellan expedition's return to Spain in 1522 marks the end of the Age of European Discovery. An unpleasant era of exploitation of the human and natural resources of the Americas followed. Native empires were destroyed, and

Figure 2.8 Ferdinand Magellan, a Portuguese explorer in service to Spain whose expedition was first to circumnavigate the world. Magellan himself did not survive the voyage; only 34 out of 260 sailors managed to return after three years of arduous voyaging.

objects of priceless religious and archaeological value were melted into coin to fund European warfare and intrigue.

Voyaging for Science

British sea power arose after the Age of Discovery to compete with the colonial aspirations of France and Spain. Sailing ships require dependable supply and repair stations, especially in remote areas. The great powers sent out expeditions to claim appropriate locations, preferably inhabited by friendly natives eager to help provision ships half a globe from home. The French sent Admiral de Bougainville into the South Pacific in the mid-1760s. His 1768 claim for France of what is now called French Polynesia opened the area to the powerful European nations. The British followed immediately.

James Cook

Scientific oceanography begins with the departure from Plymouth Harbor in 1768 of HMS *Endeavour* under command of **James Cook** of the British Royal Navy (**Figure 2.9**). An intelligent and patient leader, Cook was also a skillful navigator, cartographer, writer, artist, diplomat, sailor, scientist, and dietitian. The primary reason for the voyage was to assert the British presence in the South Seas, but the expedition had numerous scientific goals as well. First, Cook conveyed several members of the Royal Society (a scientific research group) to Tahiti to observe the transit of Venus across the disk of the sun. Their measurements verified calculations of planetary orbits made earlier by Edmund Halley (later of comet fame) and others. Then, Cook turned south into unknown territory to search for a hypothetical southern continent, which some philosophers believed had to exist to balance the landmass of the Northern Hemisphere. Using the newly invented chronometer to calculate their longitude, Cook and his men found and charted New Zealand, mapped Australia's Great Barrier Reef, marked the positions of numerous small islands, made notes on the natural history and human habitation of these distant places, and initiated friendly relations with many chiefs. Cook survived an epidemic of dysentery contracted by the ship's company while ashore in Batavia (Djakarta) and sailed home to England around the world in 1771. Because of his insistence on cleanliness and ventilation, and because his provisions included cress, sauerkraut, and citrus extracts, his sailors avoided scurvy—a vitamin-C deficiency disease that for centuries had decimated crews on long voyages.

The Admiralty was deeply impressed. Cook was promoted to the rank of commander and in 1772 was given command of the ships *Resolution* and *Adventure*, in which he embarked on one of the great voyages in scientific history. On this second voyage, he charted Tonga and Easter Island, discovered New Caledonia in the Pacific and South Georgia in the Atlantic. He was first to circumnavigate the world at high latitudes. Though he sailed to 71°S latitude, he never sighted Antarctica. He returned home again in 1775.

© National Maritime Museum, Greenwich

Figure 2.9 Captain James Cook, Royal Navy, painted in 1776 by Nathaniel Dance, shortly before embarking on his third, and fatal, voyage. Cook is a fully matured, self-confident captain who has twice circled the globe, penetrated into the Antarctic, and charted coastlines from Newfoundland to New Zealand.

Posted to the rank of captain, Cook set off in 1776 on his third, and last, expedition in *Resolution* and *Discovery*. His commission was to find a northwest passage around Canada and Alaska, or a northeast passage above Siberia. He "discovered" the Hawaiian Islands (Hawaiians were there to greet him, of course, as shown in **Figure 2.10**) and charted the west coast of North America. After searching unsuccessfully for a passage across the top of the world, Cook retraced his steps to Hawaii to provision for departure home. On 14 February 1779, after an elaborate farewell dinner with the chief of the island of Hawaii, Cook and his officers prepared to return to *Resolution* anchored in Kealakekua Bay. The Englishmen somehow angered the Hawaiians and were beset by the crowd. Cook, among others, was killed in the fracas.

Cook deserves to be considered a scientist as well as an explorer because of the accuracy, thoroughness, and completeness of his descriptions. He and the scientists aboard took samples of marine life, land plants and animals, the ocean floor, and geological formations; they also reported their characteristics in their logbooks and journals. Cook's navigation was outstanding, and his charts of the Pacific were accurate enough to be used by the Allies in World War II invasions of the Pacific islands. He drew accurate conclusions, did not exaggerate his findings, and opened friendly diplomatic relations with many native populations. Cook

Figure 2.10 First contact! Captain James Cook, commanding HMS *Resolution* off the Hawaiian island of Kaua'i, wrote in 1778: "It required but little address to get them to come along side, but we could not prevail upon any one to come on board; they exchanged a few fish they had in the canoes for anything we offered them, but valued nails, or iron above every other thing; the only weapons they had were a few stones in some of the canoes and these they threw overboard when they found they were not wanted."

© 1991 Herbert Kawainui Kane

recorded and successfully interpreted events in natural history, anthropology, and oceanography. Unlike most captains of his day he cared for his men. He was a thoughtful and clear writer. This first marine scientist peacefully changed the map of the world more than any explorer or scientist in history.

Captain James Cook, Royal Navy, was perhaps the first ocean observer careful enough to be considered a marine scientist.

The Sampling Problem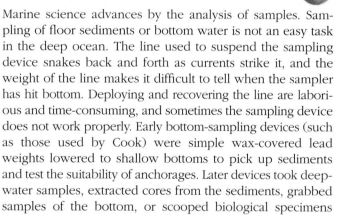

Marine science advances by the analysis of samples. Sampling of floor sediments or bottom water is not an easy task in the deep ocean. The line used to suspend the sampling device snakes back and forth as currents strike it, and the weight of the line makes it difficult to tell when the sampler has hit bottom. Deploying and recovering the line are laborious and time-consuming, and sometimes the sampling device does not work properly. Early bottom-sampling devices (such as those used by Cook) were simple wax-covered lead weights lowered to shallow bottoms to pick up sediments and test the suitability of anchorages. Later devices took deep-water samples, extracted cores from the sediments, grabbed samples of the bottom, or scooped biological specimens from the ocean floor.

The first researchers to attack the deep sampling problem successfully were British explorers Sir John Ross and his nephew Sir James Clark Ross. During an expedition to scout the Northwest Passage in 1818, Sir John Ross obtained a bot-

tom sample from 1,919 meters (3,296 feet) near Greenland by using a clamping sampler to trap the specimen. Sir James Clark Ross, discoverer of the Ross Sea and the area of Antarctica known as Victoria Land, obtained a depth **sounding** (depth measurement) of 4,893 meters (16,054 feet) in the South Atlantic.

Even today, in spite of modern advances, deep sampling remains difficult. As we will see in the next chapters, a new generation of expensive manned and remotely operated vehicles now work at great depths to return samples and pictures to the surface.

Scientific Expeditions

Great as Cook's contributions undoubtedly were, his three voyages (and those of the Rosses) were not purely scientific expeditions. These men were British naval officers engaged in Crown business, concerned with charting, "foreign relations," and natural phenomena as they applied to Royal Navy matters. The first genuine only-for-science expedition may well have been the *Challenger* expedition of 1872–1876, but the United States got into the act first with a hybrid expedition in 1838.

The United States Exploring Expedition

After a ten-year argument over its potential merits, the **United States Exploring Expedition** was launched in 1838. It was primarily a naval expedition, but its captain was somewhat more free in maneuvering orders than Cook had been. The work of the scientists aboard the flagship *Vincennes* and the

Figure 2.11 Lieutenant Charles Wilkes soon after his return from the United States Exploring Expedition in 1842.

Figure 2.12 Matthew Fontaine Maury, compiler of winds and currents. Maury was perhaps the first person for whom oceanography was a full-time occupation. This photograph was probably taken in 1853.

expedition's five other vessels helped to establish the natural sciences as reputable professions in America. Had it not been for the combative and disagreeable personality of its leader, Lieutenant Charles Wilkes (**Figure 2.11**), this expedition might have become as famous as those of Cook or the later *Challenger* voyage.

The goals of the four-year circumnavigation included showing the flag, whale scouting, mineral gathering, charting, observing, and pure exploration. One unusual objective was to disprove a peculiar theory that Earth was hollow and could be entered through huge holes at either pole.

Wilkes's team explored and charted a large sector of the east Antarctic coast and made observations that confirmed the landmass as a continent. A map of the Oregon Territory produced in 1841, one of 241 maps and charts drawn by members of the expedition, proved especially valuable when connected to the map of the Rocky Mountains prepared the following year by Captain John C. Fremont. Hawaii was thoroughly explored, and Wilkes led an ascent of Mauna Loa, one of the two peaks of Hawaii's tallest volcano. James Dwight Dana, the expedition's brilliant geologist, confirmed Charles Darwin's hypothesis of coral atoll formation (about which more will be found in Chapter 14). The expedition returned with many scientific specimens and artifacts, which formed the nucleus of the collection of the newly established Smithsonian Institution in Washington, D.C. No evidence of polar holes was found!

Upon their return in 1842, Wilkes and his "scientifics" prepared a final report totaling 19 volumes of maps, text, and illustrations. The report is a landmark in the history of American scientific achievement.

Matthew Maury

At about the time the Wilkes expedition returned, **Matthew Maury** (**Figure 2.12**), a Virginian and U.S. naval officer, was appointed director of the Navy's Bureau of Charts and Instruments. There he studied a huge and neglected treasure trove of ships' logs with their many regular readings of temperature and wind direction. By 1847 Maury had assembled much of this information into coherent wind and current charts. Maury began to issue these charts free to mariners in exchange for logs of their own new voyages.

Slowly a picture of planetary winds and currents began to emerge. Maury himself was a compiler, not a scientist, and he was vitally interested in the promotion of maritime commerce. His understanding of currents built on the work of Benjamin Franklin who, nearly a hundred years earlier, had noticed the peculiar fact that the fastest ships were not always the fastest ships; that is, hull speed did not always correlate with out-and-return time on the European run. Franklin's cousin, a Nantucket merchant named Tim Folger, noted Franklin's puzzlement and provided him with a rough chart of the "Gulph Stream" that he (Folger) had worked out. By

Figure 2.13 Benjamin Franklin's 1769 chart of the Gulf Stream system. His cousin, Timothy Folger, discovered that Yankee whalers had learned to use the Gulf Stream to their advantage. Others, especially English shipowners, were slower to learn. Folger, himself a sea captain, wrote that Nantucket whalers ". . . in crossing it have sometimes met and spoke with those packets who were in the middle of and stemming it. We have informed them that they were stemming a current that was against them to the value of three miles an hour and advised them to cross it, but they were too wise to be counseled by simple American fishermen."

Library of Congress

staying within the stream on the outbound leg and adding its speed to their own, and by avoiding it on their return, captains could traverse the Atlantic much more quickly. It was Franklin who published, in 1769, the first chart of any current (**Figure 2.13**).

But Maury was the first person to sense the worldwide pattern of surface winds and currents. His work became famous in 1849 when the California gold rush made his sailing directions essential for fast trips around Cape Horn. His crowning achievement, *The Physical Geography of the Seas*, a book explaining his discoveries, was published in 1855. Maury, considered by many to be the father of physical oceanography, was perhaps the first man to undertake the systematic study of the ocean as a full-time occupation.

The Challenger *Expedition*

The first sailing expedition devoted completely to marine science was conceived by a professor of natural history at Scotland's University of Edinburgh, Charles Wyville Thom-

son, and his Canadian-born student, John Murray. Stimulated by their own curiosity and by Charles Darwin's 1831–1836 voyage in HMS *Beagle*, they convinced the Royal Society and the British government to provide a Royal Navy ship and trained crew for a "prolonged and arduous voyage of exploration across the oceans of the world." Thomson and Murray even coined a word for their enterprise: **oceanography.** Though the term implies only marking or charting, it has come to mean the science of the ocean. The government and the Royal Society agreed to the endeavor provided a proportion of any financial gain from discoveries was handed over to the Crown. This arranged, the scientists made their plans.

HMS *Challenger*, a 2,306-ton steam corvette (**Figure 2.14**), set sail on 26 May 1872 on a four-year voyage that took them around the world and covered 127,600 kilometers (79,300 nautical miles). Although the captain was a Royal Navy officer, the six-man scientific staff directed the course of the voyage. *Challenger*'s track is shown in **Figure 2.15.**

One important mission of the *Challenger* **expedition** was to investigate Edinburgh Professor Edward Forbes's

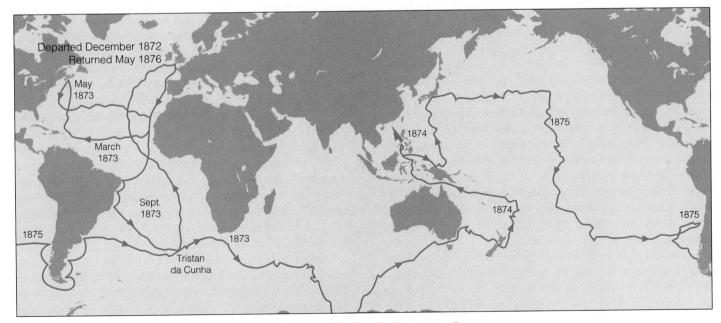

Figure 2.14 Lieutenant Pelham Aldrich, first lieutenant of HMS *Challenger*, kept a detailed journal of the *Challenger* expedition. With accuracy and humor he kept this record in good weather and bad, and he had the patience and skill to include watercolors of the most exciting events. This is part of the first page of his journal.

The Royal Geographical Society, London

Departed December 1872
Returned May 1876

May 1873

March 1873

Sept. 1873

1875

1873

Tristan da Cunha

1874

1874

1875

1875

Figure 2.15 HMS *Challenger*'s track from December 1872 to May 1876. The *Challenger* expedition remains the longest continuous oceanographic survey on record.

http://info.brookscole.com/garrison

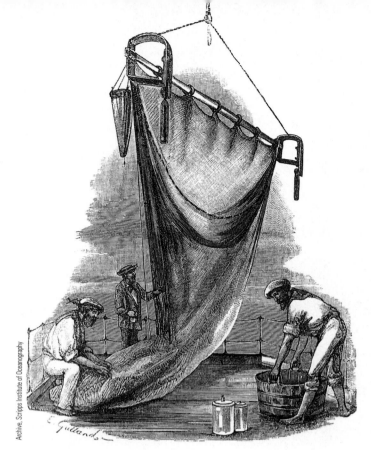

Figure 2.16 Emptying a trawl net, an engraving from the *Challenger Report*.

contention that life below 549 meters (1,800 feet) was impossible because of high pressure and lack of light. The steam winch on board made deep sampling practical, and samples from depths as great as 8,185 meters (26,850 feet) were collected off the Philippines. Through the course of 492 deep soundings with mechanical grabs and nets at 362 stations (including 133 dredgings), Forbes was proven resoundingly wrong. With each hoist, animals new to science were strewn on the deck; in all, staff biologists discovered 4,717 new species! **Figure 2.16** shows one of the large trawl nets used in making some of these discoveries.

The scientists also took salinity, temperature, and water density measurements during these soundings. Each reading contributed to a growing picture of the physical structure of the deep ocean. The crew completed at least 151 open water trawls and stored 77 samples of seawater for detailed analysis ashore. The expedition collected new information on ocean currents, meteorology, and the distribution of sediments; the locations and profiles of coral reefs were charted. Thousands of pounds of specimens were brought to British museums for study. Manganese nodules, brown lumps of mineral-rich sediments, were discovered on the seabed, sparking interest in deep-sea mining.

This first pure oceanographic investigation was an unqualified success. The discovery of life in the depths of the oceans stimulated the new science of marine biology. The scope, accuracy, thoroughness, and attractive presentation of the researchers' written reports made this expedition a high point in scientific publication. The *Challenger Report*,

the record of the expedition, was published between 1880 and 1895 by Sir John Murray in a well-written and magnificently illustrated 50-volume set. It is still used today. Indeed, it was the *Report*, rather than the voyage, that provided the foundation for the new science of oceanography. The expedition's many financial spin-offs indicated that pure research was a good investment, and the British government realized quick profits from the exploitation of newly discovered mineral deposits on islands. The *Challenger* expedition remains history's longest continuous scientific oceangraphic expedition.

The *Challenger* expedition was the first purely scientific voyage of oceanic exploration.

With successes like these, the pace of exploration accelerated. American naturalist Alexander Agassiz, sailing in 1877 on the U.S. Coast and Geodetic Survey ship *Blake*, collected data corroborating the *Challenger* material at 355 deep-sea stations. The distribution of manganese nodules was found to be widespread. Further work by Agassiz and his students on the survey ship *Albatross* helped train a generation of influential American marine biologists. In 1886 the Russians entered the field of marine exploration with the three-year cruise of *Vitiaz* under the leadership of S. O. Makarov; their main contribution was a careful analysis of the salinity and temperature of North Pacific water.

Modern Oceanography

In the twentieth century, oceanographic voyages became more technically ambitious and expensive. Scientist–explorers sought out and investigated places that once had been too difficult to reach. New electronic and optical devices aided navigation and sampling. In the last third of the century, high-speed shipboard computers made it possible for marine scientists to analyze data while still at sea.

In 1925 the German ***Meteor* expedition,** which crisscrossed the South Atlantic for two years, introduced modern optical and electronic equipment to oceanographic investigation. Its most important innovation was to use an **echo sounder,** a device that bounces sound waves off the ocean bottom, to study the depth and contour of the seafloor (**Figure 2.17**). The echo sounder revealed to *Meteor* scientists a varied and often extremely rugged bottom profile rather than the flat floor they had anticipated.

Atlantis, launched in 1931, was the first U.S. research ship built specifically for ocean studies. Investigations by her scientists confirmed Matthew Maury's findings of a mid-Atlantic ridge and helped to discover its extent. The 32-meter (104-foot) schooner *E. W. Scripps*, under the direction of Harald Sverdrup, began a wide-ranging program of chemi-

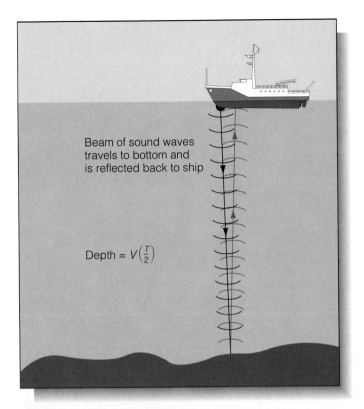

Figure 2.17 Echo sounders sense the contour of the seafloor by beaming sound waves to the bottom and measuring the time required for the sound waves to bounce back to the ship. If the round-trip travel time and wave velocity are known, then the distance to the bottom can be calculated. This technique was first used on a large scale by the German research vessel *Meteor* in the 1920s.

Beam of sound waves travels to bottom and is reflected back to ship

$$\text{Depth} = V\left(\frac{T}{2}\right)$$

Deep Sea Drilling Program, Texas A&M

Figure 2.18 *Glomar Challenger*, a research ship conceived and built in the early 1960s by an international consortium of oceanographic institutions and the U.S. National Science Foundation as part of the Deep Sea Drilling Project. Its goal: to test the then-radical hypothesis that continents are moved across Earth's surface by seafloor spreading. Operating from 1968 to 1983, the 122-meter (400-foot) ship used computers to maintain her position with the precision needed to complete the drilling of cores up to a mile long. *Glomar Challenger* was replaced in 1985 by *JOIDES Resolution* (see Figure 5.14).

cal, biological, and geophysical exploration off the coast of southern California in 1937. These voyages led to publication in 1942 of *The Oceans*, the first modern reference work on all phases of marine science.

In October 1951 a new HMS *Challenger* began a two-year voyage that would take precise depth measurements in the Atlantic, Pacific, and Indian Oceans and in the Mediterranean Sea. With echo sounders, measurements that would have taken the crew of the first *Challenger* nearly four hours to complete could be made in seconds. *Challenger II*'s scientists discovered the deepest part of the ocean's deepest trench, naming it Challenger Deep in honor of their famous predecessor. In 1960 U.S. Navy Lieutenant Don Walsh and Jacques Piccard descended into the Challenger Deep in *Trieste*, a Swiss-designed, blimplike bathyscaphe (see Chapter 4 opener).

In 1968 the drilling ship *Glomar Challenger* (**Figure 2.18**) set out to test a controversial hypothesis about the history of the ocean floor. It was capable of drilling into the ocean bottom beneath more than 6,000 meters (20,000 feet) of water and recovering samples of seafloor sediments. These long and revealing plugs of seabed provided confirming evidence for seafloor spreading and plate tectonics. (Details will be found in Chapter 3.) In 1985 deep-sea drilling duties were taken over by the much larger and more technologically advanced ship *JOIDES Resolution*. The new ship contains

equipment capable of drilling in water 8,100 meters (27,000 feet) deep, and it houses the most completely equipped geological laboratories ever put to sea (see Figure 5.14).

Polar Exploration

Polar oceanography made dramatic and dangerous advances at the turn of the century. Newly designed ships and new methods of food storage had made polar exploration possible in the last years of the nineteenth century. In 1893 **Fridtjof Nansen** began studying the north polar ocean in *Fram*, a ship designed specifically to withstand the crushing pressure of sea ice (**Figure 2.19a, b**). In the next 20 years Nansen and others probed the polar ocean depths. Researchers confirmed the feeding relationships between whales and plankton and collected much data about the whale population of the southern ocean—not out of scientific curiosity but because whales were a source of oil and baleen ("whalebone").

Modern technology has eased the burden of high-latitude travel. In 1958, under the command of Captain William Anderson, the U.S. nuclear submarine *Nautilus* sailed beneath the North Pole during a submerged transit beneath the Arctic pack from Point Barrow, Alaska, to the Norwegian Sea. As is apparent in **Figure 2.19c**, a warm, strong, and stable nuclear submarine is a nearly ideal platform for conducting oceanographic research at high latitudes.

a

b

Figure 2.19 (**a**) Fridtjof Nansen, pioneering Norwegian oceanographer and polar explorer, looking every inch the Viking. In 1908 Nansen became the first professor of oceanography, a post created for him at Christiania University. (**b**) Nansen's 123-foot schooner *Fram* ("forward"). With 13 men, *Fram* sailed on 22 June 1893 to the high Arctic with the specific purpose of being frozen into the ice. *Fram* was designed to slip up and out of the frozen ocean, and drifted with the pack ice to within about 4° of the North Pole. The whole harrowing adventure took nearly four years. The ship's 1,650-kilometer (1,025-mile) drift proved that no Arctic continent existed beneath the ice. Living conditions aboard can be sensed from this recently discovered photograph. (**c**) A very fast, immensely strong, silent nuclear submarine makes an ideal platform for oceanographic research. Scientists aboard USS *Hawkbill* prepare to take water and ice samples at the North Pole in the summer of 1998. The 1998 and 1999 expeditions aboard USS *Hawkbill* made use of a sub-bottom profiler to provide the first images of the shallow strata of the Arctic Ocean floor.

c

The Rise of Oceanographic Institutions

2.16

The demands of scientific oceanography have become greater than the capability of any single individual or voyage. Oceanographic institutions, agencies, and consortia evolved in part to ensure continuity of effort. The first of these coordinating bodies was founded by Prince Albert I of Monaco, who endowed his country's oceanographic laboratory and museum in 1906. The most famous alumnus of Albert's Musée Océanographique is Jacques Cousteau, co-inventor in 1943 of the scuba underwater breathing system. Monaco also became the site of the International Hydrographic Bureau, founded in 1921 as an association of maritime nations. This bureau published one of the first general charts of the ocean showing bottom contours.

A consortium of Japanese industries and government agencies established the Japan Marine Science and Technology Center (JAMSTEC) in 1971. In 1989 JAMSTEC launched *Shinkai 6500*, now the deepest-diving manned submersible. *Kaiko*, a remote-controlled robot sister that became fully operational in 1995, is the deepest-diving vehicle presently in service (**Figure 2.20**).

In the United States, the three preeminent oceanographic institutions are Woods Hole Oceanographic Institution on Cape Cod, founded in 1930 (and associated with the Massachusetts Institute of Technology and the neighboring Marine Biological Laboratory, founded in 1888); Scripps Institution of Oceanography, founded in La Jolla, California, and affiliated with the University of California in 1912 (**Figure 2.21**); and Lamont–Doherty Earth Observatory of Columbia

Figure 2.20 *Kaiko*, the deepest-diving vehicle presently in operation, descended to a measured depth of 10,914 meters (35,798 feet) near the bottom of the Challenger Deep on 24 March 1995. The small remotely operated vehicle (ROV) sends information back to operators in the mother ship by fiber-optic cable. *Kaiko* is operated by JAMSTEC, a Japanese marine science consortium.

a

b

Figure 2.21 (**a**) Woods Hole Oceanographic Institution, Woods Hole, Massachusetts. Marine science has been an important part of this small Cape Cod fishing community since Spencer Fullerton Baird, then assistant secretary of the Smithsonian Institution, established the U.S. Commission of Fish and Fisheries there in 1871. The Marine Biological Laboratory was founded in 1888, the Oceanographic Institution in 1930. (**b**) Scripps Institution of Oceanography, La Jolla, California. Begun in 1892 as a portable laboratory-in-a-tent, Scripps was founded by William Ritter, a biologist at the University of California, San Diego. Its first permanent buildings were erected in 1905 on a site purchased with funds donated by philanthropic newspaper owner E. W. Scripps and his sister, Ellen.

Figure 2.22 The joint U.S.–French satellite *Jason-1* orbits 1,337 kilometers (830 miles) above Earth in an orbit that allows coverage of 95% of the ice-free ocean every ten days. The satellite was launched in 2001 and is supplied with a positioning device that allows researchers to determine its position to within 2.5 centimeters (1 inch) of Earth's center. Such accuracy makes possible very accurate determination of sea-surface height by radar transmitters on board.

University, founded in 1949.[3] In 1976 these academic institutions joined with others to form a nonprofit corporation known as the Joint Oceanographic Institutions (JOI). This consortium manages the integration of facilities and specialists to undertake large research projects. JOI currently manages the Ocean Drilling Program discussed in Chapter 5.

The U.S. government has been active in oceanographic research. Within the Department of the Navy are the Office of Naval Research, the Office of the Oceanographer of the Navy, the Naval Oceanic and Atmospheric Research Laboratory, and the Naval Ocean Systems Command. These agencies are responsible for oceanographic research related to national defense. The National Oceanic and Atmospheric Administration (**NOAA**), founded within the Department of Commerce in 1970, includes the National Ocean Service, the National Weather Service, the National Marine Fisheries Service, and the Office of Sea Grant. NOAA seeks to facilitate commercial uses of the ocean.

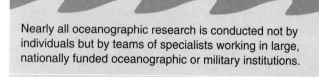

Nearly all oceanographic research is conducted not by individuals but by teams of specialists working in large, nationally funded oceanographic or military institutions.

---

[3] There are, of course, other prominent institutions involved in studying the ocean. Some thoughts on how a student might enter the field may be found in Appendix V, "Working in Marine Science."

Satellite Oceanography 2.17

The National Aeronautics and Space Administration (NASA), organized in 1958, has become an important institutional contributor to marine science. For four months in 1978, NASA's **Seasat**, the first oceanographic satellite, beamed oceanographic data to Earth. More recent contributions have been made by satellites beaming radar signals off the sea surface to determine wave height, variations in sea surface contour and temperature, and other information of interest to marine scientists.

The first of a new generation of oceanographic satellites was launched in 1992 as a joint effort of NASA and the Centre National d'Études Spatiales (the French space agency). The centerpiece of **TOPEX/Poseidon,** as the project is known, is a satellite orbiting 1,336 kilometers (835 miles) above Earth in an orbit that allows coverage of 95% of the ice-free ocean every ten days. The *TOPEX/Poseidon* satellite is supplied with a positioning device that allows researchers to determine its position to within 10 centimeters (4 inches) of Earth's center! The radars aboard can then determine the height of the sea surface with unprecedented accuracy. Other experiments in this five-year program include sensing water vapor over the ocean, determining the precise location of ocean currents, and determining wind speed and direction.

SEASTAR, launched by NASA in 1997, carries a color scanner called SeaWiFS (sea-viewing wide-field-of-view sensor). This device measures the distribution of chlorophyll at the ocean surface, a measure of marine productivity.

NASA's ambitious **Jason-1** (**Figure 2.22**), launched in December 2001, is designed to operate in tandem with

TOPEX/Poseidon. Jason-1 uses a scatterometer to measure ocean surface winds, a radiometer to sense water vapor, and an even more accurate radar altimeter to report sea-surface height.

A satellite system you can use every day? The U.S. Department of Defense has built the **Global Positioning System (GPS),** a "constellation" of 24 satellites (21 active and 3 spare) in orbit 10,600 miles above Earth. The satellites are spaced so that at least four of them are above the horizon from any point on Earth. Each satellite contains a computer, an atomic clock, and a radio transmitter. On the ground, every GPS receiver contains a computer that calculates its own position with information from at least three of the satellites. The result is provided in the form of a geographic position—longitude and latitude—that is accurate to less than 1 meter (39.37 inches), depending on the type of equipment used. The use of the GPS in marine navigation and positioning has revolutionized data collection at sea.

QUESTIONS FROM STUDENTS

1. You noted that the Chinese invented the compass and built an extensive canal system for trade. Did they make contributions to oceanic understanding?

Yes, indeed, and the story has a surprising ending. The Chinese appear to have invented the compass around A.D. 1000 and were first to use it in long-distance navigation. Around the year 1100, the Chinese began to engineer an extensive system of inland waterways, some of which connected with the Pacific Ocean to make long-distance transport of goods more convenient.

As the Dark Ages distracted Europeans, Chinese navigators became more skilled, and their vessels grew larger and more seaworthy. Then they set out to explore their world. Between 1405 and 1433, Admiral Zheng He commanded the greatest fleet the world had ever known. At least 317 ships and 37,000 men undertook seven missions to explore the Indian Ocean, Indonesia, and around the tip of Africa into the Atlantic. Their aim: to display the wealth and power of the young Ming dynasty and to "show kindness to people of distant places." The largest ship in the fleet, with nine masts and a length of 134 meters (440 feet), was a huge treasure ship carrying objects of the finest materials and craftsmanship. The mission of the fleet was not to accumulate such treasure, but to give it away! Indeed, the primary purpose of these expeditions was to convince all nations with which the fleet had contact that China was the only truly civilized state and beyond any imaginable need for knowledge or assistance!

Many technical innovations had been required to make such an ambitious undertaking possible. In addition to the compass, the Chinese invented the central rudder, watertight compartments, and sophisticated sails on multiple masts, all of which were critically important for the successful operation of large sailing vessels.

Despite having these technical advances, the Chinese intentionally abandoned ocean exploration in 1433. The political winds had changed, and the cost of the "reverse tribute" system was judged too great. In all, until recently, the Chinese made very few contributions to our understanding of the ocean. Still, their voyaging technology filtered into the West and made subsequent discoveries possible.

2. Since Columbus was unsuccessful in his attempt to sail around the world and Magellan died without completing his circumnavigation, who was the first captain to complete the trip?

Sir Francis Drake of England was the first captain to sail his own ship around the world. His expedition, begun in 1577, lasted three years. In the eastern Pacific he raided Spanish merchantmen and captured a fortune in gold, silver, coins, and precious stones. He was the first European to sight the west coast of what is now Canada, and he claimed California for Queen Elizabeth I. His transit of the Pacific lasted 68 days, and on his voyage he concluded spice trade negotiations with various heads of state; some of the agreements remain in effect to this day! On 26 September 1580 he returned to England a wealthy man, was knighted by the Queen, and went on in 1588 to help defeat the Spanish Armada.

Little was known of Drake's explorations until relatively recently. The trade and geographic information he brought home to England was considered so valuable that it was given the highest security classification—thus few people saw it or benefited from it!

3. What was James Cook's motivation for those extraordinary voyages?

One could say simply that he was a serving Royal Navy Officer and was ordered to go. But Beaglehole, Hough, and other biographers suggest the story is much more complex. How did a relatively unschooled man become leader of one of the first scientific oceanographic expeditions? Cook had the usual attributes of a successful person—intelligence, strength of character, meeting the right people at the right time, health, focus, luck—but he also had a driving intellectual curiosity and rare (for that era) tolerance and respect for alien cultures. As Hough (1994) writes, "Cook stood out like a diamond amidst junk jewelery. . . ." It was no surprise that the Lords of the Admiralty settled upon this unique man to lead the adventure.

4. How do modern navigators find their position at sea?

Very dull story. They push a few buttons on a small box and read their latitude and longitude directly on a screen. This is accomplished by analysis of radio transmissions from satellites. For about $70, you can now buy a small, handheld portable receiver capable of receiving Global Positioning System satellite signals. The GPS system is accurate to about 20 meters (66 feet) and can even tell you which direction to go to get home (or anywhere else you want to go)! The modern methods of navigation are not nearly as much fun as the old-fashioned sextant-and-chronometer method, but I suspect that any of the explorers mentioned in this chapter would be very impressed by our new tools.

CHAPTER SUMMARY

The early history of marine science is closely associated with the history of voyaging. The first marine studies had a practical aim: to facilitate travel, trade, and warfare. Later the search for new knowledge became a goal in itself. The first part of this chapter focused on *marine science for voyaging;* it looked at some of the voyagers and their voyages, the inventions that made their adventures possible, and some of the discoveries they made. The second part discussed *voyaging for marine science,* including the British *Challenger* expedition, the first wholly-for-science oceanographic research voyage. The contributions of a few of the founders of modern marine science were summarized, and the rise of oceanographic institutions and satellite oceanography was outlined.

ON-LINE STUDY RESOURCES

The Web site for this book contains a wealth of helpful study aids. Log on to:

http://info.brookscole.com/garrison

and select *Essentials of Oceanography,* 3rd edition. Select Chapter 2 from the drop-down menu or click on one of these resource areas:

- For study and review, **Chapter at a Glance** gives you an outline of the chapter, **Chapter Summary** allows you to review the chapter's main ideas, and **Glossary** lists concepts and terms for the chapter along with their definitions.

- To test your mastery of the Terms and Concepts to Remember for this chapter, you can use the electronic **Flash Cards,** play the **Concentration** game, or work the **Crossword Puzzle.**

- For practice quizzing, try the multiple-choice **Tutorial Quiz,** the ten-question **True/False Quiz,** or the **Image Analysis Quiz,** which poses questions based on art and photos from the chapter.

TERMS AND CONCEPTS TO REMEMBER

cartographers
Challenger expedition
charts
Columbus, Christopher
compass
Cook, James
echo sounder
Eratosthenes of Cyrene
Global Positioning System (GPS)
Jason-1
latitude
Library of Alexandria

longitude
Magellan, Ferdinand
Maury, Matthew
Meteor expedition
Nansen, Fridtjof
NOAA (National Oceanic and Atmospheric Administration)
oceanography
oceanus
Polynesia
Prince Henry the Navigator
SEASTAR

Seasat
sounding
TOPEX/Poseidon

United States Exploring Expedition
voyaging

STUDY QUESTIONS

1. How did the Library of Alexandria contribute to the development of marine science? What happened to most of the information accumulated there? Why do you suppose the residents of Alexandria became hostile to the librarian and the many achievements of the library?

2. What were the stimuli to Polynesian colonization? How were the long voyages accomplished?

3. Prince Henry the Navigator made only two sea voyages yet is regarded as an important figure in the history of oceanography. Why?

4. What were the main stimuli to European voyages of exploration during the Age of Discovery? Why did that era end?

5. Did Columbus discover North America? If not, who did?

6. What were the contributions of Captain James Cook? Does he deserve to be remembered more as an explorer or as a marine scientist?

7. What was the first purely scientific oceanographic expedition, and what were some of its accomplishments?

8. Who was probably the first person to undertake the systematic study of the ocean as a full-time occupation? Are his contributions considered important today?

9. What famous American is also famous for publishing the first image of an ocean current? What was his motivation for studying currents?

10. Sketch briefly the major developments in marine science since 1900. Do individuals, separate voyages, or institutions figure most prominently in this history?

11. What is an echo sounder?

12. In your opinion, where does the future of marine science lie?

RESOURCES FOR FURTHER READING AND RESEARCH

The Web site for this book contains many ideas for further reading and research. Log on to:

http://info.brookscole.com/garrison

and select *Essentials of Oceanography,* 3rd edition. Select Chapter 2 from the drop-down menu or click on one of these resource areas:

- **Additional Readings** lists the major books and articles consulted in writing this chapter, along with comments from the author about their content and reading level.

- **Hypercontents** takes you to an extensive list of current links to Internet sites with news, research, and images related to individual subjects in the chapter. Just click on the icon that corresponds to a numbered section to see the links for that subject.
- **Internet Exercises** are critical thinking questions that involve research on the Internet with starter URLs provided.
- **InfoTrac College Edition Exercises** leads you to Critical Thinking Projects that use InfoTrac College Edition as a research tool.

- **Regional InfoTrac College Edition** articles are organized into East Coast, West Coast, and Gulf Coast regions, allowing you to study oceanography on a more local level.

 For more readings, go to InfoTrac College Edition, your on-line research library, at:

http://infotrac.thomsonlearning.com

Wilbur Garrett/NGS Image Collection

Figure 3.2 Much of the Anchorage suburb of Turnagain Heights was destroyed by landslides and ground liquefaction in the 27 March 1964 Alaska earthquake. The earthquake's magnitude was 9.2—the highest ever recorded in the United States. Seismic waves associated with the earthquake passed through Earth and emerged at distant locations, carrying information about Earth's interior.

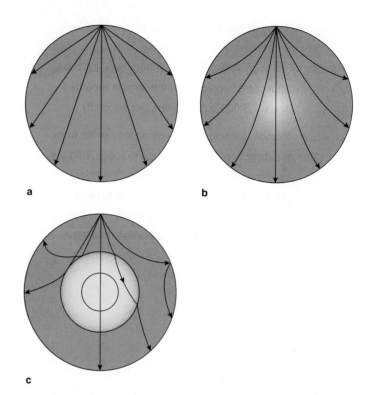

a   b

c

Figure 3.3 Possible paths of seismic waves through Earth. (**a**) If Earth were uniform (homogeneous) throughout, seismic waves would radiate out from the site of an earthquake in straight lines. (**b**) If the density, or rigidity, of Earth increased evenly with depth, seismic wave velocity would increase with depth, and the waves would bend smoothly upward toward the surface. (**c**) If Earth were layered inside, some seismic waves would be reflected at the boundaries between layers while others were bent. Seismic evidence shows that Earth is layered.

violence of the earthquake, it's a wonder that only 115 lives were lost.

Geological research stations over much of the world saw extraordinarily large seismic waves arrive from Alaska. Though they passed through Earth, the waves were so powerful that many **seismographs**—instruments that sense and record earthquakes—were physically damaged.

You might think that seismic waves would travel at constant speeds from an earthquake and that their paths through the interior would be straight lines (as in **Figure 3.3a**). This would be true if Earth were of uniform (homogeneous) composition. In fact, the speed of seismic waves increases with depth because the density of the materials they encounter increases in Earth's interior. Waves descending at shallow angles tend to bend up toward the surface as they travel into regions of faster velocity (**Figure 3.3b**).[1] The sensitive seismographs in service in the 1960s enabled researchers to study the ways seismic waves bounce off abrupt transitions between Earth's inner layers (**Figure 3.3c**). Technological advances have continued, and recent computer-assisted analysis of seismic waves has contributed to an understanding of the composition, thickness, and structure of the layers inside Earth. The shaken citizens of Anchorage could not have known at the time, but "their" earthquake helped to solidify our present model of Earth's interior.

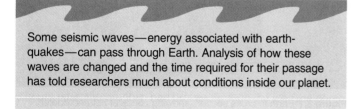

Some seismic waves—energy associated with earthquakes—can pass through Earth. Analysis of how these waves are changed and the time required for their passage has told researchers much about conditions inside our planet.

---

[1] This phenomenon, known as *refraction,* is explained in Figure 6.11.

A Layered Earth

Each layer inside Earth has different chemical and physical characteristics. One classification of Earth's interior emphasizes chemical composition. The uppermost layer is the lightweight, brittle, aptly named **crust.** The crust beneath the ocean differs in thickness, composition, and age from the crust of the continents. The thin **oceanic crust** is primarily **basalt,** a heavy, dark-colored rock composed mostly of oxygen, silicon, magnesium, and iron. By contrast, the most common material in the thicker **continental crust** is **granite,** a familiar speckled rock composed mainly of oxygen, silicon, and aluminum. The **mantle,** the layer beneath the crust, is thought to consist mainly of oxygen, iron, magnesium, and silicon. Most of Earth is mantle, which accounts for 68% of Earth's mass and 83% of its volume. The outer and inner **cores,** which consist mainly of iron and nickel, lie beneath the mantle at Earth's center.

Chemical makeup is not the only important distinction between layers. Different conditions of temperature and pressure occur at different depths, and these conditions influence the physical properties of the materials. The behavior of a rock is determined by three factors: temperature, pressure, and the rate at which a deforming force (stress) is applied. Geologists have therefore devised another classification of Earth's interior based on *physical* rather than *chemical* properties:

- The **lithosphere** (*lithos* = rock) is Earth's cool, rigid outer layer, 100–200 kilometers (60–125 miles) thick. It is composed of the continental and oceanic crusts *and* the uppermost cool and rigid portion of the mantle.

- The **asthenosphere** (*asthenes* = weak) is the hot, partially melted, slowly flowing layer of upper mantle below the lithosphere extending to a depth of 360–650 kilometers (220–400 miles).

- The **mantle** extends to the core. The asthenosphere and the mantle below the asthenosphere (the **lower mantle**) have a similar chemical composition. Although it is hotter, the mantle below the asthenosphere does not melt because of rapidly increasing pressure. As a result it is more dense and flows much more slowly.

- The **core** has two parts. The outer core is a dense, viscous liquid. The inner core is a solid with a maximum density of about 16 grams per cubic centimeter (g/cm^3), nearly six times the density of granite rock. Both parts are extremely hot, with an average temperature of about 5,500°C (9,900°F). Recent evidence indicates that the inner core may be as hot as 6,600°C (12,000°F) at its center, hotter than the surface of the sun!

Figure 3.4 shows the lithosphere and asthenosphere in detail. *Note that the rigid sandwich of crust and upper mantle—the lithosphere—floats on (and is supported by) the*

Figure 3.4 A cross section through Earth showing the internal layers. Note in the expanded section the relationship between lithosphere and asthenosphere and between crust and mantle. This representation is not to scale.

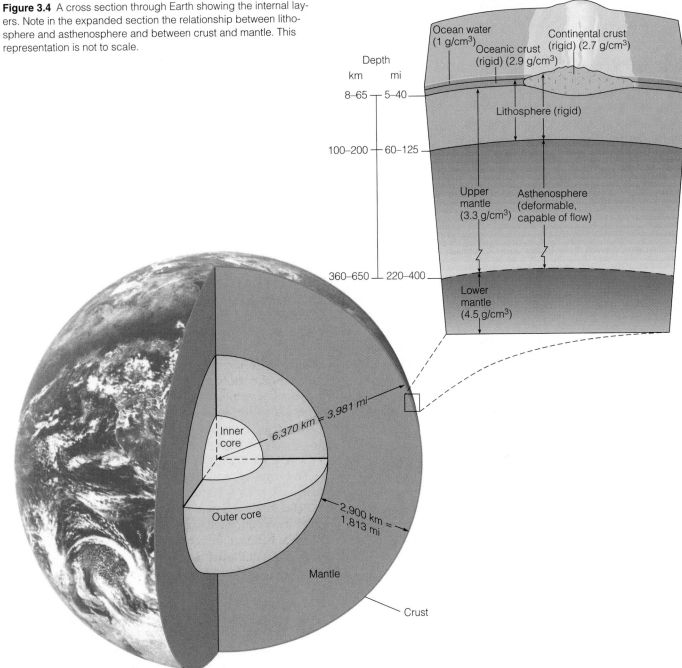

denser deformable asthenosphere. Note also that the structure of oceanic lithosphere differs from that of continental lithosphere. Because the thick granitic continental crust is not exceptionally dense, it can project above sea level. In contrast, the thin, dense, basaltic oceanic crust is almost always submerged. Research has shown that slabs of Earth's relatively cool and solid surface—its lithosphere—float and move independently of one another over the hotter, partially molten asthenosphere layer directly below.

Earth is composed of concentric spherical layers, with the least dense layer on the outside and the most dense as the core. The lithosphere, the outermost solid shell that includes the crust, floats on the hot, deformable asthenosphere.

Internal Heat

The interior of Earth is hot. The main source of that heat is **radioactive decay,** a process that generates heat when unstable forms of elements are transformed into new elements. As you read in Chapter 1, radioactive decay within the newly formed Earth released heat that contributed to the melting of the original mass. Most of the melted iron sank toward the core, releasing huge amounts of energy. By now almost all of the heat generated by the formation of the core has dissipated, but radioactive elements within Earth's core and mantle continue to decay and produce new heat. Today most of the radioactive heating takes place in the crust and upper mantle rather than in the deeper layers. Some of this heat journeys toward the surface by **conduction,** a process analogous to the slow migration of heat along a skillet's handle. Some heat also rises by **convection** in the asthenosphere. Convection occurs when a fluid is heated, expands and becomes less dense, and rises. (It is convection that causes air to rise over a warm radiator, as in Figure 7.4.)

Even after 4.6 billion years, heat continues to flow from within Earth. This heat powers the construction of mountains and volcanoes, causes earthquakes, moves continents, and shapes ocean basins.

Isostatic Equilibrium

Why do large regions of continental crust stand high above sea level? If the asthenosphere is nonrigid and deformable, why don't mountains sink because of their mass and disappear? Another look at Figure 3.4 will help to explain the situation. The mountainous parts of continents have "roots" extending into the asthenosphere. The continental crust and the rest of the lithosphere "float" on the denser asthenosphere. The situation involves buoyancy, the principle that explains why ships float.

Buoyancy is the ability of an object to float in a fluid by displacing a volume of that fluid equal in weight to the floating object's own weight. *A steel ship floats because it displaces*

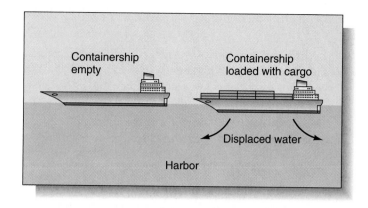

Figure 3.5 The principle of buoyancy. A ship sinks until it displaces a volume of water equal in weight to the weight of the ship and its cargo. If cargo is removed, the ship will rise.

a volume of water equal in weight to its own weight plus the weight of its cargo. An empty containership displaces a smaller volume of water than the same ship when fully loaded (**Figure 3.5**). The water that supports the ship is not *strong* in the mechanical sense; water does not support a ship the same way a steel bridge supports a car. Buoyancy, rather than mechanical strength, supports the ship and its cargo.

Any part of a continent that projects above sea level is supported in the same way. Consider the continent that contains Mt. Everest, the highest of Earth's mountains at 8.84 kilometers (29,007 feet) above sea level. Mt. Everest and its neighboring peaks are not supported by the *mechanical* strength of the materials within Earth—nothing in our world is that strong. Over a long period, and under the tremendous weight of the overlying crust, the asthenosphere behaves like a dense, viscous, slowly moving fluid. *The continent's mountains float high above sea level because the lithosphere gradually sinks into the deformable asthenosphere until it has displaced a volume of asthenosphere equal in mass to the mountains' mass.* The mountains stand at great height, nearly in balance with their subterranean underpinnings but susceptible to rising or falling as erosion or crustal stresses dictate. Lower regions are supported by shallower roots. In a slow-motion version of a ship floating in water, the entire continent stands in **isostatic equilibrium.**

What happens when a mountain erodes? In much the same way a ship rises when cargo is removed, Earth's crust will rise in response to the reduced load. Ancient mountains that have undergone millions of years of erosion often expose rocks that were once embedded deep within their roots. This kind of isostatic readjustment results in the thinning of the continental crust beneath the mountains, and subsidence beneath areas of deposited sediments. This process is shown in **Figure 3.6.**

Unlike the asthenosphere on which the lithosphere floats, crustal rock does not slowly flow at normal surface temperatures. A ship reacts to any small change in weight with a correspondingly small change in vertical position in the water, but an area of continent or ocean floor cannot react to every small weight change because the underlying rock is *not* liquid, the deformation does not occur rapidly,

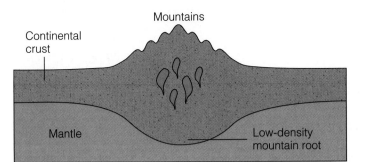

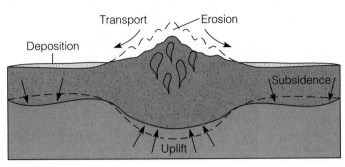

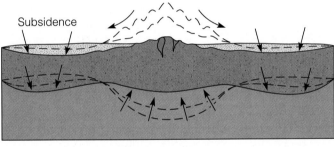

Figure 3.6 Erosion and isostatic readjustment can cause continental crust to become thinner in mountainous regions. As mountains are eroded (**a**–**c**), isostatic uplift causes their roots to rise. The same thing happens when a ship is unloaded. Further erosion exposes rocks that were once embedded deep within the peaks. Deposition of sediments away from the mountains often causes nearby crust to sink.

and the edges of the continent or seabed are mechanically bound to adjacent crustal masses. When the force of uplift or downbending exceeds the mechanical strength of the adjacent rock, the rock will fracture along a plane of weakness—a **fault.** The adjacent crustal fragments will move vertically in relation to each other. This sudden adjustment of the crust to isostatic forces by fracturing, or faulting, is one cause of earthquakes.

Large regions of Earth's continents are held above sea level by isostatic equilibrium, a process analogous to a ship floating in water.

Curious Coincidences?

Look for the site of the 1964 Alaska earthquake on **Figure 3.7.** Great as it was, this earthquake is not easy to distinguish; it blends with the many thousands of significant earthquakes that have occurred on Earth in recent times. Notice that earthquakes *do not occur at random places.* Seismic activity appears concentrated in lines and arcs across Earth's surface. Sometimes these lines coincide with the edge of a continent (as along the western edge of the Americas), and sometimes the lines trace the middle or sides of ocean basins.

Other features appear curious. As Leonardo da Vinci noticed on early charts, in some regions the continents look as if they would fit together like jigsaw puzzle pieces if the intervening ocean were removed. In 1620 Francis Bacon also wrote of a "certain correspondence" between shorelines on either side of the South Atlantic. In 1885 Edward Suess, a respected German scientist, suggested that the Southern Hemisphere's continents might once have been a single large landmass. He based his belief in part on the similarities of fossils found on these continents, especially fossils of the fern *Glossopteris.* Suess was not taken seriously by his colleagues because he could not explain how the continents had moved.

As they probed the submerged edges of the continents, researchers found that the ocean bottom nearly always sloped gradually out to sea for some distance and then dropped steeply to the deep-ocean floor. They realized that these shelflike continental edges were extensions of the continents themselves. In the few locations where they had measurements, researchers found that the "fit" between South America and Africa, impressive at the shoreline, was even better along the submerged edges of the continents.

Though so accurate a fit almost certainly could not have occurred by chance, no one had yet proposed a mechanism that could separate whole continents into moving pieces.

Continental Drift

Into the fray stepped **Alfred Wegener,** a busy German meteorologist and polar explorer (**Figure 3.8**). In a lecture in 1912 he proposed the startling and original theory of **continental drift.** Wegener suggested that all Earth's land had once been joined into a single supercontinent surrounded by an ocean. He called the land mass **Pangaea** (meaning "all Earth") and the surrounding ocean **Panthalassa** ("all ocean"). Wegener thought Pangaea had broken into pieces about 200 million years ago. Since then, he said, the pieces had moved to their present positions and were still moving.

The greatest block to the acceptance of continental drift was geology's view of Earth's mantle. The available evidence seemed to suggest that a deep, solid mantle supported the crust and mountains mechanically (not isostatically) from below. Drift would be impossible with this kind of subterranean construction. A few perceptive seismic researchers then noticed that the upper mantle reacted to earthquake waves as if it were a deformable mass, not a rigid solid. Perhaps such a layer would resemble a slug of iron heated in a

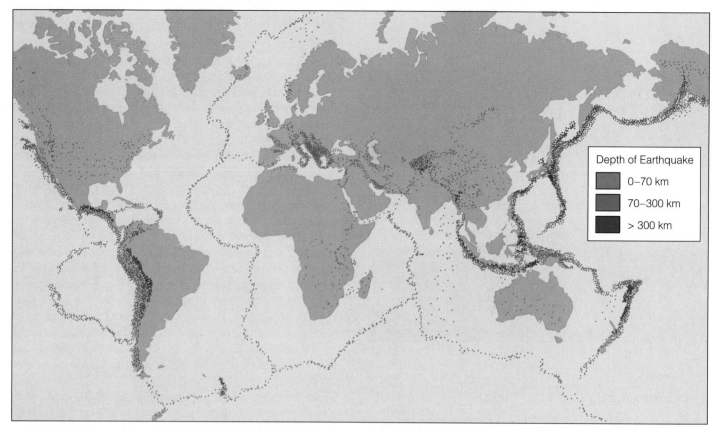

Depth of Earthquake
- 0–70 km
- 70–300 km
- > 300 km

Figure 3.7 Seismic events worldwide, January 1977 through December 1986. The locations of about 10,000 earthquakes are colored red, green, and blue to represent event depths of 0–70 kilometers, 70–300 kilometers, and below 300 kilometers, respectively.

Figure 3.8 Alfred Lothar Wegener writes in his journal at the beginning of what would be his last expedition in Greenland in 1930. His remarkable book *The Origin of Continents and Oceans* was published in 1915. In it, he outlined interdisciplinary evidence for his theory of continental drift.

blacksmith's forge; it would deform with pressure and even flow slowly. Established geologists dismissed this interpretation, however, saying that the mountains would simply fall over or sink without rigid underpinnings. By 1926 the "drifters" were in full retreat. When Wegener died on an expedition across Greenland in 1930, his theory was already in eclipse.

The Idea Transformed

The theory of continental drift refused to die, however; those neatly fitted continents provided a haunting reminder of Wegener to anyone looking at an Atlantic chart. In 1935 a Japanese scientist, Kiyoo Wadati, speculated that earthquakes and volcanoes near Japan might be associated with continental drift. In 1940, seismologist Hugo Benioff plotted the locations of deep earthquakes at the edges of the Pacific. His charts revealed the true extent of the **"Pacific Ring of Fire,"** a circle of violent geological activity surrounding much of the Pacific Ocean. Seismographs were now beginning to reveal a worldwide pattern of earthquakes and volcanoes. Deep earthquakes did not occur randomly over Earth's surface but were concentrated in zones that extended in lines along Earth's crust. Benioff, Wadati, and others wondered what could cause such an orderly pattern of deep earthquakes. Figure 3.7 is based on a plot of about 30,000 earthquakes.

Notice the odd pattern they form—almost as if Earth's lithosphere is divided into sections! Benioff's sensitive seismographs also began to gather strong evidence for a partially molten, nonrigid layer in the upper mantle. Could the continents somehow be sliding on that?

Other seemingly unrelated bits of information were accumulating. **Radiometric dating** of sediments and rocks was perfected after World War II. This technique is based on the discovery that unstable, naturally radioactive elements lose particles from their nuclei and change into new stable elements. The radioactive decay occurs at a constant rate, and the ratio of radioactive to stable atoms in a sample provides its age. To the surprise of many geologists, the maximum age of the ocean floor and its overlying sediments was radiometrically dated at less than 200 million years, only about 4% of the age of Earth.[2] The centers of the continents are much older; some parts of the continental crust are more than 3.9 billion years old, about 85% of the age of Earth. *Why was oceanic crust so young?*

Attention quickly turned to the ocean floors, the complex profiles of which were now being revealed by **echo sounders,** devices that measure depth by bouncing high-frequency sound waves off the bottom (see Figure 2.17). Researchers on ships compiled more complete, accurate charts of the submerged edges of continents. At England's Cambridge University, Sir Edward Bullard used an early computer to process these data to achieve the best possible fit of the continental jigsaw puzzle pieces around the Atlantic. The fit was astonishingly good (**Figure 3.9**).

Mantle studies were keeping pace. In 1957 the first links in the Worldwide Standardized Seismograph Network began to report data from seismic waves reflected and refracted through Earth's inner layers. This information verified the existence of a layer in the upper mantle that caused a decrease in the velocity of seismic waves. This finding strongly suggested that the layer was not solid. Perhaps the lithosphere was isostatically balanced in this deformable layer, and perhaps continents could move around in it *if a suitable power source existed.*

The Breakthrough: From Seafloor Spreading 3.9 to Plate Tectonics

In 1960 Harry Hess of Princeton University and Robert Dietz of Scripps Institution of Oceanography proposed a radical idea to explain the features of the ocean floor and the "fit" of the continents. They suggested that new seafloor develops at the Mid-Atlantic Ridge (and the other newly discov-

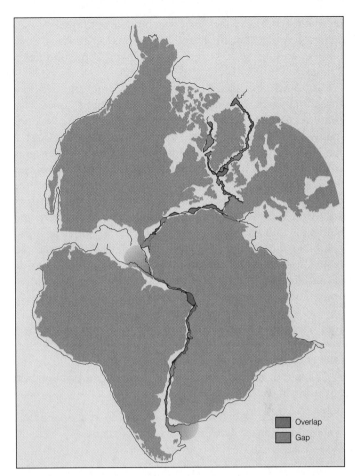

Figure 3.9 The fit of all the continents around the Atlantic at a water depth of about 137 meters (450 feet), as calculated by Sir Edward Bullard at Cambridge University in the early 1960s.

Legend: Overlap, Gap

ered ocean ridges) and then spreads outward from this line of origin. Continents would be pushed aside by the same forces that cause the ocean to grow. This motion could be powered by **convection currents,** slow-flowing circuits of material within the mantle.

Seafloor spreading, as the new theory was called, pulled many loose ends together. If the mid-ocean ridges were **spreading centers** and sources of new ocean floor rising from the asthenosphere, then they should be hot. They were. If the new oceanic crust cooled as it moved from the spreading center, then it should shrink in volume and become more dense, and the ocean should be deeper farther from the spreading center. It was. Sediments at the edges of the ocean basin should be thicker than those near the spreading centers. They were, and they were also older.

Did this mean that Earth was continuously expanding? Since there was no evidence for a growing Earth, the creation of new crust at spreading centers would have to be balanced by the destruction of crust somewhere else. Then researchers discovered that the crust plunges down into the mantle along the periphery of the Pacific. The process is known as **subduction,** and these areas are called **subduction zones** (or *Wadati–Benioff zones* in honor of their

---

[2] For a discussion of geological time, please see Figure 1.12 and Appendix II.

discoverers). The zones of concentrated earthquakes (see again Figure 3.7) were found in regions of crustal formation (spreading centers) and crustal destruction (subduction zones).

In 1965 the ideas of continental drift and seafloor spreading were integrated into the overriding concept of **plate tectonics** (our word *architect* has the same root), primarily by the work of **John Tuzo Wilson,** a geophysicist at the University of Toronto. In this theory, Earth's outer layer consists of about a dozen separate major lithospheric **plates** floating on the asthenosphere. When heated from below, the deformable asthenosphere expands, becomes less dense, and rises. It turns aside when it reaches the lithosphere and then drags the plates laterally until it turns under again to complete the circuit. The large plates include both continental and oceanic crust. The major plates, which jostle about like huge flats of ice on a warming lake, are shown and named in **Figure 3.10.** Plate movement is slow in human terms, averaging about 5 centimeters (2 inches) a year. The plates interact at converging, diverging, or slipping boundaries, sometimes forcing one another below the surface or wrinkling into mountains.

Plate movement appears to be caused by a combination of three forces:

- The outward push of new seabed formed at spreading centers. Plates slide off the raised ridges.

- Friction of mantle convection currents against the bottom of a plate.

- The downward pull of a descending plate's dense leading edge.

We now know that *through the great expanse of geological time, this slow movement remakes the surface of Earth, expands and splits continents, and forms and destroys ocean basins.* The less dense, ancient granitic continents ride high in the lithospheric plates, rafting on the slowly moving asthenosphere below. This process has progressed since Earth's crust first cooled and solidified.

Figure 3.11 presents an overview of the tectonic system. Literally and figuratively, it all fits; a cooling, shrinking, raisinlike wrinkling is no longer needed to explain Earth's surface features. This twentieth century understanding of the ever-changing nature of the Earth has given fresh meaning to historian Will Durant's warning: "Civilization exists by geological consent, subject to change without notice."

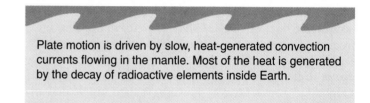

Plate motion is driven by slow, heat-generated convection currents flowing in the mantle. Most of the heat is generated by the decay of radioactive elements inside Earth.

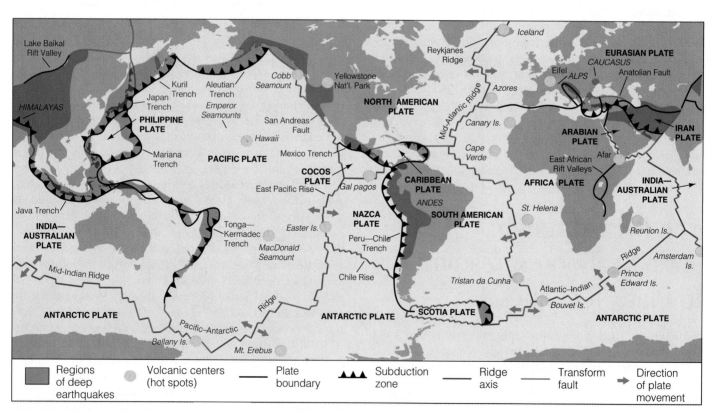

Figure 3.10 The major lithospheric plates, with their directions of movement and locations of the principal hot spots. Note the correspondence of plate boundaries and earthquake locations: Compare this figure to Figure 3.7. Most of the million or so earthquakes and volcanic events each year occur along plate boundaries.

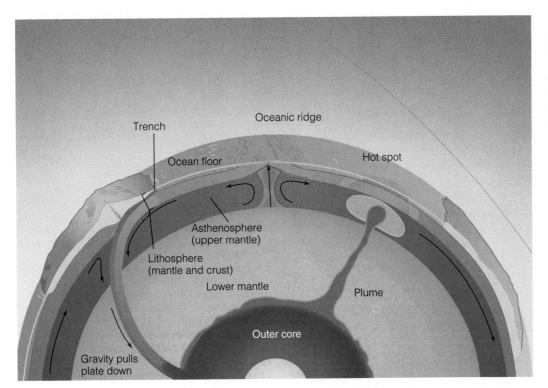

Figure 3.11 The tectonic system is powered by heat. Some parts of the mantle are warmer than others, and convection currents form when warm mantle material rises and cool material falls. Above the mantle floats the cool, rigid, lithosphere, which is fragmented into plates. Large convection currents in the partially melted, deformable asthenosphere drag the plates away from one another (along the ocean ridges), toward one another (at subduction zones or areas of mountain building), or past one another (as at California's San Andreas Fault). Smaller localized convection currents form cylindrical plumes that rise to the surface to form hot spots (like the Hawaiian Islands). Note that the whole mantle is involved in thermal convection currents (see again Figure 3.1). (From *The Earth's Dynamic Systems*, 9/e by Hamblin & Christiansen, Fig. 2.8. Reprinted by permission of Pearson Education, Inc., Upper Saddle River, NJ.)

Plate Boundaries   3.10

The lithospheric plates shown in Figure 3.10 float on a dense, deformable asthenosphere and are free to move relative to one another. Plates interact with neighboring plates along their mutual boundaries. In **Figure 3.12,** movement of Plate A to the left (west) requires it to slide along its north and south margins. An overlap is produced in front (to the west), and a gap is created behind (to the east). Different places on the margins of Plate A experience separation and extension, convergence and compression, and transverse movement (shear).

The three types of plate boundaries that result from these interactions are called *divergent, convergent,* and *transform boundaries,* depending on their sense of movement.

Divergent Plate Boundaries— Forming Ocean Basins   3.11

The spreading center at the Mid-Atlantic Ridge is a **divergent plate boundary,** a line along which two plates are moving apart. Oceanic crust forms along divergent plate boundaries. The formation of the Atlantic is shown in **Figure 3.13.** Heat from the lower crust and mantle accumulates beneath the continents. This heat caused the asthenosphere to expand and rise, thus lifting and fracturing the lighter, solid lithosphere above. The plate and its embedded continent split in two, and a new plate boundary—a rift valley—formed between the pieces. As the broken plate separated at this new spreading center, molten rock called **magma** rose into the crustal fractures. (Magma is called *lava* when found

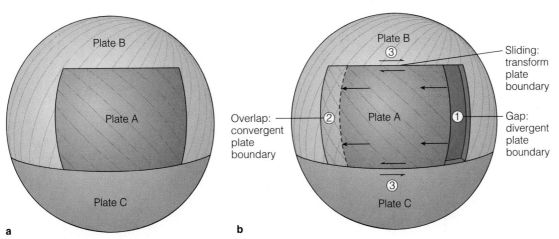

a   b

Figure 3.12 Plate boundaries in action. As Plate A moves to the left (west), a gap forms behind it ①, and an overlap with Plate B forms in front ②. Sliding occurs along the top and bottom sides ③. The margins of Plate A experience the three types of interactions: at ①, extension characteristic of divergent boundaries; at ②, compression characteristic of convergent boundaries; and at ③, the shear characteristic of transform plate boundaries.

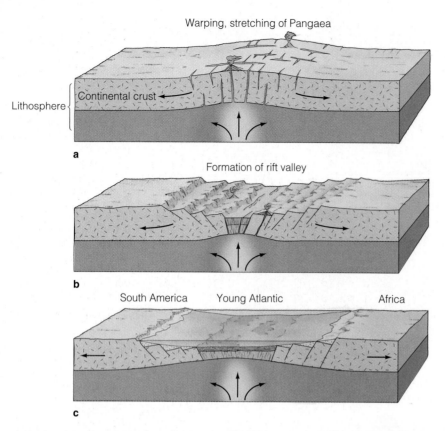

Warping, stretching of Pangaea

Lithosphere { Continental crust

a

Formation of rift valley

b

South America   Young Atlantic   Africa

c

Figure 3.13 A model for the formation of a new plate boundary: the breakup of Pangaea and the formation of the Atlantic. (**a**) As the lithosphere began to crack, a rift formed beneath the continent and molten basalt from the asthenosphere began to rise. (**b**) As the rift continued to open, the two new continents were separated by a growing ocean basin. Volcanoes and earthquakes occur along the active rift area, which is the mid-ocean ridge. The East African Rift Valley currently resembles this stage. (**c**) A new ocean basin (shown in green) forms beneath a new ocean. (**d**) The Red Sea currently resembles this stage. Note the remarkable sawtooth configuration of the peaks on the horizon and their similarity to the diagram. (**e**) The South Atlantic in cross section, showing the mid-ocean ridge in the middle of the growing basin.

d

V. Courtillot

|←—Mid-ocean ridge—→|

South America   Rift   Atlantic   Africa

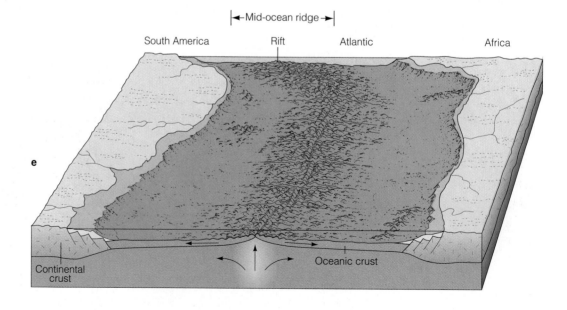

Continental crust   Oceanic crust

e

aboveground.) Some of the magma solidified in the fractures; some erupted from volcanoes on the seafloor. Together these processes produced new oceanic crust. The South Atlantic, a large new ocean basin formed between the diverging plates, is shown in **Figure 3.13e.** A long mid-ocean ridge divided by a central rift valley traverses the ocean floor roughly equidistant from the shorelines in both the North and South Atlantic, terminating north of Iceland.[3]

Plate divergence is not confined to the Atlantic, nor has it been limited to the last 200 million years. As may also be seen in Figure 3.10, the Mid-Atlantic Ridge has counterparts in the Pacific and Indian Oceans. The Pacific floor, for example, diverges along the East Pacific Rise and the Pacific Antarctic Ridge, spreading centers that form the eastern and southern boundaries of the great Pacific Plate. In East Africa, rift valleys have formed relatively recently as plate divergence begins to separate another continent. As happened in the Red Sea (**Figure 3.13d**), the ocean will invade when the rift becomes deep enough.

Figure 3.14 shows how divergence has formed other ocean basins. About 20 cubic kilometers (4.8 cubic miles) of new ocean crust forms each year.

Convergent Plate Boundaries— Recycling Crust, Building Island Arcs and Continents

Since Earth is not getting larger, divergence in one place must be offset by convergence in another. Crust is destroyed at **convergent plate boundaries,** regions of violent geological activity where plates are pushing together. South America, embedded in the westward-moving South American Plate, encounters the Pacific's Nazca Plate as it moves eastward. The relatively thick and light continental lithosphere of South America rides up and over the heavier oceanic lithosphere of the Nazca Plate, which is subducted along the deep trench that parallels the west coast of South America. **Figure 3.15** is a cross section through these plates.

The subducting plate's periodic downward lurches cause earthquakes. Some of the oceanic crust and its sediments will melt as the plate plunges downward, forming magma rich in water and carbon dioxide. In places this magma rises through overlying layers to the surface and causes volcanic eruptions. The active volcanoes of Central America and South America's Andes Mountains are a product of this activity, as are the area's numerous earthquakes. The North American Cascade volcanoes, including Mount St. Helens, result from similar processes. Most of the subducted crust mixes with the mantle. As shown in **Figure 3.16,** some of it continues downward through the mantle, eventually reaching the mantle–core boundary! Subduction at converging oceanic plates was responsible for the great Alaska earthquake of 1964. Plate convergence (and divergence) is

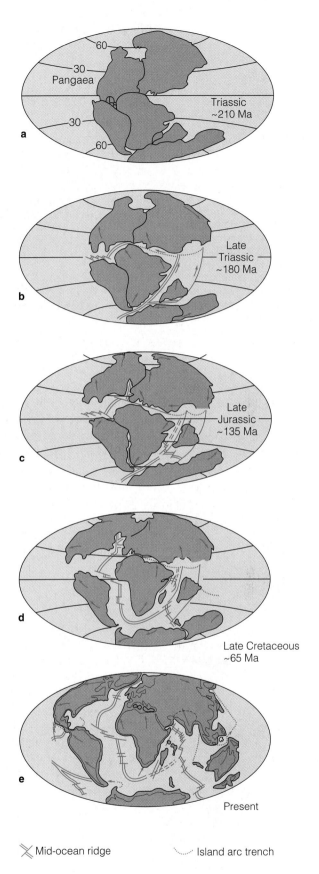

Ma = *mega-annum,* indicating millions of years ago

Figure 3.14 The breakup of Pangaea shown in five stages beginning about 210 million years ago. Inferred motion of lithospheric plates is indicated by arrows. Spreading centers (mid-ocean ridges) are shown in red.

---

[3] The Atlantic's rate of spreading is about 5 centimeters (2 inches) a year. This young ocean was about 25 meters (82 feet) narrower when Columbus sailed than it is today.

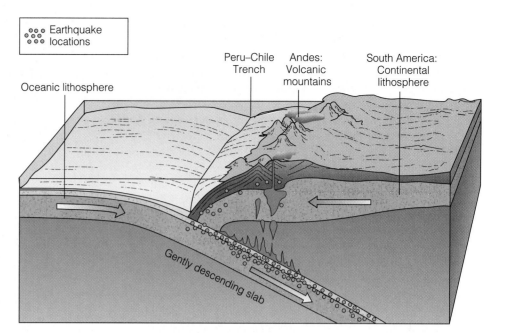

Peru–Chile
Trench

Andes:
Volcanic
mountains

South America:
Continental
lithosphere

Oceanic lithosphere

Gently descending slab

Figure 3.15 A cross section through the west coast of South America, showing the convergence of a continental plate and an oceanic plate. The subducting oceanic plate becomes more dense as it descends, its downward slide propelled by gravity. Starting at a depth of about 100 kilometers (60 miles), heat drives water and other volatile components from the subducted sediments into the overlying mantle, lowering its melting point. Masses of the melted material, rich in water and carbon dioxide, rise to power Andean volcanoes.

faster in the Pacific than in the Atlantic, in a few places reaching a rate of 18 centimeters (7 inches) a year. You can now clearly see the source of the Pacific Ring of Fire.

In the previous example, continental crust met oceanic crust. What happens when two *oceanic* plates converge? One of the colliding plates will usually be older, and therefore

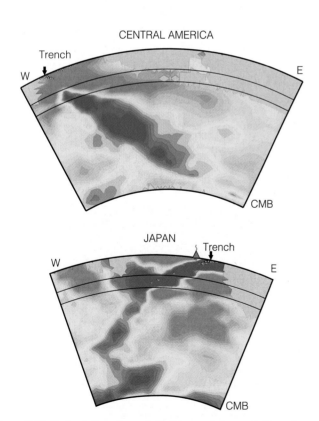

CENTRAL AMERICA

Trench

W

E

CMB

JAPAN

Trench

W

E

CMB

Figure 3.16 Vertical slices through Earth's mantle beneath Central America and Japan. Colder material is shown in blue, warmer in red. The distribution of colder material suggests that the subducting slabs beneath both areas have penetrated to the core–mantle boundary (CMB) at a depth of 2,900 kilometers (1,800 miles).

cooler and denser, than the other. Pulled by gravity, this heavier plate will slip steeply below the lighter one into the asthenosphere. The ocean bottom is distorted in these areas to form deep trenches, the ocean's greatest depths. Water and carbon dioxide trapped with the melting rock of the subducting plate rise into the overlying mantle, lowering its melting point. This fluid mix of magma and subducted material forms a relatively light magma that powers vigorous volcanoes, but the volcanoes emerge from the seafloor rather than from a continent. These volcanoes appear in patterns of curves on the overriding oceanic crust; when they emerge above sea level, they form curving arcs of islands (**Figure 3.17**).

Convergent margins are vast "continent factories" where materials from the surface descend and are heated, compressed, partially liquefied, separated, mixed with surrounding materials, and recycled to the surface. Relatively light continental crust is the main product, and it is produced at a rate of about 1 cubic kilometer (0.24 cubic mile) per year. Some geophysicists believe all of Earth's continental crust may have originated from granitic rock produced in this way. The island arcs may have coalesced to form larger and larger continental masses.

Two plates bearing continental crust can also converge (**Figure 3.18**). The most spectacular example of such a collision, between the India–Australian and Eurasian Plates some 45 million years ago, formed the Himalayas. Neither plate edge is being subducted; instead, both are compressed, folded, and uplifted. The lofty top of Mt. Everest is made of rock formed from sediments deposited long ago in a shallow sea!

Transform Plate Boundaries— Fracturing Crust

In some places, crustal plates shear laterally past one another. These areas are called **transform plate boundaries.** Crust is neither produced nor destroyed at this type of boundary,

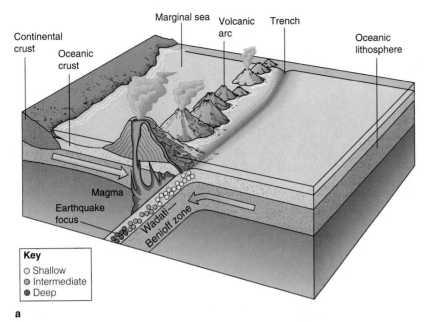

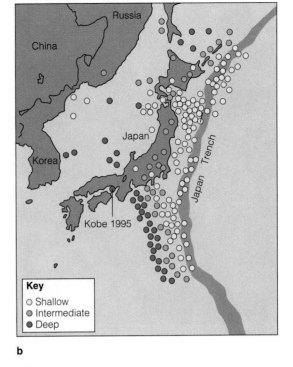

Figure 3.17 (**a**) The formation of an island arc along a trench as two oceanic plates converge. The volcanic islands form as masses of magma reach the seafloor. The Japanese islands were formed in this way. (**b**) The distribution of shallow, intermediate, and deep earthquakes for part of the Pacific Ring of Fire in the vicinity of the Japan Trench. Note that earthquakes occur on only one side of the trench, the side on which the plate subducts. The site of the catastrophic 1995 Kobe subduction earthquake is marked.

but the potential for earthquakes can be great as the plate edges slip past one another. The eastern boundary of the Pacific Plate is a long transform fault system. California's San Andreas Fault (**Figure 3.19**) is merely the most famous of the many faults that mark the junction between the Pacific and North American Plates. The Pacific Plate moves steadily, but its movement is stored elastically at the North American Plate boundary until friction is overcome. Then the Pacific Plate lurches in abrupt jerks to the northwest along much of its shared border with the North American Plate, an area that includes the major population centers of California. These jerks cause California's famous earthquakes. Because of this

movement, western California is gradually sliding north along the rest of North America; some 50 million years from now, it will encounter the Aleutian Trench.

Plate Interactions— A Summary

There are, then, two kinds of plate divergences:

- Divergent oceanic crust (such as in the Mid-Atlantic)
- Divergent continental crust (as in the Rift Valley of East Africa)

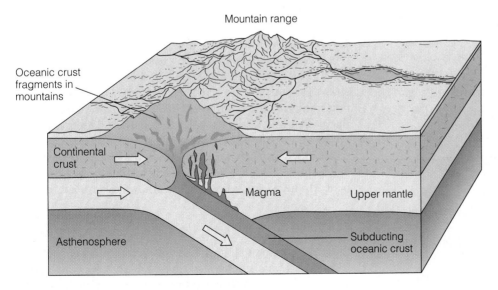

Figure 3.18 A cross section showing the convergence of two continental plates. Neither plate is dense enough or thin enough to subduct; instead, compression and folding uplift the plate edges to form mountains. Notice the supporting "root" beneath the emergent mountain needed to maintain isostatic equilibrium.

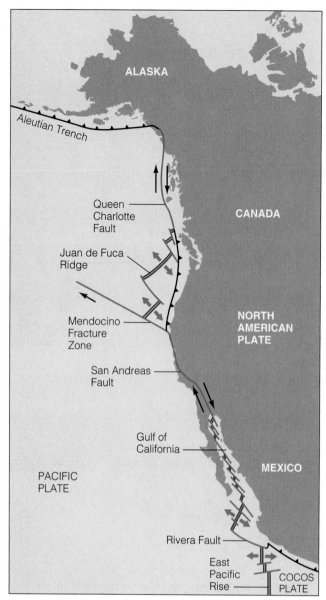

Figure 3.19 A long transform plate boundary, which includes California's San Andreas Fault.

And there are three kinds of plate convergences:

- Oceanic crust toward continental crust (west coast of South America)
- Oceanic crust toward oceanic crust (northern Pacific)
- Continental crust toward continental crust (Himalaya Mountains)

> Plate tectonics theory suggests that Earth's surface is not a static arrangement of continents and ocean, but a dynamic mosaic of jostling segments of lithosphere called lithospheric plates. Plates have collided, moved apart, and slipped past one another since Earth's crust first solidified.

Transform boundaries mark the locations at which crustal plates move past one another (San Andreas Fault).

Each of these movements produces a distinct topography, and each zone contains potential dangers for its human inhabitants.

The Confirmation of Plate Tectonics

The theory of plate tectonics has had the same effect on geology that the theory of evolution has had on biology. In each case a catalog of seemingly unrelated facts was unified by a powerful central idea. Many discoveries contributed to our present understanding of plate tectonics. Some of the compelling evidence for plate tectonics is outlined next.

Hot Spots

Hot spots are relatively stationary sources of heat in the mantle. Hot spots are not always located at plate boundaries, and no one yet knows why their source of heat is localized or what anchors them in place. Some hot spots probably originate near the core–mantle boundary. As lithospheric plates slide over these fixed locations, they are weakened from below by rising heat and magma. A volcano can form over the hot spot, but because the plate is moving, the volcano is carried away from its source of magma after a few million years and becomes inactive. It is replaced at the hot spot by a new volcano a short distance away. A chain of volcanoes and volcanic islands results. **Figure 3.20** shows the sequence.

Figure 3.21 shows the most famous of these "assembly line" chains, which extends from the old eroded volcanoes of the Emperor Seamounts to the still-growing island of Hawaii. In fact, the abrupt bend in the chain was caused by a change in direction of the Pacific Plate from largely northward to a more westward movement about 40 million years ago. The next Hawaiian island—already named Loihi—is building on the ocean floor to the southeast. Now about 1,000 meters (3,200 feet) beneath the surface, Loihi will break the surface about 30,000 years from now.

There are other hot spots in the Pacific. The island chains formed by their activity "jog" in the Hawaiian pattern, indicating that they are positioned on the same lithospheric plate. Chains of undersea volcanoes in the Atlantic, centered on the Mid-Atlantic Ridge, suggest a similar process is at work there. Hot spots can exist beneath continental crust as well; Yellowstone National Park is believed to be over a hot spot beneath the westward-moving North American Plate.

Terranes

Buoyant continental and oceanic plateaus, fragments of granitic rock and sediments, can be rafted along with a plate and scraped off onto a continent when the plate is sub-

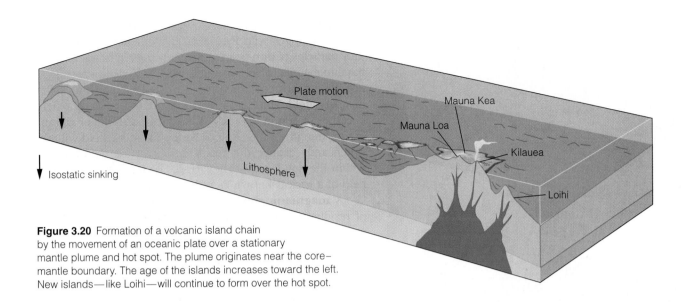

Figure 3.20 Formation of a volcanic island chain by the movement of an oceanic plate over a stationary mantle plume and hot spot. The plume originates near the core–mantle boundary. The age of the islands increases toward the left. New islands—like Loihi—will continue to form over the hot spot.

ducted. This process is similar to what happens when a knife is scraped across a table top to remove pieces of cool candle wax. The wax accumulates and wrinkles on the knife blade in the same way landmasses and ocean sediments accumulate against the face of a continent as the lithosphere in which they are embedded reaches a plate boundary. Plateaus, isolated segments of seafloor, ocean ridges, ancient island arcs, and parts of continental crust that collect on the face of a continent are called **terranes.** The thickness and low density of terranes prevent their subduction. A simplified account of terrane accumulation is diagrammed in **Figure 3.22.**

Terranes are surprisingly common. New England, much of North America west of the Rocky Mountains, and all of Alaska appear to be composed of this sort of crazy-quilt assemblage of material, some of which has evidently arrived from thousands of miles away in the Southern Hemisphere!

Paleomagnetism

The young ocean basins hold the most convincing evidence for plate tectonics. This evidence was obtained through the magnetic analysis of rocks formed during the last 200 million years.

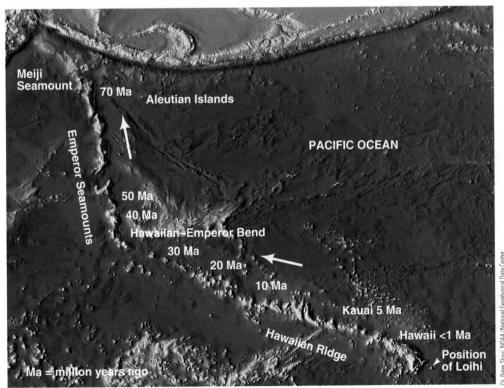

Figure 3.21 The Hawaiian chain, islands formed one by one as the Pacific Plate slid over a hot spot. The oldest known member of the chain, the Meiji Seamount, formed about 70 million years ago (70 Ma), and the bend in the chain shows that the plate changed direction about 40 million years ago. The island of Hawaii still has active volcanism, and the next island in the chain, Loihi, has begun building on the ocean floor. The upper arrow shows the initial direction of plate motion; the lower arrow shows the present direction.

4. How do geologists determine the location and magnitude of an earthquake?

Geologists use the time difference in the arrival of the seismic waves at their instruments to determine the distance to an earthquake. At least three seismographs in widely separated locations are needed to get a fix on the location.

The strength of the waves, adjusted for the distance, is used to calculate the earthquake's magnitude. Earthquake magnitude is often expressed on the **Richter scale.** Each full step on the Richter scale represents a 10-fold change in surface wave amplitude and a 32-fold change in energy release. Thus an earthquake with a Richter magnitude of 6.5 releases about 32 times more energy than an earthquake with a magnitude of 5.5, and about 1,000 times that of a 4.5-magnitude quake. People rarely notice an earthquake unless the Richter magnitude is 3.2 or higher, but the energy associated with a magnitude-6 quake may cause significant destruction.

The energy released by the 1964 Alaska earthquake was more than a billion times greater than the energy released by the smallest earthquakes felt by humans. The energy release was equal to about twice the energy content of world coal and oil production for an entire year. Very low or very high Richter magnitudes are not easy to measure accurately. The Alaska earthquake's magnitude was initially calculated between 8.3 and 8.6 on the Richter scale, but recent reassessment has yielded an extraordinary magnitude of 9.2. The Earth rang like a great silent bell for ten days after the earthquake.

5. How common are large earthquakes?

About every two days, somewhere in the world, there's an earthquake of from 6 to 6.9 on the Richter scale—roughly equivalent to the quake that shook Northridge and the rest of southern California in January 1994 or Kobe, Japan, in January 1995. Once or twice a month, on average, there's a 7 to 7.9 quake somewhere. There is about one 8 to 8.9 earthquake—similar in magnitude to the 1964 earthquake in Alaska—each year.

Large losses of life and property can occur when these earthquakes occur in populated areas, however. Damage estimates from the Northridge earthquake exceeded $40 billion. In Kobe, more than 5,000 people died and more than 26,000 were injured. Some 56,000 buildings were destroyed; estimates of the cost of reconstruction exceeded $400 billion.

CHAPTER SUMMARY

Earth is composed of concentric spherical layers, with the least dense layer on the outside and the most dense as the core. The layers may be classified by chemical composition into crust, mantle, and core; or by physical properties into lithosphere, asthenosphere, mantle, and core. Geologists have confirmed the existence and basic properties of the layers by analysis of seismic waves, which are generated by the forces that cause large earthquakes.

The theory of plate tectonics explains the nonrandom distribution of earthquake locations, the curious jigsaw-puzzle fit of the continents, and the patterns of magnetism in surface rocks. Plate tectonics theory suggests that Earth's surface is not a static arrangement of continents and ocean, but a dynamic mosaic of jostling lithospheric plates. The plates have converged, diverged, and slipped past one another since Earth's crust first solidified and cooled, driven by slow, heat-generated currents flowing in the asthenosphere. Continental and oceanic crusts are generated by tectonic forces, and most major continental and seafloor features are shaped by plate movement. Plate tectonics explains why our ancient planet has surprisingly young seafloors; the oldest is only as old as the dinosaurs—that is, about ⅓ the age of Earth.

ON-LINE STUDY RESOURCES

The Web site for this book contains a wealth of helpful study aids. Log on to:

http://info.brookscole.com/garrison

and select *Essentials of Oceanography*, 3rd edition. Select Chapter 3 from the drop-down menu or click on one of these resource areas:

- For study and review, **Chapter at a Glance** gives you an outline of the chapter, **Chapter Summary** allows you to review the chapter's main ideas, and **Glossary** lists concepts and terms for the chapter along with their definitions.

- To test your mastery of the Terms and Concepts to Remember for this chapter, you can use the electronic **Flash Cards,** play the **Concentration** game, or work the **Crossword Puzzle.**

- For practice quizzing, try the multiple-choice **Tutorial Quiz,** the ten-question **True/False Quiz,** or the **Image Analysis Quiz,** which poses questions based on art and photos from the chapter.

TERMS AND CONCEPTS TO REMEMBER

asthenosphere	lithosphere
basalt	lower mantle
buoyancy	magma
conduction	magnetometer
continental crust	mantle
continental drift	oceanic crust
convection	"Pacific Ring of Fire"
convection current	paleomagnetism
convergent plate boundary	Pangaea
core	Panthalassa
crust	plate tectonics
divergent plate boundary	plates
echo sounders	radioactive decay
fault	radiometric dating
granite	Richter scale
hot spot	seafloor spreading
isostatic equilibrium	seismic waves

seismograph
spreading center
subduction
subduction zone

terrane
transform plate boundary
Wegener, Alfred
Wilson, John Tuzo

STUDY QUESTIONS

1. How are Earth's internal layers classified?

2. How is crust different from lithosphere?

3. Where are the youngest rocks in the seabed? The oldest? Why?

4. On what points was Wegener correct? Wrong?

5. Would the most violent earthquakes be associated with spreading centers or with subduction zones? Why?

6. Describe the mechanism that powers the movement of the lithospheric plates.

7. Why is paleomagnetic evidence thought to be the linchpin in the plate tectonics argument? Can you think of any objections to the Matthews/Vine/Morley interpretation of the paleomagnetic data?

8. What biological evidence supports plate tectonic theory?

9. What evidence can you cite to support the theory of plate tectonics? What questions remain unanswered? Which side would you take in a debate?

10. Why are the continents about 20 times older than the oldest ocean basins?

 Earth Systems Today CD Questions

1. Go to "Geologic Time" in the CD-ROM's main menu, and click on "Changing Earth." Begin the animation of plate motion. When it arrives at "present day," predict plate movement 250 million years into the future. Sketch your answer; then play the rest of the animation to see if you are correct.

2. Go to "Geologic Time," then "Relative Dating" and "Absolute Dating." After reviewing the contents of Box 3.2 (Absolute and Relative Dating), try your hand at some of the experiments. Pay particular attention to the concept of half-life. See if you can correctly answer the questions on the age of layers in "Layer Study."

3. Go to "Earth's Layers," then to "P and S Waves." Before activating the animation, can you predict the type of movement for each wave?

4. Go to "Earth's Layers," then to "Core Studies." Put yourself in the place of Inge Lehmann and Richard Oldham (Box 3.1), and perform some experiments in all three sections of "Core Studies." Would you have reached the same conclusions they did?

5. Go to "Plate Tectonics," then "Plate Locations." With the "Tectonic Activity" boxes toggled off, predict the positions of recent earthquakes and volcanoes based on the plate locations shown. Toggle the "Tectonic Activity" boxes back on, and gauge your success.

6. Go to "Plate Tectonics," then "Plate Boundaries." Note the three types of boundaries: divergent, convergent, and transform. Note also the three settings for convergent boundaries: ocean–ocean, ocean–continent, and continent–continent. Before animating the frames, write a brief prediction of the resulting geological activity for each of the combinations of boundaries and settings. Now animate the frames, and see if your choices were appropriate.

7. Go to "Plate Tectonics," then "Triple Junctions and Seafloor Study." Take the VR (virtual reality) tour of a segment of the East Pacific Risa. Before you move and zoom (using your mouse), can you predict the kinds of features you will encounter?

8. Go to "Volcanism," then "Distribution of Volcanism." Toggle off the volcano types. Now looking at the upper chart showing spreading plate boundaries, directions of plate motion, and locations of subduction zones, predict where subduction zone volcanoes, rifting site volcanoes, and hot spots might be expected. Now toggle the volcanoes on to see how accurate your predictions were.

RESOURCES FOR FURTHER READING AND RESEARCH

The Web site for this book contains many ideas for further reading and research. Log on to:

http://info.brookscole.com/garrison

and select *Essentials of Oceanography*, 3rd edition. Select Chapter 3 from the drop-down menu or click on one of these resource areas:

- **Additional Readings** lists the major books and articles consulted in writing this chapter, along with comments from the author about their content and reading level.

- **Hypercontents** takes you to an extensive list of current links to Internet sites with news, research, and images related to individual subjects in the chapter. Just click on the icon that corresponds to a numbered section to see the links for that subject.

- **Internet Exercises** are critical thinking questions that involve research on the Internet with starter URLs provided.

- **InfoTrac College Edition Exercises** leads you to Critical Thinking Projects that use InfoTrac College Edition as a research tool.

- **Regional InfoTrac College Edition** articles are organized into East Coast, West Coast, and Gulf Coast regions, allowing you to study oceanography on a more local level.

 For more readings, go to InfoTrac College Edition, your on-line research library, at:

http://infotrac.thomsonlearning.com

(Left) The most famous and accomplished of small research submarines, *Alvin*. (Right) *Alvin's* "standby" titanium pressure sphere. One sphere is installed in the submarine and houses a pilot and two researchers. This identical sphere, on display at the Washington, DC, naval yard, was to be tested to destruction, but the test equipment failed before the sphere did!

GOING DEEP

In this chapter you will read about ocean bottom features. Our knowledge of the seabed has largely been gained by people working at a distance from their goal. They use remote sensors to measure sound waves, radar beams, and differences in the pull of gravity; then they combine this information to "see" details of the seafloor. Sometimes, though, there is no substitute for actually seeing—focusing a well-trained set of eyes on the ocean floor. Here's where research submarines come in handy.

Alvin, the best-known and oldest of the deep-diving, manned research submarines now in operation, has made more than 4,200 dives. *Alvin* carries three people in a sealed titanium sphere and is capable of diving to 4,000 meters (13,120 feet). The 6.7-meter (22-foot) sub was commissioned in 1964 and is operated jointly by the Woods Hole Oceanographic Institution, National Science Foundation, Office of Naval Research, and National Oceanic and Atmospheric Administration. The most famous of the research submersibles, *Alvin* has explored the Mid-Atlantic Ridge near the Azores at 2,700 meters (9,000 feet), measuring rock temperatures, collecting water samples for chemical analysis, and taking photographs. Many of the features discussed here were first observed by researchers in this trusty little sub.

4.1

Ocean Basins

Bathymetry

The discovery and study of ocean floor contours is called **bathymetry.** The earliest known bathymetric studies were carried out in the Mediterranean by a Greek named Posidonius in 85 B.C. He and his crew let out nearly 2 kilometers (1.25 miles) of rope until a stone tied to the end of the line touched bottom. Bathymetric technology had not improved by the time Sir James Clark Ross obtained soundings (depth readings) of 4,893 meters (16,054 feet) in the South Atlantic in 1818. In the 1870s the researchers aboard HMS *Challenger* added the innovation of a steam-powered winch to raise the line and weight, but the method was the same (**Figure 4.1**). The *Challenger* crew made 492 bottom soundings and confirmed Matthew Maury's earlier discovery of the Mid-Atlantic Ridge.

Echo Sounding

The sinking of RMS *Titanic* in 1912 stimulated research that finally ended slow, laborious weight-on-a-line efforts. By April 1914, Reginald A. Fessenden, a former employee of Thomas Edison, had developed an "Iceberg Detector and

Figure 4.1 An illustration from the *Challenger Report* (1880). Seamen are handling the steam winch used to lower a weight on the end of a line to the seabed to find the ocean depth.

Archive, Scripps Institute of Oceanography

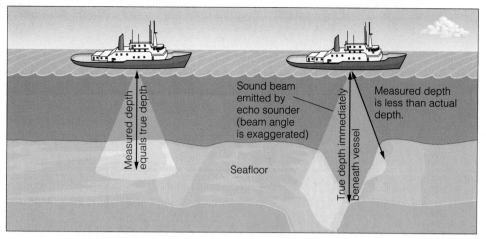

a

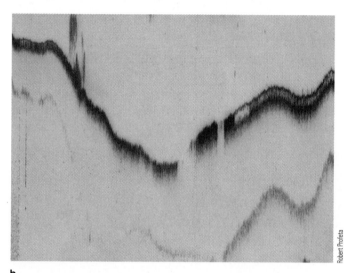

b

Figure 4.2 (**a**) The accuracy of an echo sounder can be affected by water conditions and bottom contours. The pulses of sound energy, or "pings," from the sounder spread out in a narrow cone as they travel from the ship. When depth is great, the sounds reflect from a large area of seabed. Because the first sound of the returning echo is used to sense depth, measurements over deep depressions are often inaccurate. (**b**) An echo sounder trace. A sound pulse from a ship is reflected off the seabed and returns to the ship. Transit time provides a measure of depth. For example, it takes about 2 seconds for a sound pulse to strike the bottom and return to the ship when the water depth is 1,500 meters (4,900 feet). Bottom contours are revealed as the ship sails a steady course. In this trace, the horizontal axis represents the course of the ship, and the vertical axis represents the water depth. The ship has sailed over a small submarine canyon.

Echo Depth Sounder." The detector worked by directing a powerful sound pulse ahead of a ship and then listening for an echo from the submerged portion of an iceberg. It was easy to direct the beam downward to sense the distance to the bottom.[1] It might take most of a day to lower and raise a weighted line, but echo sounders could take many bottom recordings in a minute. In June 1922 an echo sounder made the first continuous profile across an ocean basin aboard USS *Stewart*, a U.S. Navy vessel. Using an improved echo sounder based on Fessenden's design, the German research vessel *Meteor* made 14 profiles across the Atlantic from 1925 to 1927. The wandering path of the Mid-Atlantic Ridge was revealed, and its obvious coincidence with coastlines on both sides of the Atlantic stimulated the discussions that culminated in our present understanding of plate tectonics.

Echo sounding wasn't perfect. The ship's exact position was sometimes uncertain. The speed of sound through seawater varies with temperature, pressure, and salinity, and those variations made depth readings slightly inaccurate.

Simple depth sounder images (such as that shown in **Figure 4.2**) were also unable to resolve the fine detail that oceanographers needed to explore seabed features. Even so, researchers using depth sounder tracks had painstakingly compiled the first comprehensive charts of the ocean floor by 1959. (A portion of one of those hand-drawn charts is shown in Figure 4.18a.)

Two new techniques—made possible by improved sensors and high-speed computers—have been perfected to minimize inaccuracies and speed the process of bathymetry. The features discussed in this chapter have been studied using these (and other) systems, any of which is surely an improvement over lowering rocks into the ocean!

Multibeam Systems

Like an echo sounder, a multibeam system bounces sound off the seafloor to measure ocean depth. Unlike a simple echo sounder, a multibeam system may have as many as 121 beams radiating from a ship's hull. Fanning out at right angles to the direction of travel, these beams can cover a 120° arc (**Figure 4.3a**). Typically, a pulse of sound energy is

---

[1] Figure 2.17 shows this method of depth detection.

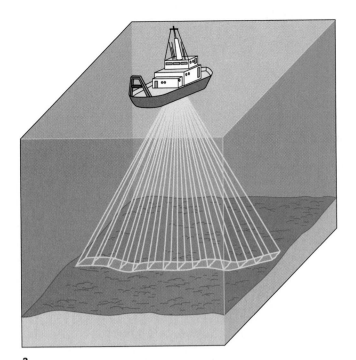

a

Andrew Goodwillie, SIO

b

Figure 4.3 (**a**) A multibeam echo sounder uses as many as 121 beams radiating from a ship's hull. Fanning out at right angles to the direction of travel, these beams can cover a 120° arc and measure a swath of bottom about 3.4 times as wide as the water is deep. Typically, a "ping" is sent toward the seabed every 10 seconds. Listening devices record sounds reflected from the bottom, but only from the narrow corridors corresponding to the outgoing pulse. Thus a multibeam system is much less susceptible to contour error than the single-beam device shown in Figure 4.2a. (**b**) A multibeam reading of a fragment of seafloor near the East Pacific Rise south of the tip of Baja California, Mexico. The uneven coverage reflects the path of the ship across the surface. Detailed analysis requires sailing a careful pattern, rather like one would use to mow a lawn evenly. Computer processing provides extraordinarily detailed images, such as those in Figures 4.11, 4.12, and 4.14.

sent toward the seabed every 10 seconds. Listening devices record sounds reflected from the bottom, but only from the narrow corridors corresponding to the outgoing pulse. Successive observations build a continuous swath of coverage beneath the ship. By "mowing the lawn"—moving the ship in a coverage pattern similar to one you would follow in cutting grass—researchers can build a complete map of an area (**Figure 4.3b**). (Further processing can yield remarkably detailed images like those of Figures 4.11 and 4.12.) Fewer than 200 research vessels are equipped with multibeam sys-

tems. At the present rate, charting the entire seafloor in this way would require more than 125 years.

Satellite Altimetry

Satellites cannot measure ocean depths directly, but they can measure small variations in the elevation of surface water. Using about a thousand radar pulses each second, the U.S. Navy's *Geosat* satellite (**Figure 4.4a**) measured its distance from the ocean surface to within 0.03 meter (1 inch)! Because the precise position of the satellite could be calculated, the average height of the ocean surface could be known with great accuracy.

Disregarding waves or tides or currents, researchers have found the ocean surface can vary from the ideal smooth (ellipsoid) shape by as much as 200 meters (660 feet). This is because the pull of gravity varies across the Earth's surface depending on the nearness (or distance) of massive parts of Earth. The mountain or ridge "pulls" water toward it from the sides, forming a mount of water over itself (**Figure 4.4b**). For example, a typical undersea volcano with a height of 2000 meters (6600 feet) above the seabed and a radius of 20 kilometers (32 miles) would produce a 2-meter (6.6-foot) rise in the ocean surface. (This mound cannot be seen with the unaided eye because the slope of the surface is very gradual.) The large features of the seabed are amazingly and accurately reproduced in the subtle standing irregularities of the sea surface (**Figure 4.4c**)!

Geosat and its successors, *TOPEX/Poseidon* and *Jason-1*, have allowed the rapid mapping of the world ocean floor from space. Hundreds of previously unknown features have been discovered through the data they have provided.

New bathymetric devices used to study seabed features include side-scan sonar and satellites that use sensitive radar for altimetry. Their use has revolutionized our understanding of ocean floor formation and topography.

The Shape of Ocean Floors

Most people think an ocean basin looks like a giant bathtub. They imagine that the continents drop off steeply just beyond the surf zone and that the ocean is deepest somewhere out in the middle. As is clear in **Figure 4.5,** bathymetric studies have shown this impression is wrong.

Why? As you read in the last chapter, plate tectonics theory suggests that Earth's surface is not a static arrangement of continents and ocean, but a dynamic mosaic of jostling lithospheric plates. The lighter continental lithosphere floats in isostatic equilibrium above the level of the heavier lithosphere of the ocean basins. The great density of the seabed partly

a

Figure 4.4 (**a**) *Geosat*, a U.S. Navy satellite that operated from 1985 through 1990, provided measurements of sea surface height from orbit. Moving above the ocean surface at 7 kilometers (4 miles) a second, *Geosat* bounced a thousand pulses of radar energy off the ocean every second. Height accuracy was within 0.03 meter (1 inch)! (**b**) Distortion of the sea surface above a seabed feature occurs when the extra gravitational attraction of the feature "pulls" water toward it from the sides, forming a mound of water over itself. (**c**) This view of the complex system of ridges, trenches, and fracture zones on the South Atlantic seafloor east of the tips of South America and Antarctica's Palmer Peninsula was derived in large part from *Geosat* data declassified by the U.S. Navy in 1995.

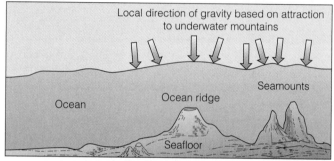

Local direction of gravity based on attraction to underwater mountains

Seamounts

Ocean

Ocean ridge

Seafloor

b

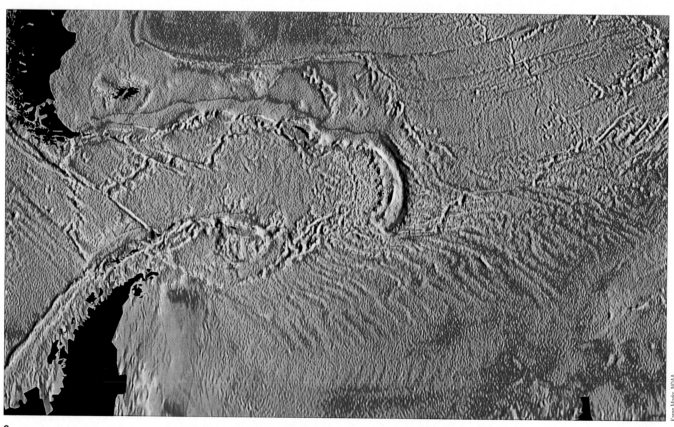

c

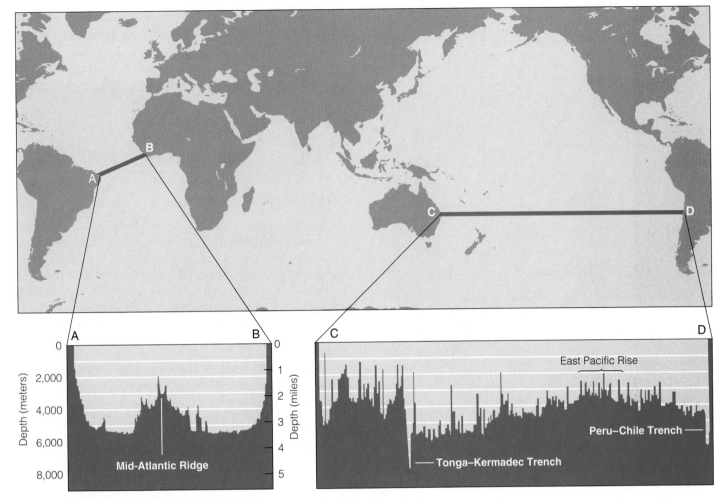

Figure 4.5 Typical cross sections of the Atlantic and Pacific ocean basins. The vertical exaggeration is 100:1.

explains why more than half of Earth's solid surface is at least 3,000 meters (10,000 feet) below sea level (**Figure 4.6**).

Notice in **Figure 4.7** the transition between the thick (and less dense) granitic rock of the continents and the relatively thin (and more dense) basalt of the deep-ocean floor. Near shore the features of the ocean floor are similar to those of the adjacent continents because they share the same granitic basement. The transition to basalt marks the *true* edge of the continent and divides ocean floors into two major provinces. The submerged outer edge of a continent is called the **continental margin.** The deep seafloor beyond the continental margin is properly called the **ocean basin.**

Seafloor features result from a combination of tectonic activity and the processes of erosion and deposition. The seabed is divided into continental margins (which resemble the adjacent continents) and deep-ocean basins (which do not).

Continental Margins

You learned in Chapter 3 that lithospheric plates converge, diverge, or slip past one another. As you might expect, the submerged edges of continents—continental margins—are greatly influenced by this tectonic activity. Continental margins that face the edges of *diverging* plates are called **passive margins** because relatively little earthquake or volcanic activity is now associated with them. Because they surround the Atlantic, passive margins are sometimes referred to as *Atlantic-type* margins. Continental margins near the edges of *converging* plates (or near places where plates are slipping past one another) are called **active margins** because of their earthquake and volcanic activity. Because of their prevalence in the Pacific, active margins are sometimes referred to as *Pacific-type* margins.

Active and passive margins west and east of South America are shown in **Figure 4.8.** Note that active margins coincide with plate boundaries but passive margins do not. Passive margins are also found outside the Atlantic, but active margins are confined mostly to the Pacific.

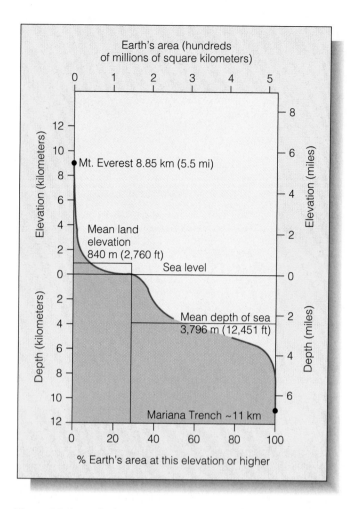

Figure 4.6 A graph showing the distribution of elevations and depths on Earth. This curve is not a land-to-sea profile of Earth, but rather a plot of the area of Earth's surface above any given elevation or depth below sea level. Note that more than half of Earth's solid surface is at least 3,000 meters (10,000 feet) below sea level. The average depth of the world ocean (3,796 meters or 12,451 feet) is much greater than the average height of the continents (840 meters or 2,760 feet). There are two dominant levels on our planet, one exposed and the other submerged.

Continental margins have two main divisions: a shallow, nearly flat continental *shelf* close to shore and a more steeply sloped continental *slope* to seaward.

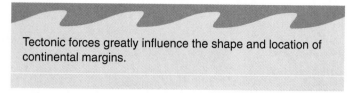

Tectonic forces greatly influence the shape and location of continental margins.

Continental Shelves

The shallow, submerged extension of a continent is called the **continental shelf.** Continental shelves, an extension of the adjacent continents, are underlain by granitic continental crust. They are much more like the continent than like the deep-ocean floor, and they may have hills, depressions, sedimentary rocks, and mineral and oil deposits similar to those on the dry land nearby. Earth's continental shelves are shown in **Figure 4.9.** Taken together, the area of the continental shelves is 7.4% of Earth's ocean area.

Figure 4.10 shows a passive-margin continental shelf characteristic of Atlantic Ocean edges. The broad shelf extends far from shore in a gentle incline, typically 1.7 meters per kilometer (0.1°, or about 9 feet per mile), much more gradual than the slope of a well-drained parking lot. Shelves along the margin of the Atlantic Ocean often reach 350 kilometers (220 miles) in width, and end at a depth of about 140 meters (460 feet) where a steeper drop-off begins.

The passive-margin shelves of the Atlantic Ocean formed as the fragments of Pangaea were carried away from each other by seafloor spreading. The continental lithosphere, thinned during initial rifting, cooled and contracted as it moved away from the spreading center, submerging the trailing edges of the continents and forming the shelves.

Most of the material comprising a shelf comes from erosion of the adjacent continental mass. Rivers assist in passive

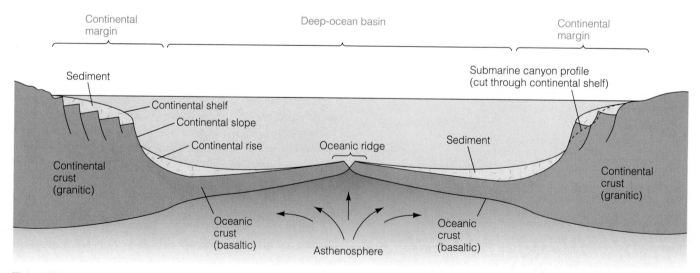

Figure 4.7 A cross section of a typical ocean basin flanked by passive continental margins. (The vertical scale has been greatly exaggerated to emphasize the basin contours.)

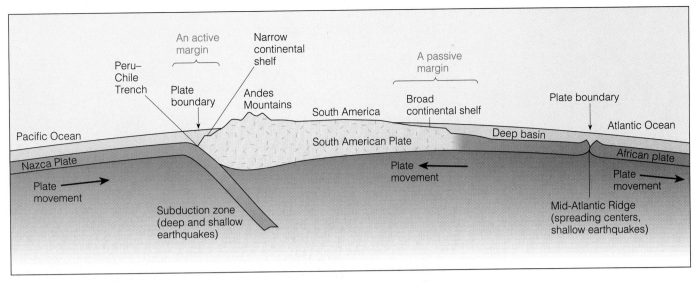

Figure 4.8 Typical continental margins bordering the leading (tectonically active) and trailing (passive) edges of a moving continent. (The vertical scale has been exaggerated.)

shelf building by transporting huge amounts of sediments to the shore from far inland. In some places the sediments accumulate behind natural dams formed by ancient reefs or ridges of granitic crust (see again Figure 4.7). The weight of the sediment isostatically depresses the continental edges and allows the sediment load to grow even thicker. Sediment at the outer edge of a shelf can be up to 15 kilometers (9 miles) thick and 150 million years old.

The width of a shelf is usually determined by its proximity to a plate boundary. You can see in Figure 4.8 that the shelf at the *passive* margin (east of South America) is broad, but the shelf at the *active* margin (west of South America) is very narrow. The widest shelf—1,280 kilometers (800 miles) across—lies north of Siberia in the tectonically quiet Arctic Sea. Shelf width depends not only on tectonics but also on

marine processes: Fast-moving ocean currents can sometimes prevent sediments from accumulating. For example, the east coast of Florida has a very narrow shelf because there is no natural offshore dam formed by ridges of granitic crust, and because the swift current of the nearby Gulf Stream scours surface sediment away. The west coast, however, has a broad shelf with a steep terminating slope (**Figure 4.11**).

The shelves of the active Pacific margins are generally not as broad and flat as the Atlantic shelves. An example is the abbreviated shelf off the west coast of South America, where the steep western slope of the Andes Mountains continues nearly uninterrupted beneath the sea into the depths of the Peru–Chile Trench (see again Figure 4.8). Active-margin shelves have more varied topography than passive-margin shelves; the character of continental shelves at an active

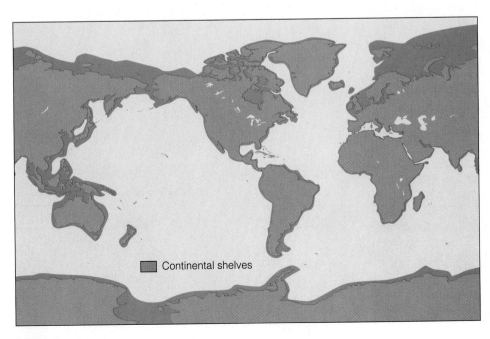

Figure 4.9 The worldwide distribution of continental shelves. The continental islands of Ireland and Great Britain are part of the continent of Europe, and Alaska connects to Siberia.

Continental shelves

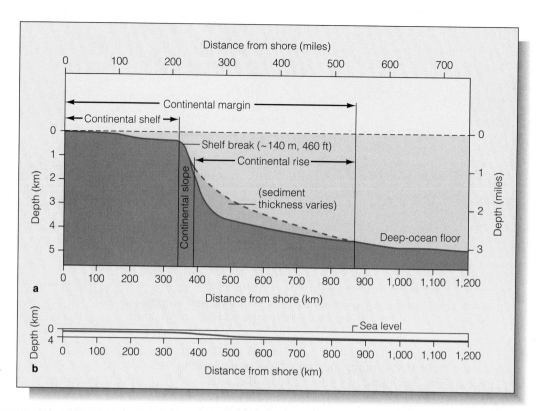

Figure 4.10 The features of a passive continental margin. (**a**) Vertical exaggeration 50 : 1. (**b**) No vertical exaggeration.

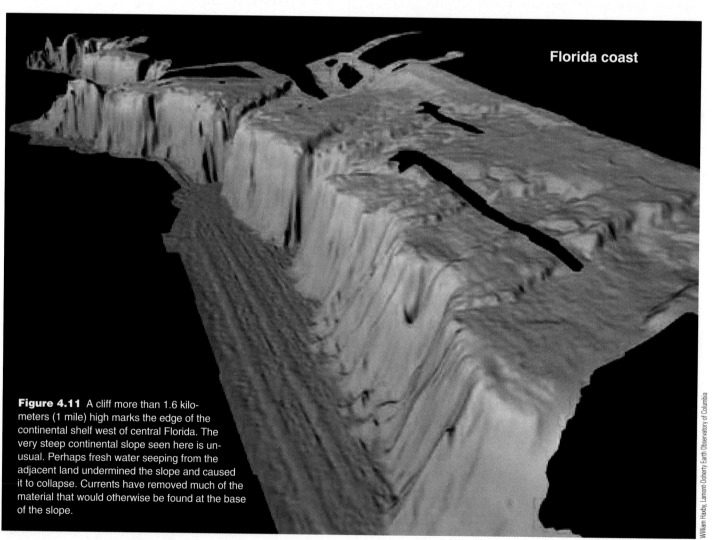

Figure 4.11 A cliff more than 1.6 kilometers (1 mile) high marks the edge of the continental shelf west of central Florida. The very steep continental slope seen here is unusual. Perhaps fresh water seeping from the adjacent land undermined the slope and caused it to collapse. Currents have removed much of the material that would otherwise be found at the base of the slope.

margin may be determined more by faulting, volcanism, and tectonic deformation than by sedimentation.

Because of their gentle slope, continental shelves are greatly influenced by changes in sea level. Around 18,000 years ago—at the height of the last **ice age** (a period of extensive glaciation)—massive ice caps covered huge regions of the continent. The water that formed the thick ice sheets was taken from the ocean, and sea level fell about 125 meters (410 feet) below its present position. The continental shelves were almost completely exposed, and the surface area of the continents was about 18% greater than it is today. Rivers and waves cut into the sediments that had accumulated during periods of higher sea level, and they transported coarse sediments to their present locations at the shelves' outer edges. Sea level began to rise again when the ice caps melted, and sediments again began to accumulate on the shelves. (More on the history and effects of sea level change will be found in the discussion of coasts in Chapter 11.)

The continental shelves have been the focus of intense exploration for natural resources. Because shelves are the submerged margins of continents, any deposits of oil or minerals along a coast are likely to continue offshore. Water depth over shelves averages only about 75 meters (250 feet), so large areas of the shelves are accessible to mining and drilling activities. Many of the techniques used to find and exploit natural resources on land can also be used on the continental shelves. Resource development requires intense scientific investigations, and our understanding of the geology of the shelves has benefited greatly from the search for offshore oil and natural gas.

Near shore, the features of the ocean floor are similar to those of the adjacent continents because they share the same granitic basement. The transition to basalt marks the true edge of the continent

Continental Slopes   4.9

The **continental slope** is the transition between the gently descending continental shelf and the deep-ocean floor. Continental slopes are formed of sediments that reach the built-out edge of the shelf and are transported over the side. At active margins, a slope may also include marine sediments scraped off a descending plate during subduction (**Figure 4.12**). The inclination of a typical continental slope is about 4° (70 meters per kilometer; 370 feet per mile), slightly steeper than the steepest road slope allowed on the interstate highway system. As Figure 4.10b implies, even the

Figure 4.12 Folded ridges of sediment cover the ocean floor west of Oregon. The crinkles result from the collision between the North American Plate and the Juan de Fuca Plate. Like a bulldozer, the North American Plate scrapes sediment from the subducting Juan de Fuca Plate and piles it into folds. To the south (the upper right), part of the Juan de Fuca Plate breaks through the sediment.

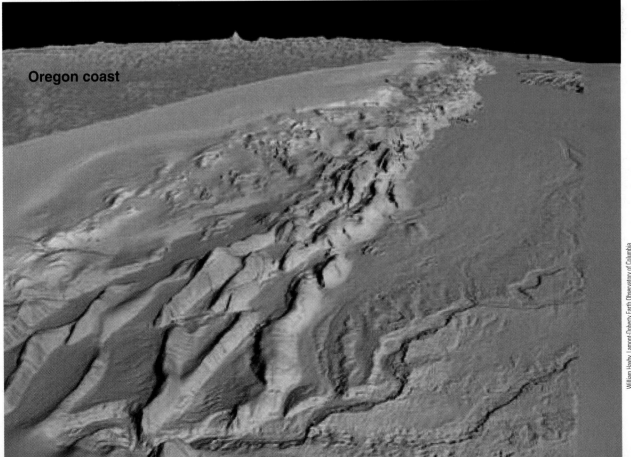

Oregon coast

William Haxby, Lamont-Doherty Earth Observatory of Columbia

steepest of these slopes is not precipitous: A 25° slope is the greatest incline yet discovered. In general, continental slopes at active margins are steeper than those at passive margins. Continental slopes average about 20 kilometers (12 miles) wide and end at the continental rise, usually at a depth of about 3,700 meters (12,000 feet). The bottom of the continental slope is the true edge of a continent.

The **shelf break** marks the abrupt transition from continental shelf to continental slope. The depth of water at the shelf break is surprisingly constant—about 140 meters (460 feet) worldwide—but there are exceptions. The great weight of ice on Antarctica, for example, has isostatically depressed that continent, and the depth at the shelf break is 300–400 meters (1,000–1,300 feet). The shelf break in Greenland is similarly depressed.

Submarine Canyons

Submarine canyons cut into the continental shelf and slope, often terminating on the deep-ocean floor in a fan-shaped wedge of sediment (**Figure 4.13**). More than a hundred submarine canyons nick the edge of nearly all of Earth's continental shelves. The canyons generally trend at right angles to the shoreline (and shelf edge), sometimes beginning very close to shore. Congo Canyon actually extends into the African continent as a deep estuary at the mouth of the Congo River. These enigmatic features can be quite large. In fact, submarine canyons similar in size and profile to Arizona's Grand Canyon have been discovered!

Hudson Canyon, a typical large canyon on a passive margin, is shown in **Figure 4.14.** Like many submarine canyons, Hudson Canyon is located just offshore of the mouth of a river or stream—in this case, New York's Hudson River. Because of their similarity to canyons on land, submarine canyons appear to have been created by erosion; so marine geologists initially thought the canyons were carved into the shelves by stream erosion at times of lower sea level. But most researchers agree that sea level has never fallen more than 200 meters (660 feet) below its present level in the

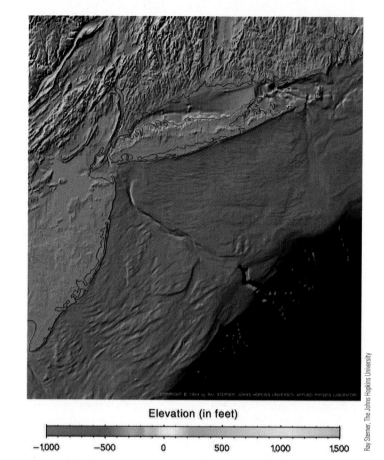

Ray Sterner, The Johns Hopkins University

Elevation (in feet)

| −1,000 | −500 | 0 | 500 | 1,000 | 1,500 |

Figure 4.14 A side-scan image of Hudson Canyon east of New Jersey. The shelf in this area is broad. The canyon can be seen nicking the shelf–slope junction and then continuing toward the abyssal plain to the southwest. The underwater topology has been exaggerated by a factor of 5 relative to the land topography.

last 600 million years. Stream erosion could account for the shape of the uppermost parts of the canyons. However, since submarine canyons can be traced to depths in excess of 3,000 meters (10,000 feet) below sea level, stream erosion could not have played a direct role in cutting their lower depths.

What, then, caused the submarine canyons to form? Local landslides or sediment liquefaction triggered by earthquakes sometimes causes an abrasive underwater "avalanche" of sediments. These mass movements of sediment, called **turbidity currents,** occur when turbulence mixes sediments into water above a sloping bottom. The sediment-filled water is denser than the surrounding water, so the thick, muddy fluid runs down the slope at speeds up to 27 kilometers (17 miles) per hour. **Figure 4.15** is a rare photograph of a turbidity current.

What is the connection between turbidity currents and submarine canyons? Most geologists believe that the canyons have been formed by abrasive turbidity currents plunging down the canyons. Small amounts of debris may cascade continuously down the canyons (**Figure 4.16**), but earthquakes can shake loose

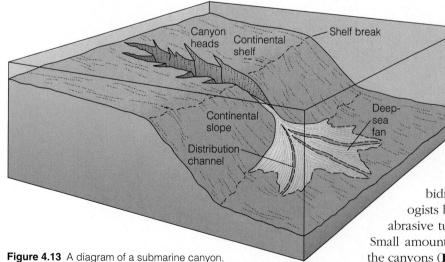

Figure 4.13 A diagram of a submarine canyon.

Figure 4.15 A turbidity current flowing down a submerged slope off the island of Jamaica. The propeller of a submarine caused the turbidity current by disturbing sediment along the slope.

huge masses of sediment that rush down the edge of the shelf, scouring the canyon deeper as they go. In this way the canyons can be cut to depths far below the reach of streams even during the low sea levels of the ice ages.

Continental Rises

Along passive margins, the oceanic crust at the base of the continental slope is covered by an apron of accumulated sediment called the **continental rise** (see again Figure 4.7). Sediments from the shelf slowly descend to the ocean floor along the whole continental slope, but most of the sediments that form the continental rise are transported to the area by turbidity currents. The width of the rise varies from 100 to 1,000 kilometers (63 to 630 miles), and its slope is gradual—about one-eighth that of the continental slope. One of the widest and thickest continental rises has formed in the Bay of Bengal at the mouths of the Ganges–Brahmaputra River, the most sediment-laden of the world's great rivers.

The submerged outer edge of a continent is called the continental margin. Features of continental margins include continental shelves, continental slopes, submarine canyons, and continental rises.

Deep-Ocean Basins

Away from the margins of continents, the structure of the ocean floor is quite different. Here the seafloor is a blanket of sediments up to 5 kilometers (3 miles) thick overlying basaltic rocks. Deep-ocean basins constitute more than half of Earth's surface.

The deep-ocean floor consists mainly of oceanic ridge systems and the adjacent sediment-covered plains. Deep basins may be rimmed by trenches or by masses of sediment. Flat expanses are interrupted by islands, hills, active and extinct volcanoes, and active zones of seafloor spreading. The sediments on the deep-ocean floor reflect the history of the surrounding continents, the biological productivity of the overlying water, and the ages of the basins themselves.

Oceanic Ridges

If the ocean evaporated, the oceanic ridges would be Earth's most remarkable and obvious feature. An **oceanic ridge** is a mountainous chain of young basaltic rock at the active

Figure 4.16 A continuous cascade of sediment at the head of San Lucas submarine canyon (off the coast of Baja California, Mexico), which may be eroding the narrow gorge in conjunction with occasional turbidity currents.

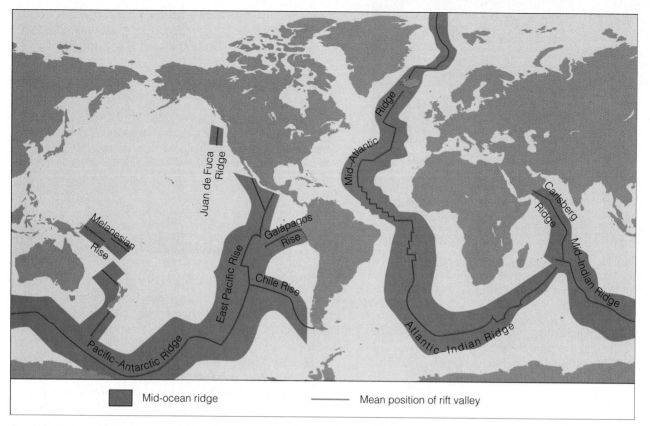

Figure 4.17 The oceanic ridge system stretches some 65,000 kilometers (40,000 miles) around Earth. If the ocean evaporated, the ridge system would be Earth's most remarkable and obvious feature.

spreading center of an ocean. Stretching 65,000 kilometers (40,000 miles), more than 1½ times Earth's circumference, oceanic ridges girdle the globe like seams surrounding a softball (**Figure 4.17**). The rugged ridges, which often are devoid of sediment, rise about 2 kilometers (1.25 miles) above the seafloor. In places they project above the surface to form islands such as Iceland, the Azores, and Easter Island. Oceanic ridges and their associated structures account for 22% of the world's solid surface area (all the land above sea level accounts for 29%). Although these features are often called mid-ocean ridges, less than 60% of their length actually exists along the centers of ocean basins.

As we saw in our discussion of plate tectonics, the rift zones associated with oceanic ridges are sources of new ocean floor where lithospheric plates diverge. The oceanic ridges are widest where they are most active. The youngest rock is located at the active ridge center, and rock becomes older with distance from the center. As the lithosphere cools, it shrinks and subsides. Slowly spreading ridges have a steeper profile than rapidly spreading ones because slowly diverging seafloor cools and shrinks closer to the spreading center. **Figure 4.18a,** a bathymetric map of the North Atlantic, clearly shows the great extent of the Mid-Atlantic Ridge, a typical oceanic ridge. **Figure 4.18b,** a multibeam image, provides a detailed look at the young central rift.

As can be seen in Figure 4.18a, the Mid-Atlantic Ridge is offset at more or less regular intervals by transform faults. A

fault is a fracture in the lithosphere along which movement has occurred, and **transform faults** are fractures along which lithospheric plates slide horizontally (**Figure 4.19**). When segments of a ridge system are offset, the fault connecting the axis of the ridge is a transform fault. Shallow earthquakes are common along transform faults. Since the ocean floor cannot expand evenly on the surface of a sphere, plate divergence on the spherical Earth can be only irregular and asymmetrical, and transform faults and fracture zones result.

Transform faults are the active part of **fracture zones.** Extending outward from the ridge axis, fracture zones are seismically inactive areas that show evidence of past transform fault activity. While segments of a lithospheric plate on either side of a transform fault move in *opposite* directions, the plate segments adjacent to the outward segments of a fracture zone move in the *same* direction, as Figure 4.19 shows.

Hydrothermal Vents

Some of the most exciting features of the ocean basins are the **hydrothermal vents.** Hot springs were discovered on oceanic ridges in 1977 by Robert Ballard and J. F. Grassle of the Woods Hole Oceanographic Institution. Diving in *Alvin* at 3 kilometers (1.9 miles) near the Galápagos Islands along the East Pacific Rise (an oceanic ridge), they came across

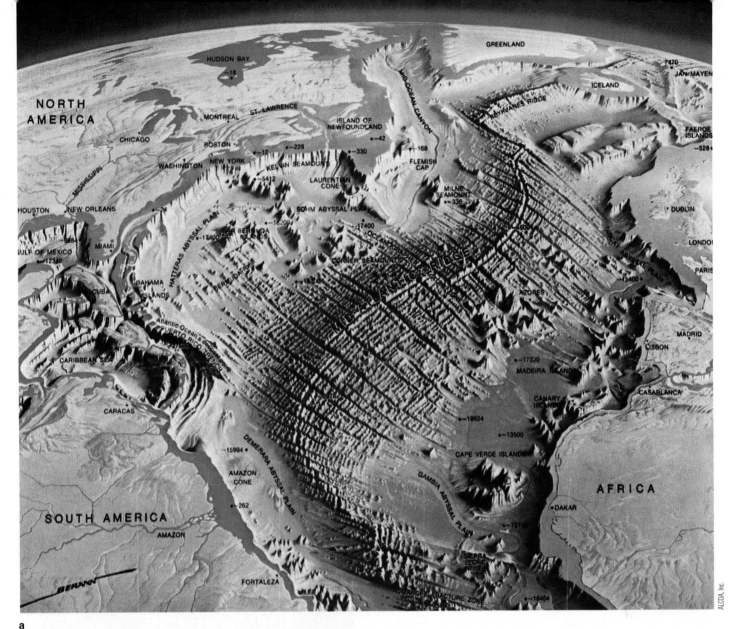

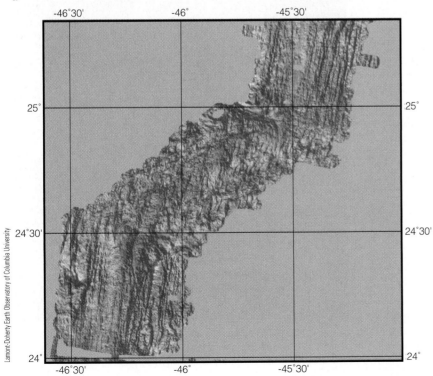

Figure 4.18 (**a**) A hand-drawn map of a portion of the Atlantic Ocean floor showing some major oceanic features: mid-ocean ridge, transform faults, fracture zones, submarine canyons, seamounts, continental rises, trenches, and abyssal plains. Depths are in feet. The map is vertically exaggerated. (**b**) The fine structure of the central portion of the Mid-Atlantic Ridge between Florida and western Africa. The depressed central valley (the spreading center, shown in blue) is clearly visible in this computer-generated multibeam image.

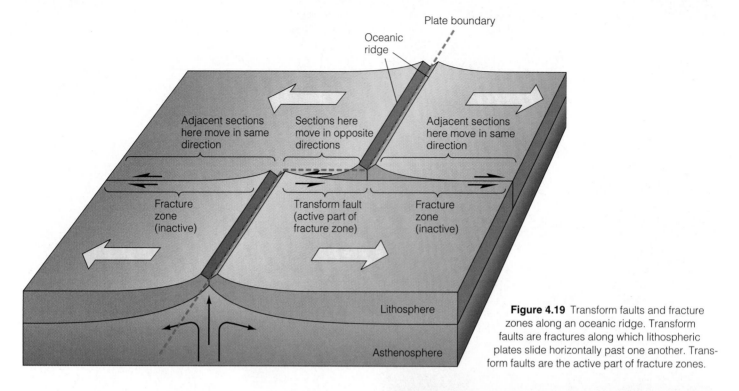

Figure 4.19 Transform faults and fracture zones along an oceanic ridge. Transform faults are fractures along which lithospheric plates slide horizontally past one another. Transform faults are the active part of fracture zones.

rocky chimneys up to 20 meters (66 feet) high, from which dark, mineral-laden water was blasting at 350°C (660°F) (**Figure 4.20**). Only the great pressure at this depth prevented the escaping water from flashing to steam. These *black smokers*, as they were nicknamed, fascinate marine geologists. It is believed that water descends through fissures and cracks in the ridge floor until it comes into contact with very hot rocks associated with active seafloor spreading. There the superheated, chemically active water dissolves minerals and gases and escapes upward through the vents by convection (**Figure 4.21**).

Since that first discovery, vents have been found on the Mid-Atlantic Ridge east of Florida, in the Sea of Cortez east and south of Baja California, and on the Juan de Fuca Ridge off the coasts of Washington and Oregon. Scientists now believe that hydrothermal vents may be very common on oceanic ridges, especially in zones of rapid seafloor spreading. In July 1990 vents were discovered in fresh water, at the bottom of Lake Baikal in southern Siberia. This discovery suggests that the world's oldest and deepest lake may someday become part of the ocean as Asia slowly breaks apart.

Not all vents form chimneys of mineral deposits—some are simply cracks in the seabed, or porous mounds, or broad segments of ocean ridge floor through which warm, mineral-laden water percolates upward. Cooler vents result when hot, rising water mixes with cold bottom water before reaching the surface. Water temperature in the vicinity of most hydrothermal vents averages 8–16°C (46–61°F), much warmer than usual for ocean bottom water, which has an average temperature of 3–4°C (37–39°F). We will study the unusual communities of animals that populate the vents in Chapter 14.

A volume of water equal to the volume of the world ocean is thought to circulate through the hot oceanic crust at spreading centers about every 10 million years! The water

Figure 4.20 A black smoker discovered at a depth of about 2,800 meters (9,200 feet) along the East Pacific Rise.

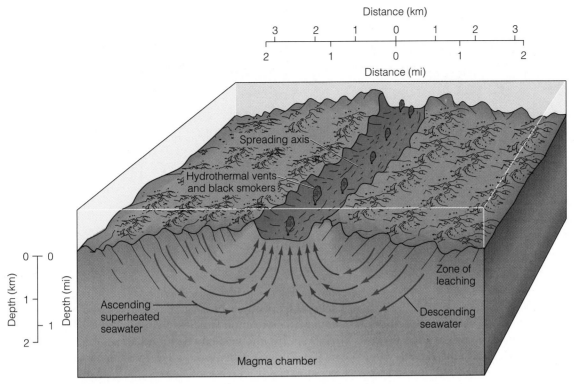

Figure 4.21 A cross section of the central part of a mid-ocean ridge—similar to that shown in Figure 4.18b—showing the origin of hydrothermal vents. Cool water (blue arrows) is heated as it descends toward the hot magma chamber, leaching sulfur, iron, copper, zinc, and other materials from the surrounding rocks. The heated water (red arrows) returning to the surface carries these elements upward, discharging them at the hydrothermal springs on the seafloor. The areas around the vents support unique communities of organisms (see Chapter 14).

coming from the vents and seeping from the floor is more acidic, is enriched with metals, and has higher concentrations of dissolved gas than seawater. The heat and chemicals issuing from these structures may play important roles in the chemical composition of seawater and the atmosphere, and in the formation of mineral deposits.

Abyssal Plains and Abyssal Hills

A quarter of Earth's surface consists of abyssal plains and abyssal hills. *Abyssal* is an adjective derived from a Greek word meaning "without bottom." While this is obviously not literally true, you can appreciate how the term came into use following the *Challenger* expedition's laborious soundings of these extremely deep areas!

Abyssal plains are flat, featureless expanses of sediment-covered ocean floor found on the periphery of all oceans. They are most common in the Atlantic, less so in the Indian Ocean, and relatively rare in the active Pacific, where peripheral trenches trap most of the sediments flowing from the continents. They lie between the continental margins and the oceanic ridges about 3,700 to 5,500 meters (12,000 to 18,000 feet) below the surface (see again Figure 4.18a). The Canary Abyssal Plain, a huge plain west of the Canary Islands in the North Atlantic, has an area of about 900,000 square kilometers (350,000 square miles).

Abyssal plains are extraordinarily flat. A 1947 survey by the Woods Hole Oceanographic Institution ship *Atlantis* found that a large Atlantic abyssal plain varies no more than a few meters in depth over its entire area. Such flatness is caused by the smoothing effect of the layers of sediment, which often exceed 1,000 meters (3,300 feet) in thickness. Most of the sediment that forms the abyssal plains appears to be of terrestrial or shallow-water origin, not derived from biological activity in the ocean above. Some of it may have been transported to the plains by winds or turbidity currents. These deep sediment layers mask irregularities in the underlying ocean crust, but a powerful type of echo sounder can "see" through this sediment to reveal the complex topography of the basaltic basin floor below (**Figure 4.22**). The broad basaltic shoulders of the Mid-Atlantic Ridge extend beneath this cloak of sediment almost as far as the bordering continental slopes.

Abyssal plain sediments may not be thick enough to cover the underlying basaltic floor near the edges bordering the oceanic ridges. Here the plains are punctuated by **abyssal hills**—small, sediment-covered extinct volcanoes or intrusions of once-molten rock, usually less than 200 meters (650 feet) high (one is seen extending above the flat surface in Figure 4.22). These abundant features are associated with

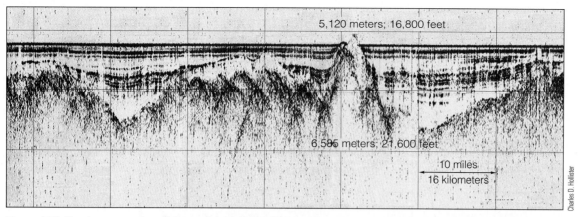

5,120 meters; 16,800 feet

6,585 meters; 21,600 feet

10 miles
16 kilometers

Charles D. Hollister

Figure 4.22 The deep, smooth sediments of the Atlantic's Northern Madeira Abyssal Plain bury 100-million-year-old mountains.

seafloor spreading; they form when newly formed crust moves away from the center of a ridge, stretches, and cracks. Some blocks of the crust drop to form valleys, and others remain higher as hills. Lava erupting from the ridge flows along the fractures, coating the hills. This helps explain why abyssal hills occur in lines parallel to the flanks of the nearby oceanic ridge, and why they occur most abundantly in places where the rate of seafloor spreading is fastest. Abyssal plains and hills account for nearly all the area of deep-ocean floor that is not part of the oceanic ridge system.

Seamounts and Guyots   4.16

The ocean floor is dotted with thousands of volcanic projections that do not rise above the surface of the sea. These projections are called **seamounts.** Seamounts are circular or elliptical, more than 1 kilometer (0.6 mile) in height, with relatively steep slopes of 20° to 25°. (Abyssal hills, in contrast, are much more abundant, less than a kilometer high, and not as steep.) Seamounts may be found alone or in groups of 10 to 100. Though many form at hot spots, most are thought to be submerged inactive volcanoes that formed at spreading centers (**Figure 4.23**). Movement of the lithosphere away from spreading centers has carried them outward and downward to their present positions. As many as 10,000 seamounts are thought to exist in the Pacific, about half the world total.

Guyots are flat-topped seamounts that once were tall enough to approach or penetrate the sea surface. Generally they are confined to the west-central Pacific. The flat top suggests that they were eroded by wave action when they were near sea level. Their plateaulike tops eventually sank too deep for wave erosion to continue wearing them down. Like the more abundant seamounts, most guyots were formed near spreading centers and transported outward and downward as the seafloor moves away from a spreading center and cools.

Trenches and Island Arcs   4.17

A **trench** is an arc-shaped depression in the deep-ocean floor. These flat-bottomed creases in the seafloor occur where a converging oceanic plate is subducted. The water temperature at the bottom of a trench is slightly cooler than the near-freezing temperatures of the adjacent flat ocean floor, reflecting the fact that trenches are underlain by old, relatively cold ocean crust sinking into the upper mantle. Trenches (and their associated island arcs topped by erupting volcanoes) are among the most active geological features on Earth. Great earthquakes and tsunami (huge waves we will discuss in Chapter 9) often originate in them. **Figure 4.24**

Stanley Hart, WHOI

169°08′W
168°04′W
169°00′W
168°56′W
168°52′W

14°12′S
14°16′S
14°20′S
14°24′S

-5000   -4000   -3000   -2000   -1000   0
Depth (m)

Figure 4.23 An undersea volcano east of the eastern-most island of the Samoan chain in the Pacific. Vailulu'u, as it has been named, rises from an ocean depth of 4,800 meters (15,700 feet) to within 590 meters (1,900 feet) of the ocean surface.

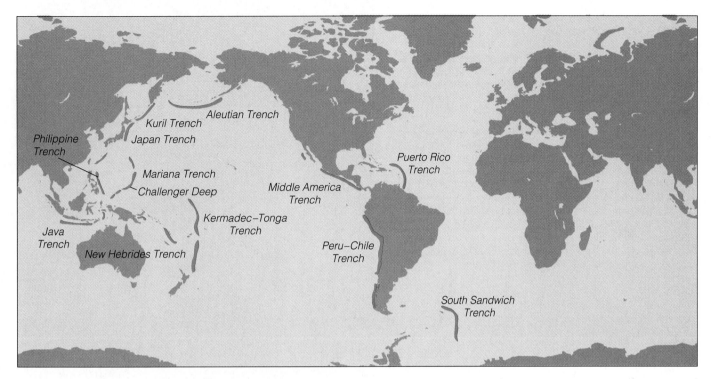

Figure 4.24 Oceanic trenches of the world.

shows the distribution of the ocean's major trenches. It is not surprising that most are around the edges of the active Pacific.

Trenches are the deepest places in Earth's crust, 3–6 kilometers (1.9–3.7 miles) deeper than the adjacent basin floor. The ocean's greatest depth is the Mariana Trench of the western Pacific, where the ocean bottom is 11,022 meters (36,163 feet) below sea level, 20% deeper than Mt. Everest is high (**Figure 4.25**). The Mariana Trench is about 70 kilome-

ters (44 miles) wide and 2,550 kilometers (1,600 miles) long, typical dimensions for these structures.

Trenches are curving chains of V-shaped indentations. The trenches are curved because of the geometry of plate interactions on a sphere. The convex sides of these curves generally face the open ocean (see again Figure 4.24). The trench walls on the island side of the depressions are steeper than those on the seaward side, indicating the direction of plate subduction. The sides of trenches become steeper with

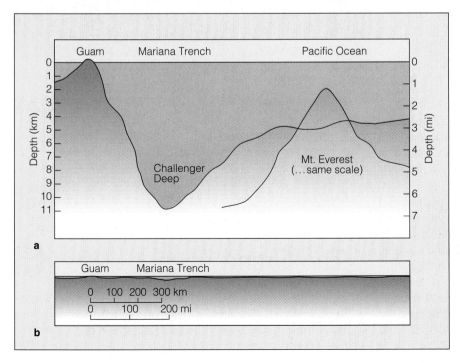

Figure 4.25 The Mariana Trench. (**a**) Comparing the Challenger Deep and Mt. Everest at the same scale shows that the deepest part of the Mariana Trench is about 20% deeper than the mountain is high. (**b**) The Mariana Trench shown without vertical exaggeration.

Figure 4.26 A technological *tour de force*, derived from data provided to the National Geophysical Data Center from satellites and shipborne sensors, shows all the features discussed in this chapter. These features—and a basic understanding of the geological reasons for their existence—will help you recall the dramatic nature and history of the seafloor that we have discussed in the past two chapters. Key to features:

(1) Aleutian Trench, (2) Hawaiian Islands, (3) Juan de Fuca Ridge, (4) Clipperton Fracture Zone, (5) Peru–Chile Trench, (6) East Pacific Rise, (7) Mariana Trench, Challenger Deep, (8) Mid-Atlantic Ridge, (9) South Sandwich Trench, (10) Red Sea, (11) Easter Island, (12) Galápagos Rift, (13) Iceland, (14) Gulf of Aden,(15) Pacific–Antarctic Ridge, (16) Azores, (17) Tristan de Cunha, (18) Kermadec Trench, (19) Tonga Trench, (20) Hatteras Abyssal Plain, (21) Grand Banks, (22) Mid-Indian Ridge, (23) Atlantic–Indian Ridge, (24) Maldive Islands, (25) Eltanin Fracture Zone, (26) Emperor Seamounts, (27) Java Trench, (28) Great Barrier Reef, (29) Kuril Trench, (30) Japan Trench.

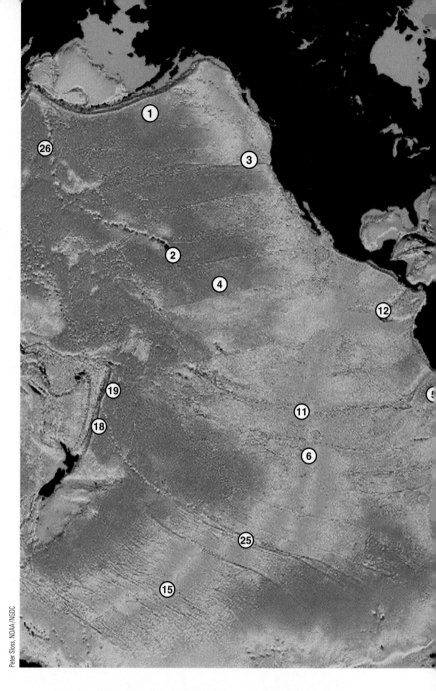

Peter Sloss, NOAA/NGDC

depth, normally reaching angles of about 10° to 16° before flattening to a floor underlain by thick sediment. (Parts of the concave wall of the Kermadec–Tonga Trench are the world's steepest at 45°.) No continental rise occurs along coasts with trenches because the sediment that would form the rise ends up at the bottom of the trench.

Island arcs, curving chains of volcanic islands and seamounts, are almost always found parallel to the concave edges of trenches. As you may remember from Chapter 3, trenches and island arcs are formed by tectonic and volcanic activity associated with subduction. The descending lithospheric plate contains some materials that melt as the plate sinks into the mantle. These materials rise to the surface as magmas and lavas that form the chain of islands behind the trench. The Aleutian Islands, most Caribbean islands, and the Marianas Islands are island arcs.

> The deep-ocean floor beyond the continental margin is called the ocean basin. Features of deep-ocean basins include oceanic ridges, hydrothermal vents, abyssal plains and hills, seamounts, guyots, trenches, and island arcs.

The Grand Tour

Researchers at the National Oceanic and Atmospheric Administration have generated a map of the world ocean floor based on satellite observations of the shape of the sea surface (**Figure 4.26**). The graphic shows all the features dis-

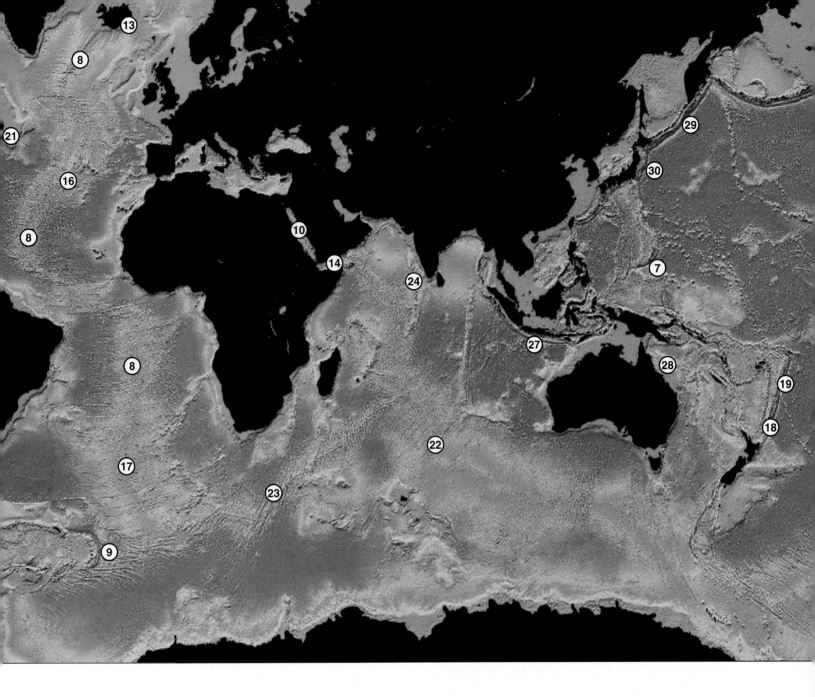

cussed in this chapter. These features—and a basic understanding of the geological reasons for their existence—will help you recall the dramatic nature and history of the seafloor that we have discussed in the past two chapters.

QUESTIONS FROM STUDENTS

1. Shouldn't the ocean be deeper in the middle? And why is Iceland one of only a few places in the world where oceanic crust is found above sea level?

Understanding *why* there are ocean basins explains their contours. A basin usually contains an expanding ridge that is higher than the surrounding bed. Oceanic crust is thin and dense. Because of isostatic equilibrium, the ocean floor lies at a lower elevation than the thicker, less dense, higher conti-

nents. Water filled the lower elevations first, submerging nearly all the basaltic basement we now call ocean floor. In areas of rapid seafloor spreading, or at hot spots, peaks are occasionally pushed toward the ocean surface by the large volume of mantle material rising in plumes from below. Large quantities of erupted magma (lava) then build the crests above sea level to form islands like Iceland. The Azores is another place on the Mid-Atlantic Ridge where this is happening.

2. Turbidity currents seem important in forming canyons and distributing deep sediments over abyssal plains. Has anybody ever seen a turbidity current in action?

Yes, surprisingly. In the late 1940s the Dutch geologist Philip Kuenen produced turbidity currents in his laboratory by pouring muddy water into a trough with a sloping bottom. His observations confirmed nineteenth-century reports that the muddy Rhône River continued to flow in a dense stream along the bottom of Lake Geneva. In the 1960s Robert Dill and Francis Shepard viewed sandfalls in Scripps Canyon from a diving saucer, and French researchers have recently photographed these currents in the Mediterranean.

3. Is Alvin *the deepest-diving, human-carrying research submersible?*

No, that honor belongs to *Shinkai 6500.* Operated by a consortium of Japanese research institutions and industries, *Shinkai 6500* is the newest deep-diving vehicle capable of carrying a human crew. Now the deepest-diving submersible in operation, *Shinkai 6500* safely descended to a depth of 6,527 meters (21,409 feet) on 11 August 1989, and it can reach all but the deepest trench floors, or about 98% of the world ocean bottom.

4. Wouldn't wave action and tides hopelessly clutter the radar signals sent from satellites to determine sea surface height?

Satellite altimetry is one of the most sophisticated oceanographic uses of high-speed computer processing. Imagine the processing power needed to reduce the data generated by more than a thousand radar pulses from orbit each second! Programmers subtract predicted tidal height from the measurements and then use algorithms to average and cancel wave crests and troughs. The remaining sea surface height—determined to perhaps 2 centimeters (slightly less than 1 inch)—is due to gravitational variations caused by submerged features. Still more processing is needed to generate graphic images like that of Figure 4.4c.

The procedures were pioneered by the U.S. Navy and the Office of Naval Research. The initial goal was to provide detailed seafloor maps for use in antisubmarine defense.

CHAPTER SUMMARY

The ocean floor can be divided into two regions: continental margins and deep-ocean basins. The continental margin, the relatively shallow ocean floor nearest the shore, consists of the continental shelf and the continental slope. The continental margin shares the structure of the adjacent continents, but the deep-ocean floor away from land has a much different origin and history. Prominent features of the deep-ocean basins include rugged oceanic ridges, flat abyssal plains, occasional deep trenches, and curving chains of volcanic islands. Most of the ocean floor is blanketed with sediment. The processes of plate tectonics, erosion, and sediment deposition have shaped the continental margins and ocean basins.

ON-LINE STUDY RESOURCES

The Web site for this book contains a wealth of helpful study aids. Log on to:

http://info.brookscole.com/garrison

and select *Essentials of Oceanography,* 3rd edition. Select Chapter 4 from the drop-down menu or click on one of these resource areas:

- For study and review, **Chapter at a Glance** gives you an outline of the chapter, **Chapter Summary** allows you to review the chapter's main ideas, and **Glossary** lists concepts and terms for the chapter along with their definitions.

- To test your mastery of the Terms and Concepts to Remember for this chapter, you can use the electronic **Flash Cards,** play the **Concentration** game, or work the **Crossword Puzzle.**

- For practice quizzing, try the multiple-choice **Tutorial Quiz,** the ten-question **True/False Quiz,** or the **Image Analysis Quiz,** which poses questions based on art and photos from the chapter.

TERMS AND CONCEPTS TO REMEMBER

abyssal hill	ice age
abyssal plain	island arc
active margin	ocean basin
bathymetry	oceanic ridge
continental margin	passive margin
continental rise	seamount
continental shelf	shelf break
continental slope	submarine canyon
fracture zone	transform fault
guyot	trench
hydrothermal vent	turbidity current

STUDY QUESTIONS

1. Why did people think an ocean was deepest at its center? What changed their minds? How is modern bathymetry accomplished?

2. Draw a rough outline of an ocean basin. Label the major parts.

3. What do the facts that granite underlies the edges of continents and basalt underlies deep ocean basins suggest?

4. The terms *leading* and *trailing* are also used to describe continental margins. How do you suppose these words relate to *active* and *passive,* or *Atlantic-type* and *Pacific-type* used in the text?

5. Why are abyssal plains relatively rare in the Pacific?

6. Answer this question if you have already read Chapter 3: Your time machine has been programmed to deliver you to Frankfurt on a chilly evening in January 1912 to hear Wegener's lectures on continental drift. What two illustrations from this chapter would you take with you to cheer him up after the lecture? Why did you select those particular illustrations?

Earth Systems Today CD Question

1. Go to "Plate Tectonics," then "Expedition." Can you find the boundaries? Using your understanding of the influence of tectonic activity on the location and characteristics of a coast, can you predict coastal features after surveying and exploring the coast in question?

RESOURCES FOR FURTHER READING AND RESEARCH

The Web site for this book contains many ideas for further reading and research. Log on to:

http://info.brookscole.com/garrison

and select *Essentials of Oceanography*, 3rd edition. Select Chapter 4 from the drop-down menu or click on one of these resource areas:

- **Additional Readings** lists the major books and articles consulted in writing this chapter, along with comments from the author about their content and reading level.

- **Hypercontents** takes you to an extensive list of current links to Internet sites with news, research, and images related to individual subjects in the chapter. Just click on the icon that corresponds to a numbered section to see the links for that subject.

- **Internet Exercises** are critical thinking questions that involve research on the Internet with starter URLs provided.

- **InfoTrac College Edition Exercises** leads you to Critical Thinking Projects that use InfoTrac College Edition as a research tool.

- **Regional InfoTrac College Edition** articles are organized into East Coast, West Coast, and Gulf Coast regions, allowing you to study oceanography on a more local level.

 For more readings, go to InfoTrac College Edition, your on-line research library, at:

http://infotrac.thomsonlearning.com

Table 5.2 Classification of Marine Sediments by Source of Particles

Sediment Type	Source	Examples	Distribution	Percent of All Ocean Floor Area Covered
Terrigenous	Erosion of land, volcanic eruptions, blown dust	Quartz sand, clays, estuarine mud	Dominant on continental margins, abyssal plains, polar ocean floors	~45%
Biogenous	Organic; accumulation of hard parts of some marine organisms	Calcareous and siliceous oozes	Dominant on deep-ocean floor (siliceous ooze below about 5 km)	~55%
Hydrogenous (authigenic)	Precipitation of dissolved minerals from water, often by bacteria	Manganese nodules, phosphorite deposits	Present with other, more dominant sediments	<1%
Cosmogenous	Dust from space, meteorite debris	Tektite spheres, glassy nodules	Mixed in very small proportion with more dominant sediments	0%

Sources: Kennett, *Marine Geology*, 1982; Weihaupt, *Exploration of the Oceans*, 1979; Sverdrup, Johnson, and Fleming, *The Oceans: Their Physics, Chemistry, and General Biology*, 1942.

Sediment is particles of organic or inorganic matter that accumulate in a loose, unconsolidated form. Sediment may be classified by grain size or by the origin of the majority of the particles.

Terrigenous Sediments

Terrigenous (*terra* = Earth, *generare* = to produce) **sediments** are the most abundant. As the name implies, terrigenous sediment originates on the continents or islands near them.

The most familiar continental igneous rock is granite, the source of quartz and clay, the two most common components of terrigenous marine sediments. Quartz, an important mineral in granite, is hard, relatively insoluble, very durable, and able to withstand lengthy weathering and transport. Quartz sands washed from the adjacent land are important components of the sediments along continental margins. Feldspar, another important mineral in granite, ultimately combines with carbonic acid (a mild acid that forms when carbon dioxide dissolves in water) and seawater to form clay. These tiny particles, which are the chief component of soils, are carried to the ocean by wind, rivers, or streams. Because of their small size, they are easily transported across the continental shelf to settle slowly to the deep-ocean floor. Although estimates vary, it appears that about 15 billion metric tons (16.5 billion tons) of terrigenous sediments are transported in rivers to the sea each year, with an additional 100 million metric tons transported annually from land to ocean as fine airborne dust and volcanic ash (**Figure 5.5**).

Biogenous Sediments

Biogenous (*bio* = life, *generare* = to produce) **sediments** are the next most abundant marine sediment. The *siliceous* (silicon-containing) and *calcareous* (calcium carbonate-

containing) compounds that make up these sediments of biological origin were originally brought to the ocean in solution by rivers or dissolved in the ocean at mid-ocean ridges. The siliceous and calcareous materials were then extracted from the seawater by the normal activity of tiny plants and animals to build protective shells and skeletons. Most of the organisms that produce biogenous sediments drift free in the water as plankton. After the death of their owners, the hard structures fall to the bottom and accumulate in layers. Biogenous sediments are most abundant where ample nutrients encourage high biological productivity, usually near continental margins and areas of upwelling. Over millions of years, organic molecules within these sediments can form oil and natural gas (see Chapter 15 for details).

Note in Table 5.2 that biogenous sediments cover a larger percent of the *area* of the ocean floor than terrigenous sediments do, but the terrigenous sediments dominate in total *volume*.

The most abundant marine sediments are broadly classified as terrigenous and biogenous. Terrigenous sediments are of geological origin and arise on the continents or islands near them. They are the more abundant. Biogenous sediments are of biological origin and may be calcareous or siliceous.

Hydrogenous Sediments

Hydrogenous (*hydro* = water, *generare* = to produce) **sediments** are minerals that have precipitated directly from seawater. The sources of the dissolved minerals include submerged rock and sediment, leaching of the fresh crust at oceanic ridges, material issuing from hydrothermal vents, and substances flowing to the ocean in river runoff. As we shall see, the most prominent hydrogenous sediments are manganese nodules, which litter some deep seabeds, and phosphorite nodules, seen along some continental margins. Hydrogenous

a

b

USGS

NASA

Figure 5.5 Sources of terrigenous sediments. (**a**) Rivers are the main source of terrigenous sediments. This photo, taken from space, shows sediment entering the Gulf of Mexico from the Mississippi River. (**b**) Wind also transports particles. A huge cloud of dust streams over the Atlantic from Africa's Sahara Desert on 26 February 2000. The particles will fall to the ocean surface and descend slowly to the bottom.

sediments are also called **authigenic** (*authis* = in place, on the spot) because they were formed in the place they now occupy. Though they usually accumulate very slowly, rapid deposition of hydrogenous sediments is possible—in a rapidly drying lake, for example.

Cosmogenous Sediments

Cosmogenous (*cosmos* = universe, *generare* = to produce) **sediments,** which are of extraterrestrial origin, are the least abundant. These sediments are typically greatly diluted by other sediment components and rarely constitute more than a few parts per million of the total sediment in any layer. Scientists believe that cosmogenous sediments come from two major sources: interplanetary dust that falls constantly into the top of the atmosphere and rare impacts by large asteroids and comets.

Interplanetary dust consists of silt- and sand-sized micrometeoroids that come from asteroids and comets or from collisions between asteroids. The silt-sized particles settle gently to Earth's surface, but larger, faster-moving dust is heated by friction with the atmosphere and melts, sometimes glowing as the meteors we see in a dark night sky. Though much of this material is vaporized, some may persist in the form of iron-rich cosmic spherules. Most of these dissolve in seawater before reaching the ocean floor. About 15,000 to 30,000 metric tons (16,500 to 33,000 tons) of interplanetary dust enter Earth's atmosphere every year.

The highest concentrations of cosmogenous sediments occur when large volumes of extraterrestrial matter arrive all at once. Fortunately, this happens only rarely, when Earth is hit by a large asteroid or comet. The debris ejected from an impact 65 million years ago (that you read about at the beginning of this chapter) was blown into space around Earth. Much of it fell back and was deposited in a layer that has now been found in more than a hundred places on Earth—mostly in marine sediments. Cosmogenous components may make up between 10% and 20% of these extraordinary sediments!

Occasionally cosmogenous sediment includes translucent, oblong particles of glass known as **tektites** (**Figure 5.6**). Tektites are thought to form from the violent impact of large meteors or small asteroids on the crust of Earth. The impact melts some of the crustal material and splashes it into space; the material melts again as it rushes through the atmosphere, producing the various raindrop shapes shown in the photo. Tektites do not dissolve easily and usually reach the ocean floor. Most are smaller than 1.5 millimeters ($\frac{1}{16}$ inch) long.

Sediment Mixtures

Sediments on the ocean floor only rarely come from a single source; most sediment deposits are a mixture of biogenous and terrigenous particles, with an occasional hydrogenous or cosmogenous supplement. The patterns and composition of sediment layers on the seabed are of great interest to

Figure 5.6 Microtektites, very rare particles that began a long journey when a large body impacted Earth and ejected material from Earth's crust. Some of this material traveled through space, re-entered Earth's atmosphere, melted, and took on a rounded or teardrop shape. These specimens of sculptured glass range from 0.2 to 0.8 millimeter in length. Glassy dust much finer in size, as well as nut-size chunks, has also fallen on Earth.

Table 5.3	The Distribution and Average Thickness of Marine Sediments		
Region	Percent of Ocean Area	Percent of Total Volume of Marine Sediments	Average Thickness
Continental shelves	9	15	2.5 km (1.6 mi)
Continental slopes	6	41	9 km (5.6 mi)
Continental rises	6	31	8 km (5 mi)
Deep-ocean floor	78	13	0.6 km (0.4 mi)

Sources: Emery in Kennett, *Marine Geology*, 1982 (Table 11-1); Weihaupt, *Exploration of the Oceans*, 1979; Sverdrup, Johnson, and Fleming, *The Oceans: Their Physics, Chemistry, and General Biology*, 1942

researchers studying conditions in the overlying ocean. Different marine environments have characteristic sediments, and these sediments preserve a record of past and present conditions within those environments.

The Distribution of Marine Sediments

The sediments on the continental margins are generally different in quantity, character, and composition from those on the deeper basin floors. Continental shelf sediments—called **neritic sediments**—consist primarily of terrigenous material. Deep-ocean floors are covered by finer sediments than those of the continental margins, and a greater proportion of deep-sea sediment is of biogenous origin. Sediments of the slope, rise, and deep-ocean floor that originate in the ocean are called **pelagic sediments.** The distribution and average thickness of the marine sediments in each oceanic region are shown in **Table 5.3.** Note that 72% of all marine sediment is associated with continental slopes and rises.

The Sediments of Continental Margins

The bulk of terrigenous sediment is eroded and carried to streams, where it is transported to the ocean. Currents distribute sand and larger particles along the coast, while wave

action carries the silts and clays to deeper water. When the water is too deep to be disturbed by wave action, the finest sediment may come to rest or continue to be transported by the turbulence of deep currents toward the deeper ocean floor. Ideally, these processes produce an orderly sorting of particles by size from relatively large grains near the coast to relatively small grains near the shelf break.

There are exceptions, however. Shelf deposits are subject to further modification and erosion as sea level fluctuates: Larger particles may be moved toward the shelf edge when sea level is low, as it was in periods of extensive glaciations—ice ages. Poorly sorted sediments are also found as glacial deposits. In polar regions, glaciers and ice shelves give rise to icebergs. These carry particles of all sizes, and when they melt they distribute their mixtures of rocks, gravel, sand, and silt onto high-latitude continental margins and deep-ocean floor. Turbidity currents also disrupt the orderly sorting of sediments on the continental margin by transporting coarse-grained particles away from coastal areas and onto the deep-ocean floor.

The rate of sediment deposition on continental shelves is variable, but it is almost always greater than the rate of sediment deposition in the deep ocean. Near the mouths of large rivers, 1 meter (about 3 feet) of sediment may accumulate every thousand years. Along the east coast of the United States, however, many large rivers terminate in estuaries, which trap most of the sediment brought to them. The continental shelf of eastern North America is therefore covered mainly by sediments laid down during the last period of glaciation, when sea level was lower.

In addition to terrigenous material, the continental margins almost always contain biogenous sediments. Biological productivity in coastal waters is often quite high, and the skeletal remains of creatures living on the bottom or in the water above mix with the terrigenous sediments and dilute them.

Sediments can build to impressive thickness on continental shelves. In some cases, shelf sediments undergo **lithification:** They are converted into sedimentary rock by pressure or by cementation. If these lithified sediments are thrust above sea level by tectonic forces, they can form mountains or plateaus.

The top of Mt. Everest, the world's tallest peak, is a shallow-water biogenic marine limestone (a calcareous rock). Much of the Colorado Plateau, with its many stacked layers, was formed by sedimentary deposition and lithification beneath a shallow continental sea beginning about 570 million years ago. The Colorado River has cut and exposed the uplifted beds to form the Grand Canyon. Hikers walking from the canyon rim down to the river pass through spectacular examples of continental shelf sedimentary deposits. Their journey takes them deep into an old ocean floor.

Though there are exceptions, the sediments of continental margins tend to be mostly terrigenous, while the generally finer sediments of the deep-ocean floor contain a larger proportion of biogenous material.

The Sediments of Deep-Ocean Basins

Deep-ocean sediment thickness is highly variable. When averaged, the Atlantic Ocean bottom is covered by sediments to a thickness of about 1 kilometer (3,300 feet), while the Pacific floor has an average sediment thickness of less than 0.5 kilometer (1,650 feet). There are three reasons for this difference: First, the Atlantic Ocean is smaller in area. Second, the Atlantic is fed by a greater number of rivers laden with sediment. Third, in the Pacific Ocean many oceanic trenches trap sediments moving toward basin centers. Beyond this, the composition and thickness of pelagic sediments also vary with location, being thickest on the abyssal plains and thinnest (or absent) on the oceanic ridges.

TURBIDITES Dilute mixtures of sediment and water periodically rush down the continental slope in turbidity currents, the erosive force of which is thought to help cut submarine canyons (see again Figure 4.15). These underwater avalanches of thick, muddy fluid can reach the continental rise and often continue moving onto an adjacent abyssal plain before eventually coming to rest. The resulting deposits are called **turbidites;** they are poorly sorted layers of terrigenous sand interbedded with the finer sediments typical of the deep-sea floor.

CLAYS About 38% of the deep seabed is covered by clays and other fine terrigenous particles. As we have seen, the finest terrigenous sediments are easily transported by wind and water currents. Microscopic waterborne particles and tiny bits of wind-borne dust and volcanic ash settle slowly to the deep-ocean floor, forming fine brown, olive-colored, or reddish clays. As Table 5.1 shows, the velocity of particle settling is related to particle size, and clay particles usually fall very slowly indeed. Terrigenous sediment accumulation on the deep-ocean floor is typically about 2 millimeters (⅛ inch) every thousand years.

OOZES Seafloor samples taken farther from land usually contain a greater proportion of biogenous sediments than those obtained near the continental margins. This is not because biological productivity is higher farther from land (the opposite is usually true), but because there is less terrigenous material far from shore, and thus the deposits contain a greater proportion of biogenous material.

Deep-ocean sediment containing at least 30% biogenous material is called an **ooze** (surely one of the most descriptive terms in the marine sciences). Oozes are named after the dominant remnant organism constituting them. The organisms that contribute their remains to deep-sea oozes are small, single-celled, drifting, plantlike organisms and the single-celled animals that feed on them. The hard shells and skeletal remains of these creatures are of relatively dense, glasslike silica or calcium carbonate (limey) substances. When these organisms die, their shells settle slowly toward the bottom, mingle with fine-grained terrigenous silts and clays, and accumulate as ooze. The silica-rich residues give rise to **siliceous ooze,** the calcium-containing material to **calcareous ooze.**

Oozes accumulate slowly, at a rate of about 1 to 6 centimeters (½ to 2½ inches) per thousand years. But they collect almost ten times more quickly than deep-ocean terrigenous clays. The accumulation of any ooze therefore depends on a delicate balance between the abundance of organisms at the surface, the rate at which they dissolve once they reach the bottom, and the rate of accumulation of terrigenous sediment.

Calcareous ooze forms mainly from shells of the amoeba-like **foraminifera (Figure 5.7a, b)**, small drifting mollusks called **pteropods,** and tiny algae known as **coccolithophores (Figure 5.7c)**. Although these creatures live in nearly all surface ocean water, calcareous ooze does not accumulate everywhere on the ocean floor because the shells are dissolved by seawater. At great depths, seawater contains more carbon dioxide and becomes slightly acid.[1] This acidity, combined with the increased solubility of calcium carbonate in cold water under pressure, dissolves the shells. At a certain depth, the **calcium carbonate compensation depth (CCD),** the rate at which calcareous sediments are supplied to the seabed, equals the rate at which those sediments dissolve. Below this depth, the tiny skeletons of calcium carbonate dissolve on the seafloor, so no calcareous oozes accumulate. Calcareous sediment dominates the deep-sea floor at depths of less than about 4,500 meters (14,800 feet). About 48% of the surface of deep-ocean basins is covered by calcareous oozes.

Siliceous (silicon-containing) ooze predominates at greater depths and in colder polar regions. Siliceous ooze is formed from the hard parts of another amoeba-like animal, the beautiful glassy **radiolarian (Figure 5.8a)**, and from single-celled algae called **diatoms (Figure 5.8b)**. After a radiolarian or diatom dies, its shell will also dissolve back into the seawater, but this dissolution occurs *much* more slowly than the dissolution of calcium carbonate. Slow dissolution, combined with

[1] A discussion of acid–base balance may be found in Chapter 6, page 117.

a

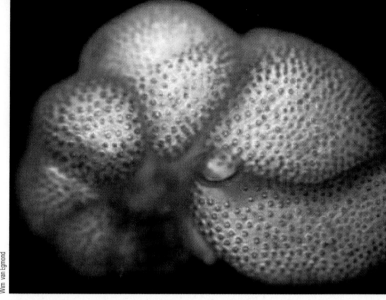

b

Figure 5.7 Organisms that contribute to calcareous ooze. (**a**) A living foraminiferan, an amoeba-like organism. The shell of this beautiful foram, genus *Hastigerina*, is surrounded by a bubble-like capsule. It is one of the largest of the planktonic species, with spines reaching nearly 5 centimeters (2 inches) in length. (**b**) The shells of two smaller foraminifera are visible in these scanning electron micrographs: the bottle-shaped, bottom-dwelling *Uvigerina* and the snail-like, planktonic *Globigerina*. (**c**) Coccoliths, individual plates of coccolithophores, a form of planktonic algae. Because of their tendency to dissolve, calcareous oozes very rarely occur at bottom depths below 4,500 meters (14,800 feet).

very high diatom productivity in some surface waters, leads to the buildup of siliceous ooze. Diatom ooze is most common in the deep ocean surrounding the Antarctic because strong ocean currents and seasonal upwelling in this area support large populations of diatoms. Radiolarian oozes occur in equatorial regions, most notably in the zone of equatorial upwelling west of South America (as will be seen in Figure 5.11). About 14% of the surface of the deep-ocean floor is covered by siliceous oozes.

The very small particles that make up most of these deep-ocean sediments would need between 20 and 50 years to sink to the bottom. By that time they would have drifted a great lateral distance from their original surface position. But researchers have noted that the composition of deep-bottom sediments is usually similar to the particle composition in the water directly above. How could such tiny particles fall quickly enough to avoid great horizontal displacement? The answer appears to involve their compression into fecal pellets (**Figure 5.9**). Though still quite small, the fecal pellets of small animals are much larger than the tiny individual skeletons of diatoms, foraminifera, and other plantlike organisms that they consumed, so they fall much faster, reaching the deep-ocean floor in about two weeks.

c

Some deep-sea oozes have been uplifted by geological processes and are now visible on land. The calcareous chalk White Cliffs of Dover in eastern England are partially lithified deposits composed largely of foraminifera and coccolithophores. Fine-grained siliceous deposits called *diatomaceous earth* are mined from other deposits. This fossil material is a valued component in flat paints, pool and spa filters, and mildly abrasive car and tooth polishes.

Deep-sea oozes—forms of biogenous sediment—contain the remains of some of the ocean's most abundant and important organisms.

HYDROGENOUS MATERIALS Hydrogenous sediments also accumulate on deep-sea floors. They are associated with terrigenous or biogenous sediments and rarely form sediments

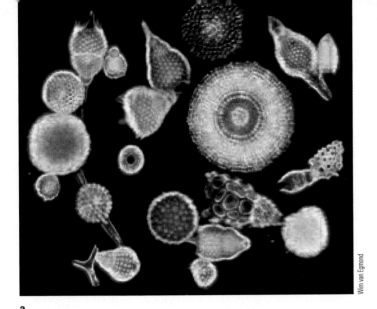

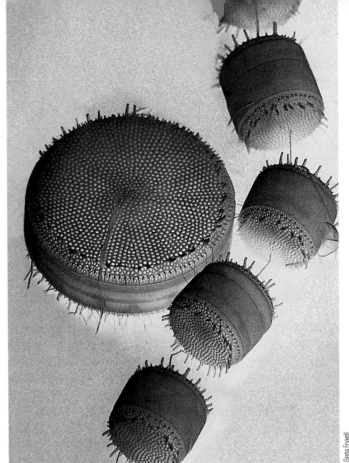

Figure 5.8 Scanning electron micrographs of siliceous oozes, which are most common at great depths. (**a**) Shells of radiolarians, amoeba-like organisms. Radiolarian oozes are found primarily in the equatorial regions. (**b**) A test (shell) of a diatom, a single-celled alga. Diatom oozes are most common at high latitudes.

by themselves. Most hydrogenous sediments originate from chemical reactions that occur on particles of the dominant sediment.

The most famous hydrogenous sediments are manganese **nodules,** which were discovered by the hard-working crew of HMS *Challenger.* The nodules consist primarily of manganese and iron oxides but also contain small amounts of cobalt, nickel, chromium, copper, molybdenum, and zinc. They form in ways not fully understood by marine chemists, "growing" at an average rate of 1 to 10 millimeters (0.04 to 0.4 inch) per *million* years, one of the slowest chemical reactions in nature. Though most are irregular lumps the size of a potato, some nodules exceed 1 meter (3.3 feet) in diameter. Manganese nodules often form around nuclei such as sharks' teeth, bits of

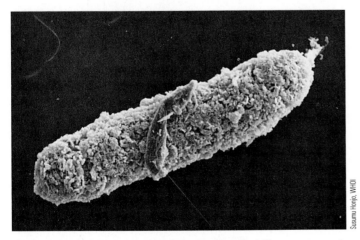

Figure 5.9 A fecal pellet of a small planktonic animal. The compressed pellet is about 80 micrometers long and consists of the indigestible remains of small microscopic plantlike organisms, mostly coccolithophores. Unaided by pellet packing, these remains might take months to reach the seabed, but compressed in this way, they can be added to the ooze in perhaps two weeks.

bone, microscopic alga and animal skeletons, and tiny crystals—as the cross section of a manganese nodule in **Figure 5.10a** shows. Bacterial activity may play a role in the development of a nodule. Between 20% and 50% of the Pacific Ocean floor may be strewn with nodules (**Figure 5.10b**).

Why don't these heavy lumps disappear beneath the constant rain of accumulating sediment? Possibly the continuous churning of the underlying sediment by creatures living there keeps the dense lumps on the surface, or perhaps slow currents in areas of nodule accumulation waft particulate sediments away. For now, the low market value of the minerals in manganese nodules makes them too expensive to recover. As techniques for deep-sea mining become more advanced and raw material prices increase, however, the nodules' concentration of valuable materials will almost certainly be exploited.

EVAPORITES **Evaporites** are an important group of hydrogenous deposits that include many salts important to humanity. These salts precipitate as water evaporates from isolated arms of the ocean or from landlocked seas or lakes. For thousands of years people have collected sea salts from evaporating pools or deposited beds. Evaporites are forming today in the Gulf of California, the Red Sea, and the Persian Gulf. The first evaporites to precipitate as water's salinity increases are the carbonates, such as calcium carbonate (from which limestone is formed). Calcium sulfate, which gives rise to gypsum, is next. Crystals of sodium chloride (table salt) will form if evaporation continues.

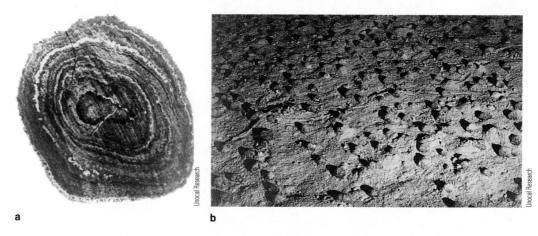

Figure 5.10 Manganese nodules. (**a**) A cross section cut through a manganese nodule, showing the concentric layers of manganese and iron oxides. This nodule is about 11 centimeters (4½ inches) long, a typical size. (**b**) Lemon-sized manganese nodules littering the abyssal Pacific.

a

b

The Distribution of Sediments

Figure 5.11 is a simplified look at the distribution of marine sediments. Notice especially the lack of radiolarian deposits in much of the deep North Pacific; the strand of siliceous oozes extending west from equatorial South America; and the broad expanses of the Atlantic, South Pacific, and Indian Ocean floors covered by calcareous oozes. The broad, deep, rela-

tively old Pacific contains extensive clay deposits, most delivered in the form of airborne dust. As you might expect, the poorly sorted glacial deposits are found only at high latitudes.

This map summarizes more than a century of effort by marine scientists. Studies of sediments will continue because of their importance to natural resource development and because of the details of Earth's history that remain locked beneath their muddy surfaces.

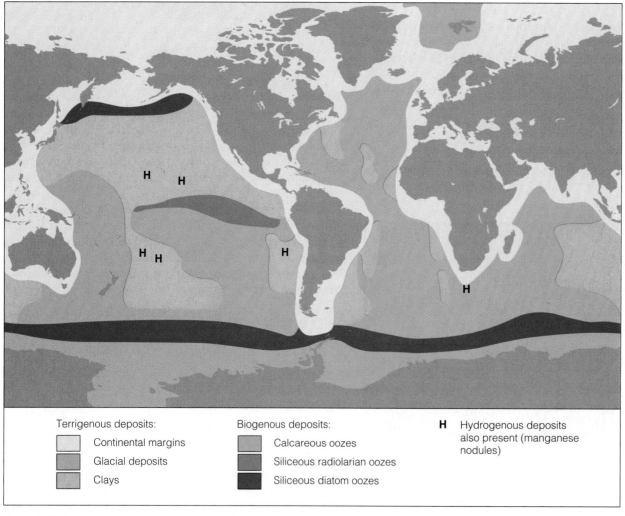

Terrigenous deposits:	Biogenous deposits:	**H** Hydrogenous deposits also present (manganese nodules)
Continental margins	Calcareous oozes	
Glacial deposits	Siliceous radiolarian oozes	
Clays	Siliceous diatom oozes	

Figure 5.11 The general pattern of sediments on the ocean floor.

Studying Sediments

Deep-water cameras have enabled researchers to photograph bottom sediments. The first of these cameras was simply lowered on a cable and triggered by a trip wire. Other more elaborate cameras have been taken to the seafloor on towed sleds or deep submersibles.

Actual samples usually provide more information than photographs do. HMS *Challenger*'s scientists used weighted, wax-tipped poles and other tools attached to long lines to obtain samples, but today's oceanographers have more sophisticated equipment. Shallow samples may be taken using a **clamshell sampler** (named because of its method of operation, not its target; see **Figure 5.12**). Deeper samples are taken by a **piston corer** (**Figure 5.13**), a device capable of punching through as much as 25 meters (82 feet) of sediment and returning an intact plug of material. Using a rotary drilling technique similar to that used to drill for oil, the drilling ship *JOIDES Resolution* (**Figure 5.14**) and its predecessors have returned long core segments, some more than 1,100 meters (3,600 feet) long from a single hole! These cores

a

a

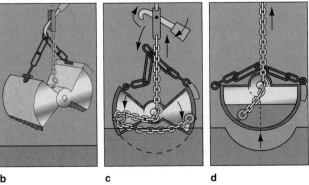

b   c   d

Figure 5.12 (**a**) On board research vessel *Robert Gordon Sproul,* a scoop of muddy ocean-bottom sediments collected with a clamshell grab is dumped onto the deck for study. (**b**) before sampling, (**c**) during sampling, and (**d**) after the sample has been taken. Note that the sample is relatively undisturbed.

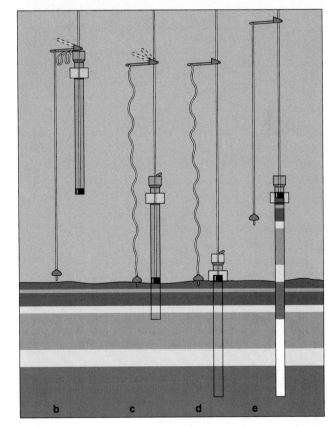

b   c   d   e

Figure 5.13 (**a**) A piston corer. (**b**) The corer is allowed to fall to the bottom. (**c**) The corer reaches the bottom and continues, forcing a sample partway into the cylinder. (**d**) Tension on the cable draws a small piston within the corer toward the top of the cylinder, and the pressure of the surrounding water forces the corer deeper into the sediment. (**e**) The corer and sample are hauled in.

Figure 5.14 *JOIDES Resolution*, the deep-sea drilling ship operated by the Joint Oceanographic Institutions for Deep Earth Sampling. The vessel is 124 meters (407 feet) long, with a displacement of more than 16,000 tons. The rig can take samples to a depth of 9,150 meters (30,000 feet) below sea level.

Deep Sea Drilling Project, Texas A&M University

Figure 5.15 Sediment cores in storage. Cores are sectioned longitudinally, placed in trays, and stored in hermetically sealed, cold rooms. The Gulf Coast Repository of the Ocean Drilling Program, located at Texas A&M University (pictured here), stores about 75,000 sections taken from more than 80 kilometers (50 miles) of cores recovered from the Pacific and Indian Oceans. Smaller core libraries are maintained at the Scripps Institution in California (Pacific and Indian Oceans) and at the Lamont–Doherty Earth Observatory in New York (Atlantic Ocean).

are stored in core libraries, a valuable scientific resource (**Figure 5.15**). Analysis of sediments and fossils from the Deep Sea Drilling Project cores helped verify the theory of plate tectonics. It has also shed light on the evolution of life-forms and helped researchers to decipher the history of changes in Earth's climate over the last 100,000 years.

Powerful new continuous seismic profilers have also been used to determine the thickness and structure of layers of sediment on the continental shelf and slope, and to assist in the search for oil and natural gas (see, for example, Figure 5.4). Recent improvements in computerized image processing of the echoes returning from the seabed now permit detailed analysis of these deeper layers.

Sediments as Historical Records

Because the deep-sea sediment record is ultimately destroyed in the subduction process, the ocean's sedimentary "memory" is not as long as early marine scientists hoped it would

be. Still, modern studies of deep-sea sediments using seafloor samples, cores obtained by deep drilling, and continuous seismic profiling have demonstrated that these deposits contain a remarkable record of relatively recent ocean history. The analysis of layered sedimentary deposits in the ocean (or on land) represents the discipline of **stratigraphy.** Deep-sea stratigraphy utilizes variations in the composition of rocks, microfossils, depositional patterns, geochemical character, and physical character (density and such) to trace or correlate distinctive sedimentary layers from place to place, establish the age of the deposits, and interpret changes in ocean and atmospheric circulation, productivity, and other aspects of past ocean behavior. In turn, these sorts of studies and the advent of deep-sea drilling have given rise to the emerging science of **paleoceanography,** the study of the ocean's past.

Early attempts to interpret ocean and climate history from evidence in deep-sea sediments occurred in the 1930s through 1950s as cores became available. These initial studies relied primarily on down-core variations in the abundance and distribution of glacial marine sediments, carbonate and siliceous oozes, and temperature-sensitive microfossils. Modern paleoceanographic studies continue to utilize these same features but with much greater understanding of their significance and aided by seismic imaging of the deposits over large areas. In addition, newer and more precise methods of dating deep-sea sediments have enabled scientists to place events in a proper time context. Finally, the appearance of instruments capable of analyzing very small variations in the relative abundances of the stable isotopes of oxygen preserved within the carbonate shells of microfossils found in deep-sea sediments has allowed scientists to interpret changes in the temperature of surface and deep water over time. These same data are also used to estimate variations in the volume of ice stored in continental ice sheets and thus to track the ice ages. Other geochemical evidence contained in the shells of marine microfossils, including variations in carbon isotopes and trace metals such as cadmium, provide insights into ancient patterns of ocean circulation, productivity of the marine biosphere, and ancient upwelling. These

sorts of data have already provided quantitative records of the glacial–interglacial climatic cycles of the past 2 million years. Future drilling and analysis of deep-sea sediments are poised to extend our paleoceanographic perspective much further back in time.

Figure 5.16 shows the age of the Pacific Ocean floor using data obtained largely from analyses of the overlying sediment. Note that sediments get older with increasing distance from the East Pacific Rise spreading center.

Global analysis of marine sediments in the modern basins can shed light on unexpected details of the last 180 million years of Earth's history. One of the most significant events is the extinction of up to 52% of known marine animal species (and the dinosaurs) at the end of the Cretaceous period 65 million years ago. As you read in the chapter opener, many researchers now believe that a sudden and violent impact of one or more asteroids or comets caused this catastrophe.

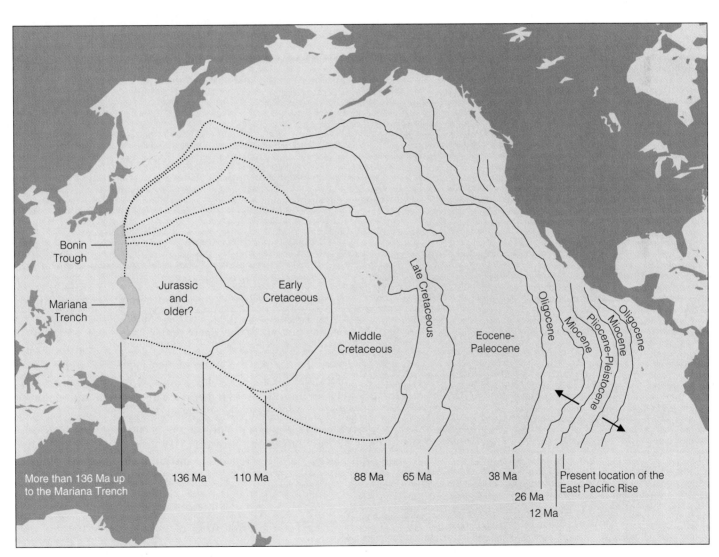

Figure 5.16 The ages of portions of the Pacific Ocean floor, based on core samples of sediments just above the basalt seabed, in millions of years ago (Ma, mega-annum). The youngest sediments are found near the East Pacific Rise and the oldest close to the eastern side of the troughs and trenches. Contrast this figure with Figure 3.25.

Sediment deposited on a quiet seabed can provide a sequential record of events in the water column above. In a sense sediments act as the recent memory of the ocean. This memory does not extend past about 200 million years because seabeds are relatively young and recycled at subduction zones.

Earth might not be the only planet where marine sediments have left historical records. As you read in Chapter 2, Mars probably had an ocean between 3.2 and 1.2 billion years ago. In May 1998 *Mars Global Surveyor* photographed sediments that look suspiciously marine near the edge of an ancient bay (**Figure 5.17**). One can only wonder what stories they will tell.

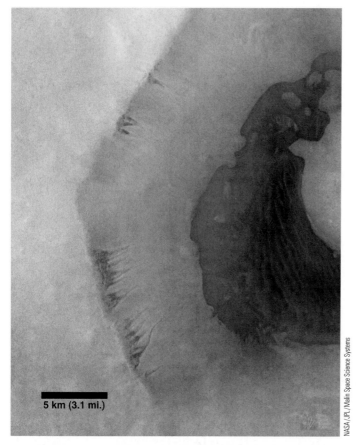

5 km (3.1 mi.)

NASA/JPL/Malin Space Science Systems

Figure 5.17 Marine sediments on Mars? Dark windblown dunes of sandy sediments rest on the floor of an ancient bay in the southern Martian lowlands. At some time in the past, water seeped out of layers within the cliffs (to the left) and flowed downhill, flooding part of the basin. The zones of contrast between the dark floor material and the lighter material of the slopes suggest the formation of bays and peninsulas. The appearance of dunes within the crater may be coincidental, or the sand may have been generated by wind and wave action. The lack of superimposed fresh impact craters suggests this process may have occurred relatively recently. Photograph by *Mars Global Surveyor*, May 1998.

QUESTIONS FROM STUDENTS

1. What economic benefit, if any, comes from studying sediments?

Study of sediments has brought practical benefits. In 2000 an estimated 34% of the world's crude oil and 28% of its natural gas were extracted from the sedimentary deposits of continental shelves and continental rises. Offshore hydrocarbons currently generate annual revenues in excess of $145 billion. Deposits within the sediments of continental margins account for about one-third of the world's estimated oil and gas reserves.

In addition to oil and gas, in 1998 sand and gravel valued at more than $450 million were taken from the ocean. This is about 1% of world needs. Commercial mining of manganese nodules has also been considered. In addition to manganese, these nodules contain substantial amounts of iron and other industrially important chemical elements. The high iron content of these nodules has prompted a proposal to rename them *ferromanganese* nodules. We will investigate these resources in more detail in Chapter 15.

2. Where are sediments thickest?

Sediment is thickest close to eroding land and beneath biologically productive neritic waters, and thinnest over the fast-spreading oceanic ridges of the eastern South Pacific. The thickest accumulations of sediment may be found along and beneath the continental margins (especially on continental rises). Some are typically more than 1,500 meters (5,000 feet) thick. Remember, much of the rocky material of the Grand Canyon was once marine sediment atop an ancient seabed. The Grand Canyon is nearly 2 kilometers (1¼ miles) deep, and the uppermost layer of sedimentary rock has already been eroded completely away!

3. What's the relationship between deep-sea animals and the sediments on which they live?

Though microscopic bacteria and benthic foraminifera may be very abundant on the seabed, visible life is not abundant on the bottom of the deep ocean. There are no plants at great depths because there is no light, but animals do live there. Some, like the brittle stars in Figure 5.2, move slowly along the surface searching for bits of organic matter to eat. Others burrow through the muck in search of food particles. Worms eat quantities of sediment to extract any nutrients that may be present and then deposit strings of fecal material as they move forward. The deeps may be uninviting places, but life is tenacious and survives even in this hostile environment.

CHAPTER SUMMARY

The ocean floor is covered in most places by layers of sediment. The sediment is composed of particles from land, from biological activity in the ocean, from chemical processes within water, and even from space. The blanket of marine sediment is thickest at the continental margins and thinnest over the active oceanic ridges.

Sediments may be classified by particle size, source, location, or color. Terrigenous sediments, the most abundant, originate on continents or islands near them. Biogenous sediments are composed of the remains of once-living organisms. Hydrogenous sediments are precipitated directly from seawater. Cosmogenous sediments, the ocean's rarest, come to the seabed from space.

The position and nature of sediments provide important clues to Earth's recent history, and valuable resources can sometimes be recovered from them.

ON-LINE STUDY RESOURCES

The Web site for this book contains a wealth of helpful study aids. Log on to:

http://info.brookscole.com/garrison

and select *Essentials of Oceanography*, 3rd edition. Select Chapter 5 from the drop-down menu or click on one of these resource areas:

- For study and review, **Chapter at a Glance** gives you an outline of the chapter, **Chapter Summary** allows you to review the chapter's main ideas, and **Glossary** lists concepts and terms for the chapter along with their definitions.

- To test your mastery of the Terms and Concepts to Remember for this chapter, you can use the electronic **Flash Cards,** play the **Concentration** game, or work the **Crossword Puzzle.**

- For practice quizzing, try the multiple-choice **Tutorial Quiz,** the ten-question **True/False Quiz,** or the **Image Analysis Quiz,** which poses questions based on art and photos from the chapter.

TERMS AND CONCEPTS TO REMEMBER

authigenic sediment	ooze
biogenous sediment	paleoceanography
calcareous ooze	pelagic sediment
calcium carbonate compensation depth (CCD)	piston corer
	poorly sorted sediment
clamshell sampler	pteropod
clay	radiolarian
coccolithophore	sand
cosmogenous sediment	sediment
diatom	siliceous ooze
evaporite	silt
foraminiferan	stratigraphy
hydrogenous sediment	tektite
lithification	terrigenous sediment
neritic sediment	turbidite
nodule	well-sorted sediment

STUDY QUESTIONS

1. In what ways are sediments classified?

2. List the four types of marine sediments. Explain the origin of each.

3. How are neritic sediments generally different from pelagic ones?

4. Is the thickness of ooze always an accurate indication of the biological productivity of surface water in a given area? (*Hint:* See the next question.)

5. What happens to the calcium carbonate skeletons of small organisms as they descend to great depths? How do the siliceous components of once-living things compare?

6. What sediments accumulate most rapidly? Least rapidly?

7. Can marine sediments tell us about the history of the ocean from the time of its origin? Why?

8. How do paleoceanographers infer water temperatures, and therefore terrestrial climate, from sediment samples?

9. Where are sediments thickest? Are any areas of the ocean floor free of sediments?

10. Are sediments commercially important? In what ways?

 Earth Systems Today CD Question

1. Go to "Rocks and the Rock Cycle," then "Rock Cycle." How do sediments fit into the rock cycle? From the animation, can you tell the origin and fate of sediments?

RESOURCES FOR FURTHER READING AND RESEARCH

The Web site for this book contains many ideas for further reading and research. Log on to:

http://info.brookscole.com/garrison

and select *Essentials of Oceanography*, 3rd edition. Select Chapter 5 from the drop-down menu or click on one of these resource areas:

- **Additional Readings** lists the major books and articles consulted in writing this chapter, along with comments from the author about their content and reading level.

- **Hypercontents** takes you to an extensive list of current links to Internet sites with news, research, and images related to individual subjects in the chapter. Just click on the icon that corresponds to a numbered section to see the links for that subject.

- **Internet Exercises** are critical thinking questions that involve research on the Internet with starter URLs provided.

- **InfoTrac College Edition Exercises** leads you to Critical Thinking Projects that use InfoTrac College Edition as a research tool.
- **Regional InfoTrac College Edition** articles are organized into East Coast, West Coast, and Gulf Coast regions, allowing you to study oceanography on a more local level.

 For more readings, go to InfoTrac College Edition, your on-line research library, at:

http://infotrac.thomsonlearning.com

http://infotrac.thomsonlearning.com

© Tom Stewart/Corbis

Tall icebergs form as glaciers reach the ocean and break. Waves and weather can sculpt them into beautiful and improbable shapes.

Water

ICEBERGS

Water occurs in three states: liquid, solid, and gas. Oceanographers are most familiar with water's liquid form, but about 6% of the world ocean is covered by ice. A small fraction of this ice is contained in the fantastic shapes of icebergs.

Icebergs in the southern ocean originate as huge ice sheets attached to the Antarctic continent; lengths of up to 8 kilometers (5 miles) are not unusual, with flat tops rising 45 meters (150 feet) above sea level. In 1927 a frozen section of 26,000 square kilometers (10,000 square miles)—about eight times the size of Rhode Island—broke from the Antarctic shore and floated north along the coast of Argentina!

Pinnacled icebergs, more characteristic of the northern polar ocean, usually begin life as glaciers in western Greenland. These slowly flowing ice rivers form at the edge of the great Greenland ice sheet, squeeze between the sharp-crested ridges of the mountains, and eventually reach the sea. Rising and falling tides at the end of the deep fjords break off huge tongues of ice, which drift from shore as icebergs. The great weight of ice in a glacier compacts these glistening masses and makes them very dense; only about one-seventh of their bulk floats above sea level.

Table 6.1	Heat Capacity of Common Substances
Substance	**Heat Capacity[a] in calories/gram/°C**
Silver	0.06
Granite	0.20
Aluminum	0.22
Alcohol (ethyl)	0.30
Gasoline	0.50
Acetone	0.51
Pure water	**1.00**
Ammonia (liquid)	1.13

[a] Heat capacity is a measure of the heat required to raise the temperature of 1 gram (0.035 ounce) of a substance by 1°C (1.8°F). Different substances have different heat capacities. *Not all substances respond to identical inputs of heat by rising in temperature the same number of degrees.* Notice how little heat is required to raise the temperature of 1 gram of silver 1 degree.

Because of the great strength and large number of the hydrogen bonds between water molecules, water can gain or lose large amounts of *heat* with very little change in *temperature*. This *thermal inertia* moderates temperatures world-wide. Of all common substances, only liquid ammonia has a higher heat capacity than liquid water.

amounts of heat while changing relatively little in temperature. Anyone who waits by a stove for water to boil knows a lot about water's heat capacity! Compared with water, ethyl alcohol has a much lower heat capacity. If both liquids absorb heat from identical stove burners at the same rate, pure ethyl alcohol, the active ingredient in alcoholic beverages, will rise in temperature about twice as fast as an equal

mass of water. Beach sand has an even lower heat capacity. Sand requires as little as 0.2 calorie to rise 1°C (1.8°F), so beaches can get too hot to stand on with bare feet while the water remains pleasantly cool.

As we will soon see, the concept of heat capacity is very important in oceanography. But for now, remember this: Water has an extraordinarily high heat capacity; it resists changing *temperature* when *heat* is added or removed.

Water Temperature and Density

The uniqueness of water becomes even more apparent when we consider the effect of a temperature change on water's **density** (its mass per unit of volume). You may recall from Chapter 3 that the density of pure water is 1 gram per cubic centimeter (1 g/cm^3). Granite rock is heavier, with a density of about 2.7 g/cm^3, and air is lighter, with a density of about 0.0012 g/cm^3. Most substances become denser (weigh more per unit of volume) as they get colder. Pure water generally becomes denser as heat is removed and its temperature falls, but water's density behaves unexpectedly as the temperature approaches the freezing point.

A **density curve** shows the relationship between the temperature or salinity of a substance and its density. Most substances become progressively denser as they cool; their temperature–density relationships are linear (that is, appear as a straight line on graphs). But **Figure 6.3** shows the unusual temperature–density relationship of pure water. Imagine heat being removed from some water in a freezer. Initially, the water is at

Figure 6.3 The relationship of density and temperature for pure water. Note that points **C** and **D** both represent 0°C (32°F) but different densities and different states of water. Ice floats because the density of ice is lower than the density of liquid water.

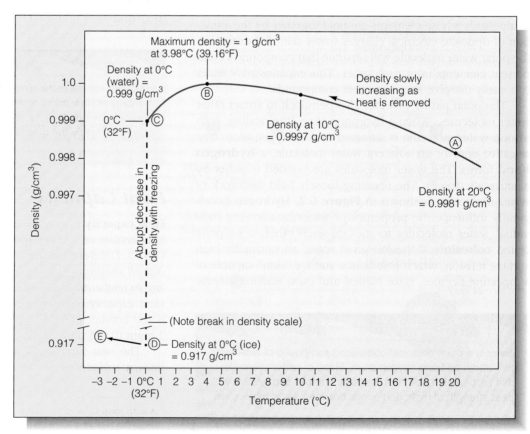

room temperature, point **A** on the graph. As expected, the density of water increases as its temperature drops along the line from point **A** toward point **B.** As the temperature approaches point **B,** the density increase slows, reaching a maximum at point **B** of 1 g/cm^3 at 3.98°C (39.16°F). As cooling continues, the water begins to set up a more rigid framework of hydrogen bonds, which causes the liquid to expand slightly. Water thus becomes slightly less dense as cooling continues, until point **C** (0°C, 32°F) is reached. At point **C** the water begins to freeze—to change state by crystallizing into ice.

State is an expression of the internal form of a substance. Changes in state are accompanied by either an input or an output of energy. Water exists on Earth in three physical states: liquid, gas (water vapor), and solid (ice). If the freezer continues to remove heat from the water at point **C** in Figure 6.3, the water will change from liquid to solid state. Through this transition from water to ice—from point **C** to point **D**—the density of the water *decreases* abruptly. Ice is therefore lighter than an equal volume of water. Ice increases in density as it gets colder than 0°C. No matter how cold it gets, however, ice never reaches the density of liquid water. Being less dense than water, ice "freezes over" as a floating layer instead of "freezing under" like the solid forms of virtually all other liquids.

As we'll see in a moment, the implications of water's high heat capacity and the ability of ice to float are vital in maintaining Earth's moderate surface temperature. First, we look at the transition from point **C** to point **D** in Figure 6.3.

Freezing Water 6.6

During the transition from liquid to solid state at the freezing point, the bond angle between the oxygen and hydrogen atoms in water expands from about 105° to slightly more than 109°. This change allows the hydrogen bonds in ice to form a crystal lattice (**Figure 6.4**). The space taken by 27 water molecules in the liquid state would be occupied by only 24 water molecules in the solid lattice, however; so water has to expand about 9% as the crystal forms. Because the molecules are packed less efficiently, ice is less dense than liquid water—and so it floats. A cubic centimeter of ice at 0°C (32°F) has a mass of only 0.917 gram, but a cubic centimeter of liquid water at 0°C has a mass of 0.999 gram.

The transition from liquid water to ice crystal (point **C** to point **D** in Figure 6.3) requires continued removal of heat energy; the change in state does not occur instantly throughout the mass when the cooling water reaches 0°C (32°F). Again, consider water in a freezer. **Figure 6.5,** a plot of heat removal versus temperature, illustrates the water's progress to ice. As in Figure 6.3, point **A** represents 20°C (68°F) water just placed in the freezer. The removal of heat does not stop when the water reaches point **C,** *but the decline in temperature stops.* Even though heat continues to be removed, the water will not get colder until all of it has changed state from liquid (water) to solid (ice). Heat may therefore be removed from water when it is changing state (that is, when it is freezing) without the water dropping in temperature. Indeed, the continued removal of heat is what makes the change in state

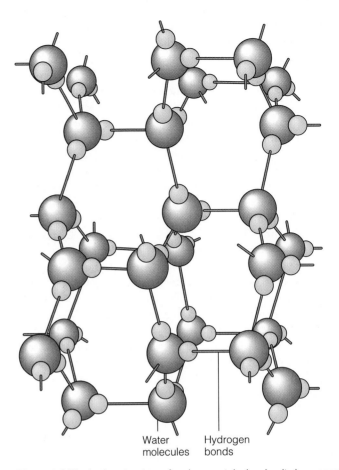

Figure 6.4 The lattice structure of an ice crystal, showing its hexagonal arrangement at the molecular level. The space taken by 24 water molecules in the solid lattice could be occupied by 27 water molecules in liquid state, so water expands about 9% as the crystal forms. Because molecules of liquid water are packed less efficiently, ice is less dense than liquid water and will float.

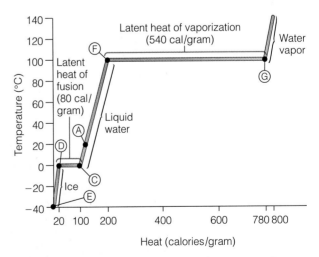

Figure 6.5 A graph of temperature versus heat as water freezes, melts, and vaporizes. The horizontal line between points **C** and **D** represents the latent heat of fusion, when heat is being added or removed but the temperature is not changing. The horizontal line between points **G** and **F** represents the latent heat of vaporization, when heat is being added or removed but the temperature is not changing. (Note that points **A–E** on this graph are the same as those in Figure 6.3.)

possible. Heat is released as bonds form to make ice, and that heat must be removed to allow more ice to form.

The removal of heat from point **A** to point **C** in Figures 6.3 and 6.5 produces a *measurable* lowering of temperature detectable by a thermometer. Removing just 1 calorie of heat from a gram of liquid water causes its temperature to drop 1°C. This detectable decrease in heat is called **sensible heat** loss. But the loss of heat as water freezes between points **C** and **D** is not measurable (that is, not sensible) by a thermometer. Removing a calorie of heat from freezing water at 0°C (32°F) won't change its temperature at all; 80 calories of heat energy must be removed per gram of pure water at 0°C (32°F) to form ice. This heat is called the **latent heat of fusion.** The straight line between points **C** and **D** in Figure 6.5 represents water's latent heat of fusion.

No more ice crystals can form when all the water in the freezer has turned to ice. If the removal of heat continues, the ice will get colder and will soon reach the temperature inside the freezer, point **E** in Figures 6.3 and 6.5.

Latent heat of fusion is also a factor during thawing. When ice melts, it *absorbs* large quantities of heat (the same 80 calories per gram), but it does not change in temperature until all the ice has turned to liquid. This is why ice is so effective in cooling drinks.

Evaporating Water

Let's reverse the process now and warm the ice. Imagine the water resting at −40°C (−40°F), point **E** at the lower left of Figure 6.5. Add heat, and the ice warms toward point **D.** It begins to melt. The horizontal line between point **D** and point **C** represents the latent heat of fusion: Heat is absorbed but temperature does not change as the ice melts. All liquid now at point **C,** the water warms past our original point **A** and arrives at point **F.** It begins to boil—it *vaporizes.*

When water vaporizes (or evaporates), individual water molecules diffuse into the air. Since each water molecule is hydrogen-bonded to adjacent molecules, heat energy is required to break those bonds and allow the molecule to fly away from the surface. Evaporation cools a moist surface

because departing molecules of water vapor carry this energy away with them. (This is how perspiring cools us when we're hot. The heat energy required to evaporate water from our skin is taken away from our bodies, cooling us.)

Hydrogen bonds are quite strong, and the amount of energy required to break them—known as the **latent heat of vaporization**—is very high. The long horizontal line between points **F** and **G** represents the latent heat of vaporization. Even though more heat is applied, the water cannot get warmer until all of it has vaporized. At 540 calories per gram at 20°C (68°F), water has the highest latent heat of vaporization of any known substance. As before, the term *latent* applies to heat input that does not cause a temperature change but does produce a change of state—in this case, from liquid to gas.

About 1 meter (3.3 feet) of water evaporates each year from the surface of the ocean, a volume of water equivalent to 334,000 cubic kilometers (80,000 cubic miles). The great quantities of solar energy that cause this evaporation are carried from the ocean by the escaping water vapor. When a gram of water vapor condenses back into liquid water, the same 540 calories is again available to do work. As we will see, winds, storms, ocean currents, and wind waves are all powered by that heat.

Why the big difference between water's latent heat of *fusion* (80 calories per gram) and its latent heat of *vaporization* (540 calories per gram)? Only a small percentage of hydrogen bonds are broken when ice melts, but *all* must be broken during evaporation. Breaking these bonds requires additional energy in proportion to their number. **Figure 6.6** summarizes this information.

Global Thermostatic Effects

The **thermostatic properties** of water are those properties that act to moderate changes in temperature. Water temperature rises as sunlight is absorbed and changed to heat, but

Figure 6.6 We must add 80 calories of heat energy to change 1 gram of ice to liquid water. After the ice is melted, about 1 calorie of heat is needed to raise each gram of water by 1°C. But 540 calories must be added to each gram of water to vaporize it—to boil it away. The process is reversed for condensation and freezing.

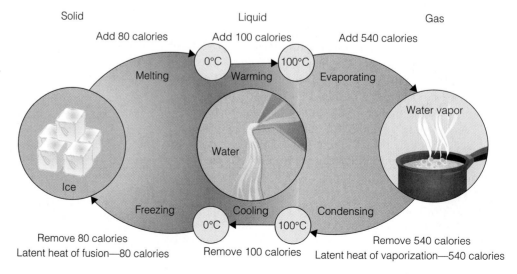

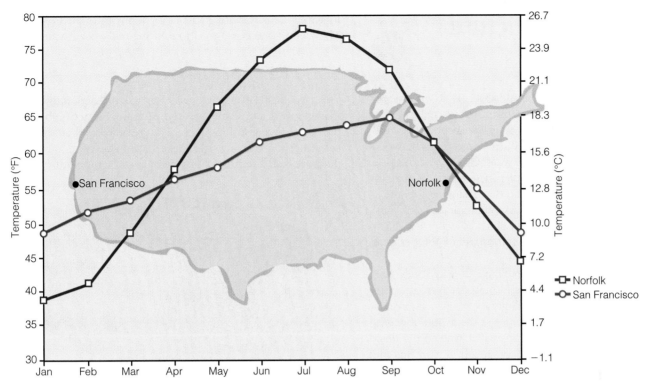

Figure 6.7 San Francisco, California, and Norfolk, Virginia, are on nearly the same line of latitude. Wind tends to flow from west to east at this latitude. Compared to Norfolk, San Francisco is warmer in the winter and cooler in the summer, in part because air in San Francisco has moved over the ocean while air in Norfolk has approached over land. Water doesn't warm as much as land in the summer, nor cool as much in winter—a demonstration of thermal inertia.

as we've seen, water has a very high heat capacity, so its temperature will not rise very much even if a large quantity of heat is added. This tendency of a substance to resist a change in temperature with the gain or loss of heat energy is called **thermal inertia.** To investigate the impact of water's thermostatic properties on conditions at Earth's surface, we need to look at the planet's overall heat balance.

Only about 1 part in 2.2 billion of the sun's radiant energy is intercepted by Earth, but that amount averages 7 million calories per square meter per day at the top of the atmosphere or, for Earth as a whole, an impressive 17 trillion kilowatts (23 trillion horsepower)! About half of this light reaches the surface, where it is converted to heat and then transferred into the atmosphere by conduction, radiation, and evaporation. The atmosphere, like the land and ocean, eventually radiates this heat back into space in the form of long-wave (infrared) radiation. As in your personal financial budget, income must eventually equal outgo. Over long periods of time the total *incoming* heat (plus that from earthly sources) equals the total *outgoing* heat, so Earth is in **thermal equilibrium;** it is growing neither significantly warmer nor colder.[2] Heat input comes mainly from the sun; heat outflow can occur only as heat radiates into the cold of space.

Liquid water's thermal characteristics prevent broad swings of temperature during day and night and, through a longer span, during winter and summer (**Figure 6.7**). Heat is stored in the ocean during the day and released at night. A much greater amount of heat is stored through the summer and given off during the winter. If our ocean were made of alcohol—or almost any other liquid—summer temperatures would be much hotter and winters bitterly cold. Sea ice in the polar regions also contributes to thermal inertia. Because water expands and floats when it freezes, ice can absorb the morning warmth of the sun, melt, and then refreeze at night, giving back to the atmosphere the heat it stored through the daylight hours. The *heat content* of the water changes through the day; its *temperature* does not.

And why doesn't the ocean freeze solid near the poles or boil away at the equator? Because heat is carried from equatorial regions to polar regions by seawater currents and atmospheric water vapor. Again, whether in liquid or vapor form, water's ability to carry heat without a dramatic rise in temperature acts to moderate surface temperatures worldwide.

Without water's unique thermal properties, temperatures on Earth's surface would change dramatically with only minor changes in atmospheric transparency or solar output. Water acts as a "global thermostat."

---

[2] Changes in heat balance do occur over short periods of geological time. Increasing amounts of carbon dioxide and methane in Earth's atmosphere may be contributing to an increase in surface temperature called the *greenhouse effect*. More on this subject may be found in Chapter 15.

The Density Structure of the Ocean

The density of water is mainly a function of its salinity and temperature. Cold, salty water is denser than warm, less salty water. The density of seawater varies between 1.020 and 1.030 g/cm³, indicating that a liter of seawater weighs between 2% and 3% more than a liter of pure water (1.000 g/cm³) at the same temperature. Seawater's density increases with increasing salinity, increasing pressure, and decreasing temperature. **Figure 6.8** shows the relationship among temperature, salinity, and density. Notice that two samples of water can have the *same* density at *different combinations* of temperature and salinity.

Much of the ocean is divided into three density zones. The **surface zone,** or **mixed layer,** is the upper layer of ocean (**Figure 6.9a**). Temperature and salinity are relatively constant with depth in the surface zone because of the action of waves and currents. The surface zone consists of water in contact with the atmosphere and exposed to sunlight; it contains the ocean's least dense water and accounts for only about 2% of total ocean volume. The surface zone (or mixed layer) typically extends to a depth of about 150 meters (500 feet), but depending on local conditions, it may reach a depth of 1,000 meters (3,300 feet) or be absent entirely.

The Pycnocline

The **pycnocline** is a zone in which density increases with increasing depth. This zone isolates surface water from the denser layer below. The pycnocline contains about 18% of all ocean water. The **deep zone** lies below the pycnocline at depths below about 1,000 meters (3,300 feet) in mid-latitudes (around 40°S to 40°N). There is little additional change

in water density with increasing depth through this zone. This deep zone contains about 80% of all ocean water.

The Thermocline

The pycnocline's rapid density increase with depth is due mainly to a decrease in water temperature. **Figure 6.9b** shows the general relationship of temperature with depth in the open sea. The surface zone is well mixed, with little decrease in temperature with depth. In the next layer, temperature drops

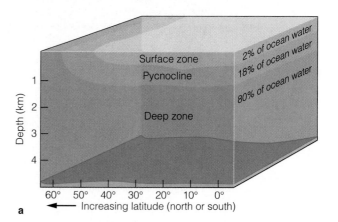

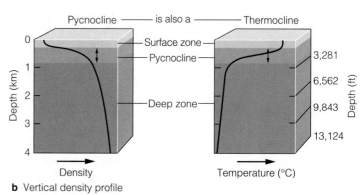

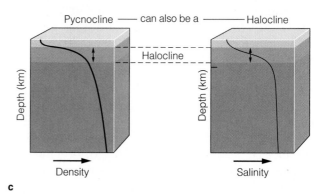

Figure 6.9 Density stratification in the ocean. (**a**) In most of the ocean, a surface zone (or mixed layer) of relatively warm, low-density water overlies a layer called the *pycnocline.* Density increases rapidly with depth in the pycnocline. Below the pycnocline lies the deep zone of cold, dense water—about 80% of total ocean volume. (**b**) The rapid density increase in the pycnocline is mainly due to a decrease in temperature with depth in this area—the *thermocline.* (**c**) In some regions, especially in shallow water near rivers, a pycnocline may develop in which the density increase with depth is due to vertical variations in salinity. In this case, the pycnocline is a *halocline.*

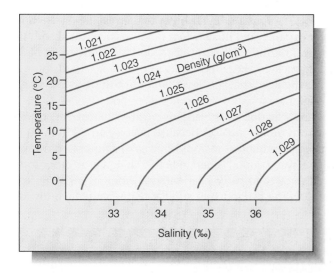

Figure 6.8 The complex relationship among temperature, salinity, and density of seawater. Note that two samples of water can have the *same* density at *different combinations* of temperature and salinity. (From G. P. Kuiper, ed., *The Earth as a Planet,* © 1954 The University of Chicago Press. Reprinted by permission.)

rapidly with depth. Beneath it lies the deep zone of cold, stable water. The middle layer, the zone in which temperature changes rapidly with depth, is called the **thermocline.**

Thermoclines are not identical in form in all areas or latitudes. Surface temperature is associated with available sunlight. More solar energy is available in the tropics than in the polar regions, so the water there is warmer. The ocean's sunlit upper layer is also thicker in the tropics, both because the solar angle there is more nearly vertical and because water in the open tropical ocean contains fewer suspended particles (and is therefore clearer than water in open temperate or polar regions). Because the ocean is heated to a greater depth, the tropical thermocline is deeper than thermoclines at higher latitudes. It is also much more pronounced: The transition to the colder, denser water below is more abrupt in the tropics than at high latitudes.

Polar waters, which receive relatively little solar warmth, are not layered by temperature and generally lack a thermocline because surface water in the polar regions is nearly as cold as water at great depths.

Figure 6.10 contrasts polar, tropical, and temperate thermal profiles, showing that the thermocline is primarily a mid- and low-latitude phenomenon. Thermocline depth and intensity also vary with season, local conditions (storms, for example), currents, and many other factors.

Below the thermocline, water is very cold, ranging from −1°C to 3°C (30.5°F to 37.5°F). Because this deep, cold layer contains the bulk of ocean water, the average temperature of the world ocean is a chilly 3.9°C (39°F).

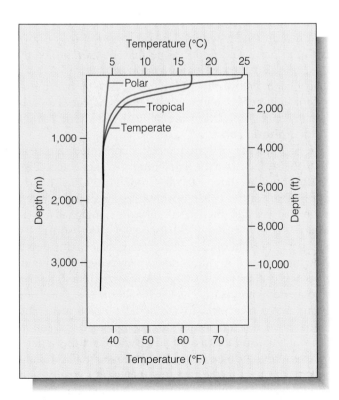

Figure 6.10 Typical temperature profiles at polar, tropical, and middle (temperate) latitudes. Note that polar waters lack a thermocline.

The Halocline

Low salinity can also contribute to the pycnocline, especially in cool regions where precipitation is high or along coasts where freshwater runoff mixes with surface water. Wherever precipitation exceeds evaporation, salinity will be low. These differences produce the **halocline,** a zone of rapid salinity increase with depth (**Figure 6.9c**). The halocline often coincides with the thermocline, and the combination produces a pronounced pycnocline.

Water density is greatly influenced by changes in temperature and salinity. Water masses are usually layered by density, with the densest (coldest and saltiest) water on or near the ocean floor.

Refraction, Light, and Sound

The ring of light sometimes seen around the moon and the safe concealment of a submarine may not seem related, but both events depend on **refraction,** the bending of waves. Light and sound are wave phenomena. When a light wave or a sound wave leaves a medium of one density—such as air—and enters a medium of a different density—such as water—at an angle other than 90°, it is bent from its original path. The reason for this bending is that light or sound waves travel at different speeds in the different media.

The situation is analogous to a line of marchers walking along a desert highway with arms intertwined. The marchers can walk faster if they stay on the street than if they walk in the sand next to the street. Their speed on the pavement, then, is greater than their speed in the sand. As long as they stay on the street, they won't change direction. But if their marching angle gradually takes them off the edge (into a medium in which their speed is *lower*), the people who reach the sand first will suddenly slow down, and the line will pivot quickly off the street. They have been refracted. Their progress is depicted in **Figure 6.11a.** Note that the transition from one medium to another must occur at an angle other than 90° for refraction to occur; our marchers will not change direction if they march straight off the asphalt into the sand. (They will still slow down, however—see **Figure 6.11b.**)

Examples of the refraction of light by water are all around you. A pencil sticking out of a glass of water looks bent because of refraction, the submerged steps of a swimming pool ladder look closer than they are because of refraction, and refraction magnifies objects and causes divers to exaggerate the size of the fish that got away.

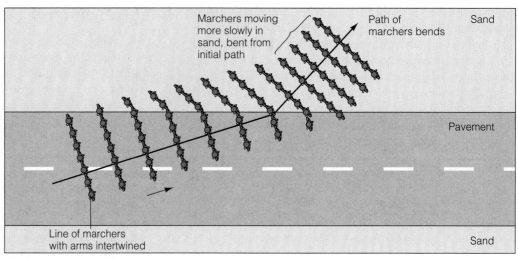

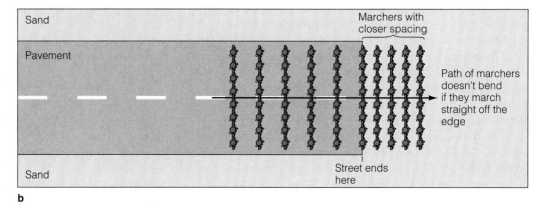

Figure 6.11 An analogy for refraction. The ranks of marchers represent light or sound waves; the pavement and sand represent different media. (**a**) If the marchers head off the pavement at an angle other than 90°, their path will bend (refract) as they hit the sand because some will be walking more slowly than others. (**b**) If they march straight off the pavement, the ranks will slow down but not bend as they hit the sand.

Light in the Ocean 6.14

As we've seen, sunlight has a difficult time reaching and penetrating the ocean; clouds and the sea surface reflect light while atmospheric gases and particles scatter and absorb it. Once past the sea surface, light is rapidly weakened by scattering and absorption. **Scattering** occurs as light is bounced between air or water molecules, dust particles, water droplets, or other objects before being absorbed. The greater density of water (along with the greater number of suspended and dissolved particles) makes scattering more prevalent in water than in air. The **absorption** of light is governed by the structure of the water molecules it happens to strike. When light is absorbed, molecules vibrate and the light's energy is converted to heat.

Even perfectly clear seawater is not perfectly transparent. If it were, the sun's rays would illuminate the greatest depths of the ocean and seaweed forests would fill its warmed basins. The thin film of lighted water at the top of the surface zone is called the **photic zone.** In clear tropical waters, the photic zone may extend to a depth of 200 meters (660 feet), but a more typical value for the open ocean is 100 meters (330 feet). All the production of food by photosynthetic marine plants occurs in this thin, warm surface. Here water is heated by the sun, heat is transferred from the ocean into the atmosphere and space, and gases are exchanged

with the atmosphere. The thermostatic effects we've discussed function largely within this zone. Most of the ocean's life is found here. The photic zone may be extraordinarily thin, but it is also extraordinarily important.

The ocean below the photic zone lies in blackness. Except for light generated by living organisms, the region is perpetually dark. This dark water beneath the photic zone is called the **aphotic zone.**

The light energy of some colors is converted into heat nearer to the surface than the light energy of other colors. **Figure 6.12** shows this differential absorption by color. The top meter (3.3 feet) of the ocean absorbs nearly all the infrared radiation that reaches the ocean surface, significantly contributing to surface warming. The top meter also absorbs 71% of red light. The dimming light becomes bluer with depth because the red, yellow, and orange wavelengths are being absorbed. By 300 meters (1,000 feet) even the blue light has been converted into heat.

From above, clear ocean water looks blue because blue light can travel through water far enough to be scattered back through the surface to our eyes. Divers near the surface see an even brighter blue color. Because nearly all red light is converted to heat in the first few meters of ocean water, red objects a short distance beneath the surface look gray. A diver working at a depth of 10 meters (33 feet) who cuts his hand will see gray blood rather than red because there is not

Color	Wavelength (nm)	% Absorbed in 1 m of Water	Depth by Which 99% Is Absorbed (m)
Infrared	800	82.0	3
Red	725	71.0	4
Orange	600	16.7	25
Yellow	575	8.7	51
Green	525	4.0	113
Blue	475	1.8	254
Violet	400	4.2	107
Ultraviolet	310	14.0	31

a

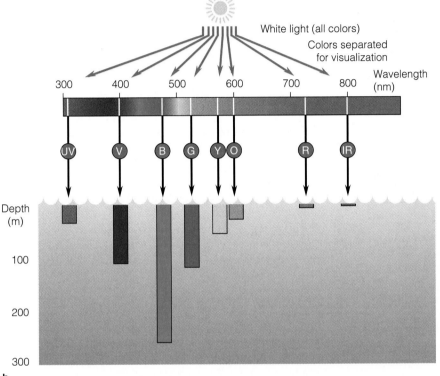

b

c

d

Figure 6.12 Only a thin surface layer of seawater is illuminated by the sun. Except for light generated by living organisms, most of the ocean lies in complete blackness. Absorption of light of different wavelengths (colors) by seawater: (**a**) The table shows the percentage of light absorbed in the uppermost meter of the ocean, and the depths at which only 1% of the light of each wavelength remains. (**b**) The bars show the depths of penetration of 1% of the light of each wavelength (as in the last column of the table). (**c**) A fish photographed in normal oceanic light—the color blue predominates. (**d**) The same fish photographed with a strobe light. The flash contains all colors, and the distance from the strobe to the fish and back to the camera is not far enough to absorb all the red light. The fish shows bright, warm colors otherwise invisible.

enough red light at that depth to reflect from blood's red pigment and stimulate his eye. The underwater pictures of red organisms you've seen are possible only because the diver has brought along a source of white light (which contains all colors). See **Figure 6.12c, d.**

Sound in the Ocean

Sound is a form of energy transmitted by rapid pressure changes in an elastic medium. Sound energy decreases as it travels through seawater because of spreading, scattering, and absorption. Energy loss due to spreading is pro-

portional to the square of the distance from the source. Scattering occurs as sound bounces off bubbles, suspended particles, organisms, the surface, the bottom, or other objects. Eventually sound is absorbed and converted by molecules into a very small amount of heat. The absorption of sound is proportional to the square of the frequency of the sound—higher frequencies are absorbed sooner. Sound waves can travel for much greater distances through water than light waves can before being absorbed. Because sound travels through water so efficiently, many marine animals use sound rather than light to "see" in the ocean.

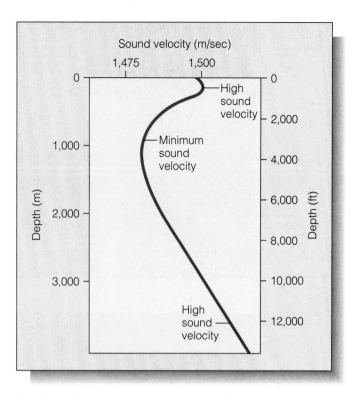

Figure 6.13 The relationship between water depth and sound velocity.

The speed of sound in seawater of average salinity is about 1,500 meters per second (3,345 miles per hour) at the surface, almost five times the speed of sound in air. The speed of sound in seawater increases as temperature and pressure increase. Sound travels faster at the warm ocean surface than it does in deeper, cooler water. Its speed decreases with increasing depth, eventually reaching a minimum at about 1,000 meters (3,300 feet). Below that depth, the effect of increasing pressure offsets the effect of decreasing temperature because temperature remains nearly constant, so speed increases again. Near the bottom of an ocean basin, the speed of sound may be higher than at the surface. Though important in the behavior of oceanic sound, these variations amount to only 2% or 3% of the average speed of sound in seawater. The relationship between depth and sound speed is shown in **Figure 6.13.**

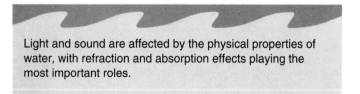

Light and sound are affected by the physical properties of water, with refraction and absorption effects playing the most important roles.

The Sofar Layer

6.16

The depth at which the speed of sound reaches its *minimum* varies with conditions but is usually near 1,200 meters (3,900 feet) in the North Atlantic or about 600 meters (2,000 feet) in the North Pacific. Transmission of

sound in this minimum-velocity layer is very efficient because refraction tends to cause sound energy to remain within the layer. The outer edges of waves escaping from this layer will enter water in which the speed of sound is higher, speed up, and cause the wave to pivot back into the minimum-velocity layer, as diagrammed in **Figure 6.14.** Upward-traveling sound waves that are generated within the minimum-velocity layer will tend to be refracted downward, and downward-traveling sound waves will tend to be refracted upward. In short, sound waves bend *toward* layers of lower sound velocity and so tend to stay within the zone. Therefore loud noises made at this depth can be heard for thousands of kilometers.

Navy depth charges detonated in the minimum-velocity layer in the Pacific have been heard 3,680 kilometers (2,280 miles) away from the explosion. In a recent test, sound generated by a U.S. Navy ship in the Indian Ocean was heard at the Oregon Coast!

In the early 1960s the U.S. Navy experimented with the use of sound transmission in the minimum-velocity layer as a lifesaving tool. Survivors in a life raft would drop a small charge into the water which was set to explode at the proper depth. A number of widely spaced listening stations ashore would compare the differences in the arrival times of the signal and then compute the position of the raft. The project—since abandoned in favor of radio beacons—was called **sofar** (for *so*und *f*ixing *a*nd *r*anging). The minimum-velocity layer has come to be known as the **sofar layer.**

Sonar

6.17

Crews aboard surface ships and submarines employ active **sonar** (*so*und *n*avigation *a*nd *r*anging), the projection and return through water of short pulses ("pings") of high-frequency sound, to search for objects in the ocean (**Figure 6.15**). In a modern system, electrical current is passed through crystals, which respond by producing powerful sound pulses pitched above the limit of human hearing. Some of the sound from the transmitter bounces off any object larger than the wavelength of sound employed and returns to a microphone-like sensor. Signal processors then amplify the echo and reduce the frequency of the sound to within the range of human hearing. An experienced sonar operator can tell the direction of the contact, its size and heading, and even something about its composition (whale or submarine or school of fish) by analyzing the characteristics of the returned ping.

Side-scan sonar is a type of active sonar. Operating with as many as 60 transmitter/receivers tuned to high sound frequencies, side-scan systems towed in the quiet water beneath a ship are sometimes capable of near-photographic resolution. Side-scan systems are used for geological investigations, archaeological studies, and locating downed ships and airplanes (**Figure 6.16**). The *multibeam system* you read about in Chapter 4 (see again Figures 4.11, 4.12, and 4.14) is a form of active sonar.

Humans are not the only organisms to use sonar. Whales and other marine mammals use clicks and whistles to find

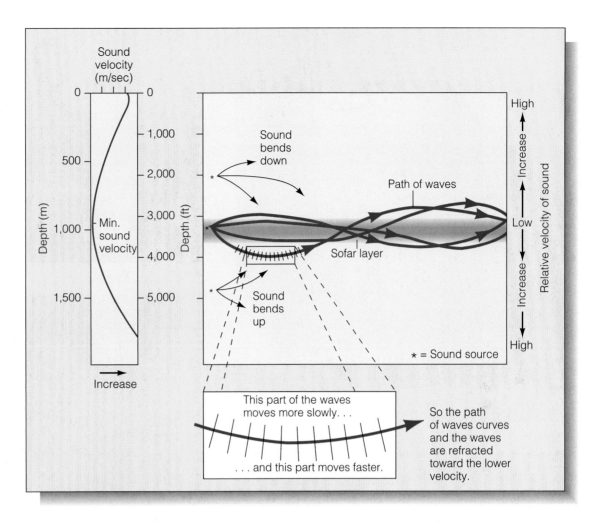

Figure 6.14 The sofar layer, in which sound waves travel at minimum speed. Sound transmission is particularly efficient—that is, sounds can be heard for great distances—because refraction tends to keep sound waves within the layer.

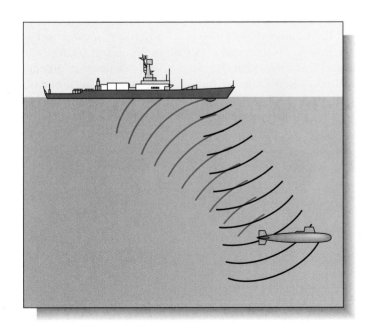

Figure 6.15 The principle of active sonar. Pulses of high-frequency sound are radiated from the sonar array of the sending vessel. Some of the energy of this ping reflects from the submerged submarine and returns to a receiver on the sending vessel. The echo is analyzed to plot the position of the submarine.

food and avoid obstacles. We'll discuss this form of active sonar in Chapter 13.

The Dissolving Power of Water

Water is a powerful solvent—it will eventually dissolve nearly any substance. No wonder, then, that seawater and most other liquids in nature are water solutions. Water's dissolving power results from the polar nature of the water molecule. Consider how water dissolves sodium chloride (or NaCl), the most common salt. In solid crystals of this salt, the sodium atoms in NaCl have lost electrons, and the chloride atoms have gained them. The resulting charged atoms are called **ions.** When liquid water is present, the polarity of water causes the sodium ion (Na^+) to separate from the chloride ion (Cl^-) (**Figure 6.17**). The ions move away from the salt crystal, permitting water to attack the next layer of NaCl. *Note that NaCl does not exist as "salt" in seawater; its components are separated when salt crystals dissolve and joined when crystals re-form.*

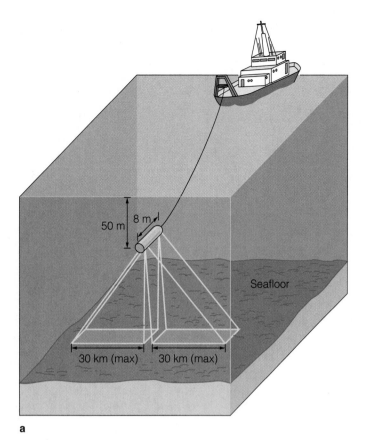

a

Figure 6.16 Side-scan sonar. (**a**) Side-scan sonar in action. Sound pulses leave the submerged towed array (**b**), bounce off the bottom, and return to the device. One of the two slits through which sound pulses depart and return is visible on its side. Computers process the impulses into images: (**c**) A side-scan sonar record showing a downed World War II–era PB4Y-2 Privateer at a depth of 50 meters (165 feet). The artificial colors enhance target detail and clarity.

b

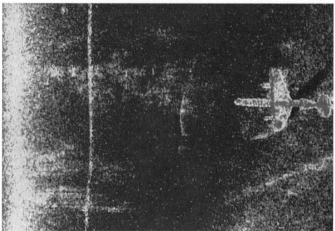

c

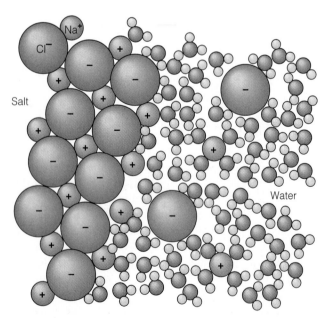

Figure 6.17 Salt in solution. When a salt such as NaCl is put in water, the positively charged hydrogen end of the polar water molecule is attracted to the negatively charged Cl^- ion, and the negatively charged oxygen end is attracted to the positively charged Na^+ ion. The ions are surrounded by water molecules that are attracted to them and become solute ions in the solvent (right side of figure).

By contrast, oil doesn't dissolve in water even if the two are thoroughly shaken together. When oil is dispersed in water, it forms a mixture because molecules of oil are non-polar. This means that oil has no positive or negative charges to attract the polar water molecule. In a way this is fortunate: Living tissues would readily dissolve in water if the oils within their membranes didn't blunt water's powerful attack.

The polar nature of the water molecule produces some unexpected chemical properties. One of the most important is water's remarkable ability to dissolve more substances than any other natural solvent.

Seawater

The total quantity (or concentration) of dissolved inorganic solids in water is its **salinity.** The ocean's salinity varies from about 3.3% to 3.7% by weight, depending on such factors as evaporation, precipitation, and freshwater runoff from the

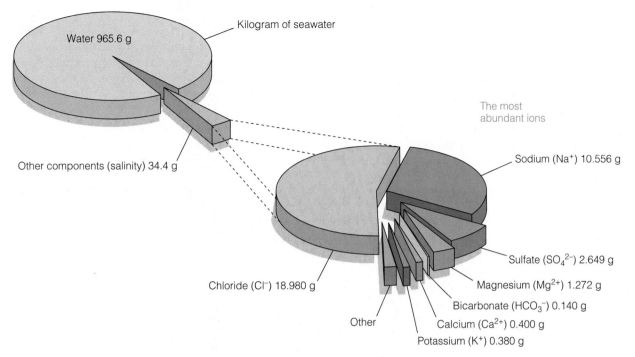

Figure 6.18 A representation of the most abundant components of a kilogram of seawater at 35‰ salinity. Note that the specific ions are represented in grams per kilogram, equivalent to parts per thousand (‰).

continents, but the average salinity is usually given as 3.5%. The world ocean contains some 5,000 trillion kilograms (5.5 trillion tons) of salt. If the ocean's water evaporated completely, leaving its salts behind, the dried residue could cover the entire planet with an even layer 45 meters (150 feet) thick! Most of the dissolved solids in seawater are salts that have been separated into ions. Chloride and sodium are the most abundant of these.

Salinity: Dissolved Solids and Water Together

About 3.5% of seawater consists of dissolved substances. Boiling away 100 kilograms of seawater theoretically produces a residue that weighs 3.5 kilograms. Variations of 0.1% are significant, and oceanographers prefer to use parts-per-thousand notation (‰) rather than percent (%, parts per hundred) in discussing these materials.[3] The seven ions shown in the pie on the right in **Figure 6.18** make up more than 99% of this residual material. When seawater evaporates, its ionic components combine in many different ways to form table salt (NaCl), epsom salts ($MgSO_4$), and other mineral salts.

Seawater also contains minor constituents. The ocean is sort of an "Earth tea": Every element present in the crust and atmosphere is also present in the ocean, though sometimes in extremely small amounts. Only 14 elements have concentrations in seawater greater than one part per million (ppm).

Elements present in amounts less than 0.001‰ (1 ppm) are known as **trace elements.**

> The most abundant ions dissolved in seawater are chloride, sodium, and sulfate.

The Source of the Ocean's Salts

Remembering the effectiveness of water as a solvent, you might think that the ocean's saltiness has resulted from the ability of rain, groundwater, or crashing surf to dissolve crustal rock. Much of the sea's dissolved material originated in that way, but is crustal rock the source of all the ocean's solutes? An easy way to find out would be to investigate the composition of salts in river water and compare these figures to those of the ocean as a whole. If crustal rock is the only source, then the salts in the ocean should be like those in concentrated river water. But they are not. River water is usually a dilute solution of calcium and bicarbonate ions, while the principal ions in seawater are chloride and sodium. The magnesium content of seawater would also be higher if seawater were simply concentrated river water. The proportions of salts in isolated salty inland lakes, such as Utah's Great Salt Lake or the Dead Sea, are indeed much different from the proportions of salts in the ocean. So, weathering and erosion of crustal rocks cannot be the sole sources of sea salts.

---

[3] Note that 3.5% = 35‰. If you began with 1,000 kilograms of seawater, you would expect 35 kilograms of residue.

The components of ocean water whose proportions are *not* accounted for by the weathering of surface rocks are called **excess volatiles.** To find the source of these excess volatiles we must look to Earth's deeper layers. The upper mantle appears to contain more of the substances found in seawater (including the water itself) than are found in surface rocks, and their proportions are about the same as found in the ocean. As you read in Chapter 3, convection currents slowly churn Earth's mantle, causing the movement of tectonic plates. This activity allows some deeply trapped volatile substances to escape to the exterior, outgassing through volcanoes and rift vents. These excess volatiles include carbon dioxide, chlorine, sulfur, hydrogen, fluorine, nitrogen, and, of course, water vapor. This material, along with residue from surface weathering, accounts for the chemical constituents of today's ocean.

Some of the salts that form from the ocean's dissolved materials are hybrids of the two processes of weathering and outgassing. Table salt, sodium chloride, is an example. The sodium ions come from the weathering of crustal rocks, while the chlorine ions come from the mantle by way of volcanic vents and outgassing from mid-ocean rifts. As for the lower-than-expected quantity of magnesium ions in the ocean, recent research at a spreading center east of the Galápagos Islands suggests that mid-ocean rifts may play a role in reducing the magnesium content and increasing the calcium content of seawater. The water that circulates through new ocean floor at these sites is apparently stripped of magnesium and a few other elements. The magnesium seems to be incorporated into mineral deposits, but calcium is added as hot water dissolves adjacent rocks.

The quantity of dissolved inorganic solids in water is its salinity. The proportion of ions in seawater is not the same as the proportion in concentrated river water, which indicates that ongoing geological and chemical processes affect the ocean's salinity.

The Principle of Constant Proportions 6.22

In 1865 the chemist Georg Forchhammer noted that although the total *amount* of dissolved solids (salinity) might vary between samples, the *ratio* of major salts was constant in samples of seawater from many locations. In other words, the percentage of various salts in seawater is the same in samples from many places regardless of how salty the water is. This constant ratio is known as **Forchhammer's principle,** or the **principle of constant proportions.** Forchhammer was also the first to observe that seawater contains fewer silica and calcium ions than concentrated river water, and the first to realize that removal of these compounds by marine animals and plants to form shells and other hard parts might account for part of the difference.

Determining Salinity 6.23

Water's salinity by weight would seem an easy property to measure. Why not simply evaporate a known weight of seawater and weigh the residue? This simple method yields imprecise results because some salts will not release all the molecules of water associated with them. If these salts are heated to drive off the water, then other salts (carbonates, for example) will decompose to form gases and solid compounds not originally present in the water sample.

Modern analysis depends instead on determining the sample's chlorinity. **Chlorinity** is a measure of the total weight of chlorine, bromine, and iodine ions in seawater. Because chlorinity is comparatively easy to measure, and because the ratio of chlorinity to salinity is constant, marine chemists have devised the following formula to determine salinity:

$$\text{Salinity in ‰} = 1.80655 \times \text{Chlorinity in ‰}$$

Chlorinity is about 19.4‰, so salinity is around 35‰.

Seawater samples can be obtained by methods ranging from tossing a clean bucket over the side of the ship to sophisticated tube-and-pump systems. Typically water samples are collected using a group of sampling bottles (as in **Figure 6.19**). The bottles are lowered from a ship and triggered to close at specific depths by an electronic signal. Later the bottles are hauled to the surface and their contents analyzed.

Until recently, marine chemists used a delicate chemical procedure involving a silver nitrate solution to measure the chlorinity of seawater. Conversion to salinity was made by a set of mathematical tables. The procedure was calibrated against a standard sample of seawater of precisely known chlorinity. Today's marine scientists use an electronic device called a **salinometer** that measures the electrical conductivity of seawater (**Figure 6.20**). Conductivity varies with the

Figure 6.19 A "rosette" of sampling bottles. Each of these 10-liter Niskin bottles may be mechanically sealed by a signal from the research ship when the array reaches predetermined depths. The bottles are hauled to the surface and their contents analyzed.

Figure 6.20 This portable salinometer reads temperature, pH, and dissolved oxygen as well as conductivity. Designed to be lowered over the side of a research vessel at the end of a line, the self-powered device contains a small pump that passes water over sensors. This model takes two readings every second and operates down to about 200 meters (660 feet). Data may be retrieved by a cable connected to the research vessel or stored in the salinometer's memory to be retrieved when it is brought back aboard ship. A microprocessor contained within the white tube converts conductivity to salinity and, when connected to a portable computer, displays the results of all readings as graphs.

concentration and mobility of ions present and with the water temperature. Circuits in the salinometer adjust for water temperature, convert conductivity to salinity, and then display salinity. Salinometers are also calibrated against a sample of known conductivity and salinity. The best salinometers can determine salinity to an accuracy of 0.001%. Some salinometers are designed for remote sensing—the electronics stay aboard ship while the sensor coil is lowered over the side.

Chemical Equilibrium and Residence Times

If outgassing and the chemical weathering of rock are continuing processes, shouldn't the ocean become progressively saltier with age? Landlocked seas and some lakes usually become saltier as they grow older, but the ocean does not. The ocean appears to be in **chemical equilibrium;** that is, the proportion *and amounts* of dissolved salts per unit volume of ocean are nearly constant. Evidently, whatever goes in must come out somewhere else.

Geologists in the 1950s developed the concept of a "steady state ocean." The idea suggests that ions are added to the ocean at the same rate as they are being removed. This theory helps explain why the ocean is not growing saltier. The idea led to the concept of **residence time**—the average length of time an element spends in the ocean. Residence time for a particular element may be calculated by this equation:

$$\text{Residence time} = \frac{\text{Amount of element in the ocean}}{\text{Rate at which the element is added to (or removed from) the ocean}}$$

Additions of salts from the mantle or from the weathering of rock are balanced by subtractions of minerals being bound into sediments. Dissolved salts precipitate out of the water and the hard parts of living organisms containing silicon and calcium carbonate drift slowly down to the seabed. Some of these sediments are removed from the ocean and drawn into the mantle at subduction zones by the cycling of crustal plates. Input from runoff and outgassing equals outfall (binding into sediments) for each dissolved component.

The residence time of an element depends on its chemical activity. Atoms (or ions) of some elements, such as aluminum and iron, remain in seawater for a relatively short time before becoming incorporated in sediments; others, such as chloride, sodium, and magnesium, remain in water for millions of years. The approximate residence times for the major constituents of seawater vary greatly. Iron has a residence time of about 200 years, but calcium stays in the ocean for about a million years, and the abundant chloride ion has a residence time of about 100 million years. Because of evaporation and precipitation, ocean water itself has a residence time of about 4,100 years.

Though most solids and gases are soluble in water, the ocean is in chemical equilibrium, and neither the proportion nor the amount of most dissolved substances changes significantly through time.

Mixing Time

If constituent minerals are added to ocean water at rates that are less than the ocean's mixing time, they will become evenly distributed throughout the ocean. Because of the vigorous activity of currents, the **mixing time** of the ocean is thought to be on the order of 1,600 years, so the ocean has been mixed hundreds of thousands of times during its long history. The relatively long residence times of seawater's major constituents assure thorough mixing, which is the basis of Forchhammer's principle of constant proportions.

Dissolved Gases

Gases in the air dissolve readily in seawater at the ocean's surface. Plants and animals living in the ocean require these dissolved gases to survive. No marine animal has the ability to break down water molecules to obtain oxygen directly, and no marine plant can manufacture enough carbon dioxide to support its own metabolism. In order of their relative abundance, the major gases found in seawater are nitrogen, oxygen, and carbon dioxide. Because of differences in their solubility in water and air, the proportions of dissolved gases in the ocean are very different from the proportions of the same gases in the atmosphere.

Unlike solids, gases dissolve most readily in *cold* water. A cubic meter of chilly polar water usually contains a greater volume of dissolved gases than a cubic meter of warm tropical water.

Nitrogen

About 48% of the dissolved gas in seawater is nitrogen. (In contrast, the atmosphere is slightly more than 78% nitrogen by volume.) The upper layers of ocean water are usually saturated with nitrogen; that is, additional nitrogen will not dissolve. Living organisms require nitrogen to build proteins and other important biochemicals, but they cannot use the free nitrogen in the atmosphere and ocean directly. It must first be "fixed" into usable chemical forms by specialized organisms. Though some species of bottom-dwelling bacteria can manufacture usable nitrates from the nitrogen dissolved in seawater, most of the nitrogen compounds needed by living organisms must be recycled among the organisms themselves.

Oxygen

About 36% of the gas dissolved in the ocean is oxygen, but there is about a hundred times more gaseous oxygen in Earth's atmosphere than is dissolved in the whole ocean. An average of 6 milligrams of oxygen is dissolved in each liter of seawater (that is, 6 parts per million parts of oxygen per liter of seawater, by weight). Yet this small amount of oxygen is a vital resource for animals that extract oxygen with gills. The sources of the ocean's dissolved oxygen are the photosynthetic activity of plants and plantlike organisms, and the diffusion of oxygen from the atmosphere (as seen in **Figure 6.21**).

Carbon Dioxide (CO_2) 6.29

The amount of carbon dioxide in the atmosphere is very small (0.03%) because CO_2 is in great demand by photosynthesizers as a source of carbon for growth. Carbon dioxide is very soluble in water, though; the proportion of dissolved CO_2 in water is about 15% of all dissolved gases. Because CO_2 combines chemically with water to form a weak acid (H_2CO_3, carbonic acid), water can hold perhaps a thousand times more carbon dioxide than either nitrogen or oxygen at saturation. Carbon dioxide is quickly used by marine plants, so dissolved quantities of CO_2 are almost always much less than this theoretical maximum. Even so, at the present time there is about 60 times as much CO_2 dissolved in the ocean as in the atmosphere. Much more CO_2 moves from atmosphere to ocean than from ocean to atmosphere, in part because some dissolved CO_2 forms carbonate ions, which are locked into sediments, minerals, and the shells and skeletons of living organisms.

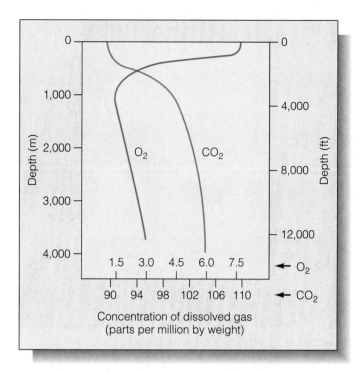

Figure 6.21 How concentrations of oxygen and carbon dioxide vary with depth. Oxygen is abundant near the surface because of the photosynthetic activity of marine plants. The oxygen concentration decreases below the sunlit layer because of the respiration of marine animals and bacteria, and because of the oxygen consumed by the decay of tiny dead organisms slowly sinking through the area. In contrast, plants use carbon dioxide during photosynthesis, so surface levels of CO_2 are low. Because photosynthesis cannot take place in the dark, CO_2 given off by animals and bacteria tends to build up at depths below the sunlit layer. Levels of CO_2 also increase with depth because its solubility increases as pressure increases and temperature decreases.

Figure 6.21 illustrates how carbon dioxide and oxygen concentrations vary with depth. Carbon dioxide concentrations increase with increasing depth, but oxygen concentrations usually decrease through the middle depths and then rise again toward the bottom. High concentrations of oxygen at the surface are usually by-products of photosynthesis in the ocean's brightly lit upper layer. Since plants and plantlike organisms require carbon dioxide for metabolism, surface CO_2 concentrations tend to be low. A decrease in oxygen below the sunlit upper layer is usually due to the respiration of bacteria and marine animals, which leads to higher concentrations of carbon dioxide. Oxygen levels are slightly higher in deeper water because fewer animals are present to take up the oxygen that reaches these depths and

Gases dissolve in water in proportions that vary with their physical properties. Nitrogen is the most abundant dissolved gas in seawater; oxygen is the second most abundant. Carbon dioxide is the most soluble gas, and one of many substances that affect the ocean's pH balance.

because oxygen-rich polar water that sinks from the surface is the greatest source of deep water.

Acid–Base Balance

Water can separate to form hydrogen ions (H$^+$) and hydroxide ions (OH$^-$). These two ions are present in equal concentrations in pure water. An imbalance in the proportion of ions produces an acidic (or basic) solution. An **acid** is a substance that *releases* a hydrogen ion in solution; a **base** is a substance that *combines with* a hydrogen ion in solution. Basic solutions are also called **alkaline** solutions.

The acidity or alkalinity of a solution is measured in terms of the **pH scale,** which measures the concentration of hydrogen ions in a solution. An excess of hydrogen ions (H$^+$) in a solution makes that solution acidic. An excess of hydroxide ions (OH$^-$) makes a solution alkaline. **Figure 6.22** shows a pH scale and the pH values of a few familiar solutions. The scale is logarithmic, which means that a change of one pH unit represents a tenfold change in the hydrogen ion concentration. A modern nonphosphate detergent is a thousand times more alkaline than seawater, and black coffee is a hundred times more acidic than pure water. Pure water, which is neutral (neither acidic nor alkaline), has a pH of 7; lower numbers indicate greater acidity (more H$^+$ ions) and higher numbers indicate greater alkalinity (fewer H$^+$ ions).

Seawater is slightly alkaline; its average pH is about 7.8. This seems odd because of the large amount of CO$_2$ dissolved in the ocean. If dissolved CO$_2$ combines with water to form carbonic acid, why is the ocean mildly alkaline and not slightly acidic? When dissolved in water, CO$_2$ is actually present in several different forms. Carbonic acid (H$_2$CO$_3$) is only one of these. In water solutions, some carbonic acid breaks down to produce the hydrogen ion (H$^+$), the bicarbonate ion (HCO$_3^-$), and the carbonate ion (CO$_3^{2-}$). This behavior acts to **buffer** seawater, preventing broad swings of pH when acids or bases are introduced.

Though seawater remains slightly alkaline, it is subject to some variation. In areas of rapid plant growth, for example, pH will rise because CO$_2$ is used by the plants for photosynthesis. Because temperatures are generally warmer at the surface, less CO$_2$ can dissolve in the first place. So, surface pH in warm productive water is usually around 8.5.

At middle depths and in deep water, more CO$_2$ may be present. Its source is the respiration of animals and bacteria. With cold temperatures, high pressure, and no photosynthetic plants to remove it, this CO$_2$ will lower the pH of water, making it more acidic with depth. Deep, cold seawater below 4,500 meters (15,000 feet) has a pH of around 7.5. This lower pH can dissolve calcium-containing marine sediments; you may recall from Chapter 5 that sediments containing calcium carbonate are rarely found in deep water. A drop to pH 7 can occur at the deep-ocean floor when bottom bacteria consume oxygen and produce hydrogen sulfide.

Seawater acts as a buffer to prevent broad swings of pH when acids or bases are introduced.

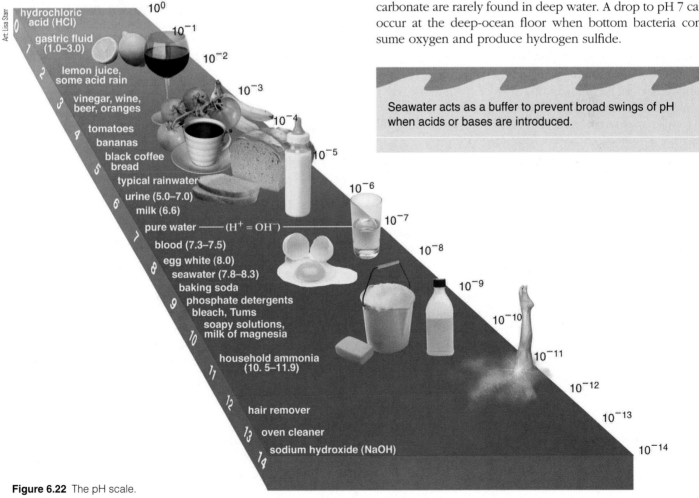

Figure 6.22 The pH scale.

QUESTIONS FROM STUDENTS

1. If outgassing continues today, why aren't the oceans getting bigger? By now shouldn't they have covered the continents?

Well, yes, outgassing does continue, but the water is being drawn from the mantle. Some marine geologists believe that the mantle shrinks a little, making the ocean basins a bit deeper, which accommodates the ocean's greater volume. Also, geological processes have increased the amount of continental crust, and water is drawn back into the mantle at subduction zones. Though some of this water eventually reappears at the surface in volcanic vapor, some is incorporated back into mantle material. It probably just about balances out.

2. I see that there is a hundred times more oxygen in the atmosphere than in the ocean. How can that be? Isn't water 86% oxygen by weight?

Yes, but the oxygen of water (H_2O) is bonded tightly to hydrogen atoms. Unlike atmospheric oxygen, the oxygen in water molecules cannot be used by organisms for respiration. Fish can't extract this oxygen with their gills. Marine animals must depend on dissolved oxygen—paired molecules of oxygen (O_2) present in the water and free to move through the gill membranes. Compared to the atmosphere, very little of this free oxygen is present in the ocean.

3. If water has the highest latent heat of vaporization, why does a drop of alcohol make my hand feel colder when it evaporates than a drop of water does?

Heat *versus* temperature again. The alcohol makes your hand *colder*, but it evaporates much *more quickly*. The water stays around longer (takes longer to evaporate), and although it may not cause your skin to become as *cold*, it will remove more *heat*.

4. Liquid methane appears to be the only other substance found in quantity in a liquid state in the solar system. If our ocean were of liquid methane rather than liquid water, how would conditions here be different?

Liquid methane freezes at $-183°C$ ($-272°F$) and boils at $-162°C$ ($-234°F$). Methane has a much lower heat capacity than liquid water because it is not dipolar—it does not form a great bonded mass as water does—and consequently it does not require a large energy input to break the maze of hydrogen bonds to release molecules. Because of methane's low heat capacity, the difference between Earth's polar and tropical daytime temperatures would be drastically greater unless the circulation rate of liquid and vapor currents accelerated to keep pace. Computer modeling yields a very unattractive vision of a planet with a boiling equatorial ocean, crushing atmospheric pressures from the greater amount of vapor in the air, torrential polar methane rainfall, cataclysmic cyclonic storms, and average wind speeds at mid-latitudes of hundreds of kilometers per hour. We miss all this excitement because our ocean is made of water.

5. Input of solar radiation in the Northern Hemisphere peaks on 22 June and reaches a minimum on 22 December because of Earth's orbital tilt (more about that in Chapter 8). Why, then, do our warmest days occur in August or September and our coldest days in January or February?

Because of thermal inertia. There is a lag between maximum sunlight and maximum warmth because of water's great heat capacity. The sun must shine on this watery planet for many weeks to raise the summer hemisphere's temperature. Of course, water also retains heat well, so the coldest days in the winter hemisphere come well after the darkest ones.

6. Why don't sounds seem to travel easily from air to ocean, or vice versa?

Sound waves can make the transition from one medium to another with little energy loss only when the speeds of sound waves in the two different media are similar. The speeds of sound in water and air, however, are too different for an efficient transition to be made. Too great a contrast in speed produces reflection (not refraction) of the sound waves at the junction. This is why you can't hear people shouting from the edge of the pool while you're underwater, even though the weak sound of a submerged pebble clicking against the side is very clear and sharp.

If you place a solid medium in which the speed of sound is intermediate between air and water, the sound can move across one junction and then the other, for a more efficient total transition. Wood works well for this, which explains why ocean noises are easy to hear in wooden boats. Even if the speed of sound in the intermediate medium is higher than in water (as in steel, for instance), some sound will be audible simply because the hard surface provides a good radiating surface for noises coming from the water.

7. Why is water blue?

The absorption of red light by hydrogen bonds is what gives pure water—and thick ice—its pale bluish hue. Some of the blue light that remains after the red is absorbed is scattered back through the surface, and that's what we see.

CHAPTER SUMMARY

Water, a chemical compound composed of two hydrogen atoms and one oxygen atom, is abundant on and within Earth. The polar nature of the water molecule and the hydrogen bonds that form between water molecules result in some unexpected physical and chemical properties.

The thermal properties of water are responsible for the mild physical conditions at Earth's surface. Liquid water is remarkably resistant to temperature change with the addition or removal of heat; ice, with its high latent heat of fusion and low density, melts and re-freezes over large areas of the ocean to absorb or release heat with no change in temperature. These thermostatic effects, combined with the mass movement of water and water vapor, prevent large swings in Earth's surface temperature.

The physical characteristics of the world ocean are largely determined by the physical properties of seawater. These properties include water's heat capacity, density, salinity, and ability to transmit light and sound.

Changes in temperature and salinity greatly influence water density. Ocean water is usually layered by density, with the most dense water on or near the bottom.

Sound and light in the sea are affected by the physical properties of water, with refraction and absorption effects playing important roles.

Water also has the remarkable ability to dissolve more substances than any other natural solvent. Though most solids and gases are soluble in water, the ocean is in chemical equilibrium, and neither the proportion nor the amount of most dissolved substances changes significantly through time. Most of the properties of seawater are different from those of pure water because of the substances dissolved in the seawater.

ON-LINE STUDY RESOURCES

The Web site for this book contains a wealth of helpful study aids. Log on to:

http://info.brookscole.com/garrison

and select *Essentials of Oceanography*, 3rd edition. Select Chapter 6 from the drop-down menu or click on one of these resource areas:

- For study and review, **Chapter at a Glance** gives you an outline of the chapter, **Chapter Summary** allows you to review the chapter's main ideas, and **Glossary** lists concepts and terms for the chapter along with their definitions.

- To test your mastery of the Terms and Concepts to Remember for this chapter, you can use the electronic **Flash Cards,** play the **Concentration** game, or work the **Crossword Puzzle.**

- For practice quizzing, try the multiple-choice **Tutorial Quiz,** the ten-question **True/False Quiz,** or the **Image Analysis Quiz,** which poses questions based on art and photos from the chapter.

TERMS AND CONCEPTS TO REMEMBER

absorption
acid
adhesion
alkaline
aphotic zone
base
buffer
calorie
chemical bond
chemical equilibrium
chlorinity
cohesion
covalent bonds
deep zone
degrees
density
density curve
electron
excess volatiles
Forchhammer's principle
halocline
heat
heat capacity
hydrogen bond
ion
latent heat of fusion
latent heat of vaporization
mixed layer (or surface zone)
mixing time
molecule
pH scale
photic zone
polar
principle of constant proportions
proton
pycnocline
refraction
residence time
salinity
salinometer
scattering
sensible heat
sofar
sofar layer
sonar
sound
state
surface zone (or mixed layer)
temperature
thermal equilibrium
thermal inertia
thermocline
thermostatic properties
trace elements

STUDY QUESTIONS

1. Why is water a polar molecule? What properties of water derive from its polar nature?

2. Other than hydrogen and oxygen, what are the most abundant elements/ions in seawater?

3. How is salinity determined? How are modern methods dependent on the principle of constant proportions?

4. Which dissolved gas is represented in the ocean in much greater proportion than in the atmosphere? Why the disparity?

5. What factors affect seawater's pH? How does the pH of seawater change with depth? Why?

6. How is heat different from temperature?

7. How does water's high heat capacity influence the ocean? Leaving aside its effect on beach parties, how do you think conditions on Earth would differ if our ocean consisted of ethyl alcohol?

8. Why does ice float? Why is this fact important to thermal conditions on Earth?

9. What factors affect the density of water? Why does cold air or water tend to sink? What is the role of salinity in water density?

10. How is the ocean stratified by density? What physical factors are involved? What names are given to the ocean's density zones?

11. If the residence time of the water in the ocean is about 4,100 years, how many times has an average water molecule evaporated from and returned to the ocean?

12. What factors influence the intensity and color of light in the sea? What factors affect the depth of the photic zone? Could there be a "photocline" in the ocean?

Atmospheric Circulation

About half of the energy radiated toward Earth from the sun is absorbed by Earth, but this energy is not distributed evenly across the planet's surface. The amount of solar energy that reaches Earth's surface per minute varies with the angle of the sun above the horizon, the transparency of the atmosphere, and the local reflectivity of the surface. The most important factors that affect the angle of the sun above the horizon are latitude and season.

Uneven Solar Heating and Latitude

As can be seen in **Figure 7.2a,** sunlight striking the polar regions spreads over a greater area, approaches the surface at a low angle favoring reflection, and filters through more atmosphere. The high angle at which sunlight strikes in the

tropics distributes the same amount of sunlight over a much smaller area; the more nearly vertical angle at which the light approaches means that it passes through less atmosphere and minimizes reflection. As you would expect, the tropics are warmer than the polar regions.

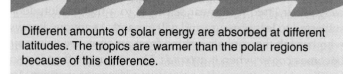

Different amounts of solar energy are absorbed at different latitudes. The tropics are warmer than the polar regions because of this difference.

Uneven Solar Heating and the Seasons

At mid-latitudes the Northern Hemisphere receives about three times as much solar energy per day in June as it does in December. This difference is due to the 23½° tilt of Earth's rotational axis relative to the plane of its orbit around the sun

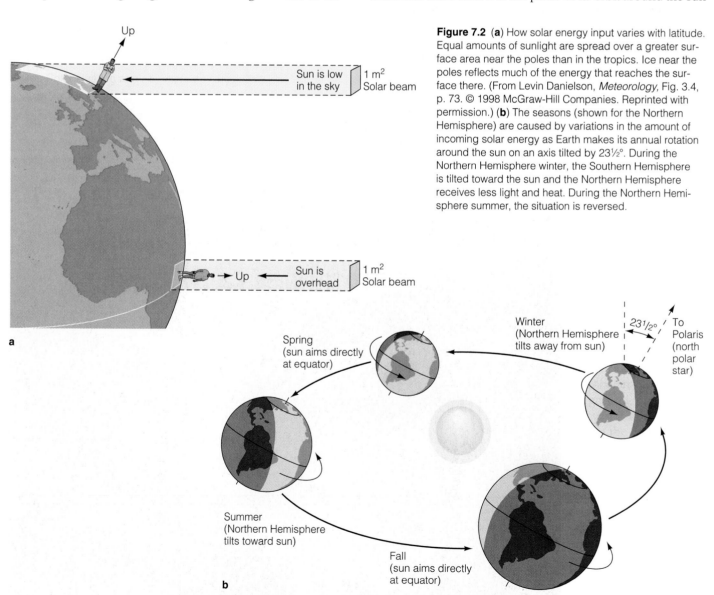

Figure 7.2 (**a**) How solar energy input varies with latitude. Equal amounts of sunlight are spread over a greater surface area near the poles than in the tropics. Ice near the poles reflects much of the energy that reaches the surface there. (From Levin Danielson, *Meteorology,* Fig. 3.4, p. 73. © 1998 McGraw-Hill Companies. Reprinted with permission.) (**b**) The seasons (shown for the Northern Hemisphere) are caused by variations in the amount of incoming solar energy as Earth makes its annual rotation around the sun on an axis tilted by 23½°. During the Northern Hemisphere winter, the Southern Hemisphere is tilted toward the sun and the Northern Hemisphere receives less light and heat. During the Northern Hemisphere summer, the situation is reversed.

(**Figure 7.2b**). As Earth revolves around the sun, the constant tilt of its rotational axis causes the Northern Hemisphere to lean *toward* the sun in June but *away* from it in December. The sun therefore appears higher in the sky in the summer but lower in winter. The inclination of Earth's axis also causes days to become longer as summer approaches but shorter with the coming of winter. Longer days mean more time for the sun to warm Earth's surface.

Uneven Solar Heating and Atmospheric Circulation

As can be seen in **Figure 7.3,** near the equator, the amount of solar energy received by Earth greatly exceeds the amount of heat radiated into space. In the polar regions the opposite

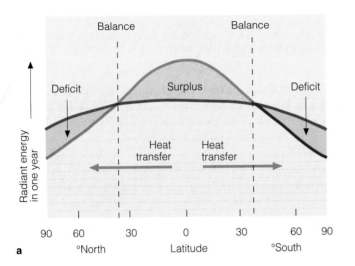

a

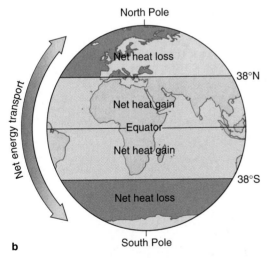

b

Figure 7.3 Areas of heat gain and loss on Earth's surface. (**a**) The average annual incoming solar radiation (red line) *absorbed* by Earth is shown along with the average annual infrared radiation (blue line) *emitted* by Earth. Note that polar latitudes lose more heat to space than they gain, and tropical latitudes gain more heat than they lose. Only at about 38°N and 38°S latitudes does the amount of radiation received equal the amount lost. Since the area of heat gained (orange area) equals the area of heat lost (blue areas), Earth's total heat budget is balanced. (**b**) The ocean does not boil away near the equator or freeze solid near the poles because heat is transferred by winds and ocean currents from equatorial to polar regions.

is true. If nothing intervened, the tropical ocean would boil and the polar regions freeze solid. Global air and water circulation moves heat and thus prevents this unpleasant outcome! Circulation of air is responsible for about two-thirds of the heat transfer from tropical to polar regions. (Ocean currents account for the other third.)

Think of air circulation in a room with a hot radiator opposite a cold window (**Figure 7.4**). Air warms, expands, becomes less dense, and *rises* over the radiator. Air cools, contracts, becomes more dense, and *falls* near the cold glass window. The circular current of air in the room—a **convection current**—is caused by the difference in temperature between the ends of the room. A similar process occurs over the surface of Earth. As we have seen, surface temperatures are higher at the equator than at the poles, and air can gain heat from warm surroundings. Since air is free to move over Earth's surface, it would be reasonable to assume that an air circulation pattern like the one shown in **Figure 7.5** would develop over Earth. In this ideal model, air heated in the tropics would expand and become less dense, rise to high altitude, turn poleward, and "pile up" as it converged near the poles. The air would then cool and contract by radiating heat into space, sink to the surface, and turn equatorward, flowing along the surface back to the tropics to complete the circuit.

But this is *not* what happens. Global circulation of air is governed by two factors: uneven solar heating *and* Earth's rotation. The eastward rotation of Earth on its axis deflects the moving air or water (or any moving object that has mass) away from its initial course. This deflection is called the **Coriolis effect** in honor of **Gaspard Gustave de Coriolis,** the French scientist who worked out its mathematics in 1835.

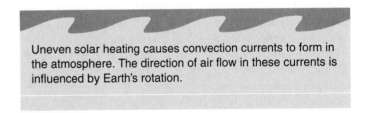

Uneven solar heating causes convection currents to form in the atmosphere. The direction of air flow in these currents is influenced by Earth's rotation.

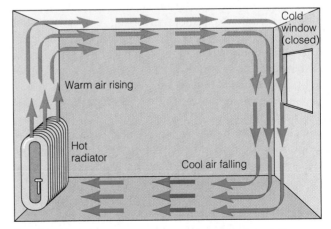

Figure 7.4 A convection current forms in a room when air flows from a hot radiator to a cold window and back. (For a practical oceanic application of this principle, look ahead to Figure 7.12.)

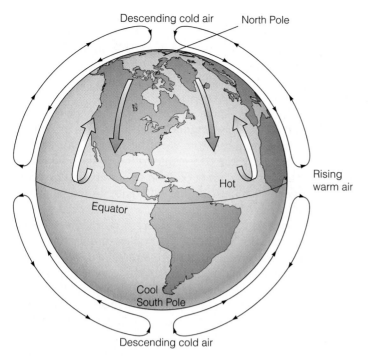

Figure 7.5 A hypothetical model of Earth's air circulation if uneven solar heating were the only factor to be considered. (The thickness of the atmosphere is greatly exaggerated in this drawing.)

An understanding of the Coriolis effect is important to an understanding of atmospheric and oceanic circulation.

The Coriolis Effect

To an earthbound observer, any object moving freely across the globe appears to curve slightly from its initial path. In the Northern Hemisphere this curve is to the right from the expected path; in the Southern Hemisphere it is to the left. To earthbound observers the deflection is very real; it isn't caused by some mysterious force, and it isn't an optical illusion or some other trick caused by the shape of the globe itself. *The observed deflection is caused by the observer's moving frame of reference on the spinning Earth.*

The influence of this deflection can be illustrated by performing a mental experiment involving objects—in this case, cities and cannonballs—and then applying the principle to atmospheric circulation. Let's pick as examples for our experiment the equatorial city of Quito, the capital of Ecuador, and Buffalo, New York. Both cities are on almost the same line of longitude (79°W), so Buffalo is almost exactly north of Quito (as **Figure 7.6** shows). Like everything else attached to the rotating Earth, both cities make one trip around the world each 24 hours. The north-south relationship of the two cities never changes—Quito is *always* due south of Buffalo.

A complete trip around the world is 360°, so each city moves eastward at an angular rate of 15° per hour (360°/24 hours = 15°/hour). Yet, even though their angular rates are the same, the two cities move eastward at different speeds. Quito is on the equator, the "fattest" part of Earth. Buffalo is

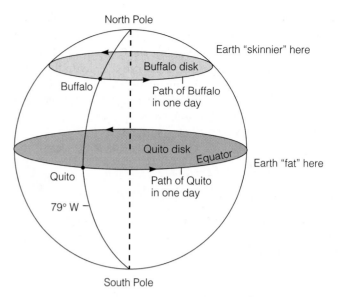

Figure 7.6 Sketch of the thought experiment in the text, showing that Buffalo travels a shorter path on the rotating Earth each day than Quito does.

farther north at a "skinnier" part. Imagine both cities isolated on flat disks, and imagine Earth's sphere being made of a great number of these disks strung together on a rod connecting North Pole to South Pole. From Figure 7.6, you can see that Buffalo's disk has a smaller circumference than Quito's. Buffalo doesn't have as far to go in one day as Quito because its disk is not as large. That means that Buffalo must move eastward *more slowly* than Quito to maintain its position due north of Quito.

Look at Earth from above the North Pole in **Figure 7.7.** The Quito disk and the Buffalo disk must turn through 15°

Quito moves at 1,668 km/hr (1,036 mi/hr). Note: Quito's longer *distance* through space in one hour is still 15°.

Buffalo moves at 1,260 km/hr (783 mi/hr). Note: Buffalo's shorter *distance* through space in one hour is still 15°.

Figure 7.7 A continuation of the thought experiment. A look at Earth from above the North Pole shows that Buffalo and Quito move at different velocities.

of longitude each hour (or Earth would rip itself apart), but the city on the equator must move faster to the east to turn its 15° each hour because its "slice of the pie" is larger. Buffalo must move at 1,260 kilometers (783 miles) per hour to go around the world in one day, while Quito must move at 1,668 kilometers (1,036 miles) per hour to do the same.

Now imagine a massive object moving between the two cities. A cannonball shot north from Quito toward Buffalo would carry Quito's eastward component as it goes; that is, regardless of its northward speed, the cannonball is also moving *east* at 1,668 kilometers (1,036 miles) per hour. The fact of being fired northward by the cannon does not change its eastward movement in the least. As the cannonball streaks north, an odd thing happens. The cannonball veers from its northward path, angling slightly to the right (east) (**Figure 7.8**). Actually, this first cannonball is moving just as an observer from space would expect it to, but to those of us on the ground the cannonball "gets ahead of Earth." As cannonball 1 moves north, the ground beneath it is no longer moving eastward at 1,668 kilometers per hour. During the ball's time of flight, Buffalo (on its smaller disk) *has not moved eastward enough to be where the ball will hit.* If the time of flight for the cannonball is one hour, a city 408 kilometers east of Buffalo (1,668 [Quito's speed] − 1,260 [Buffalo's speed] = 408) will have an unexpected surprise. Albany may be in for some excitement!

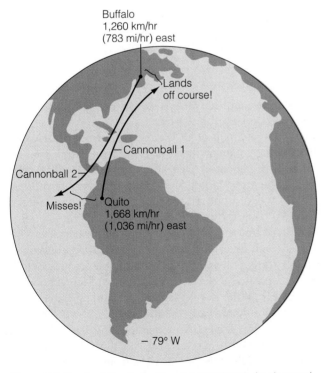

Figure 7.8 The final step in the thought experiment. As observed from space, cannonball 1 (shot northward) and cannonball 2 (shot southward) move as we might expect; that is, they travel straight away from the cannons and fall to Earth. Observed from the ground, however, cannonball 1 veers slightly east and cannonball 2 veers slightly west of their intended targets. The effect depends on the observer's frame of reference.

If a cannonball were fired south from Buffalo toward Quito, the situation would be reversed. This second cannonball has an eastward component of 1,260 kilometers (783 miles) per hour even while it sits in the muzzle. Once fired and moving southward, cannonball 2 travels over portions of Earth that are moving ever faster in an eastward direction. The ball again appears to veer off course to the right (see Figure 7.8 again), falling into the Pacific to the west of Ecuador. Don't be deceived by the word *appears* in the last sentence. The cannonballs really do veer to the right. Only to an observer in space would they appear to go straight, and points on Earth would appear to move out from underneath them.

The Coriolis effect is a real effect dependent on our rotating frame of reference. Part of that frame of reference involves the direction from which you view the problem. Thus in Figure 7.8, cannonball 2 looks to you as if it is veering left, but to the citizens of Buffalo facing south to watch the cannonball disappear, it moves to the right (west). Coriolis deflection causes moving objects to move to the left in the Southern Hemisphere because the frame of reference there is reversed. Also, except at the equator (where the Coriolis effect is nonexistent), the Coriolis effect influences the path of objects moving from east to west, or west to east.

An easy way to remember the Coriolis effect: In the Northern Hemisphere, objects move to the right; in the Southern Hemisphere, objects move to the left.

Because the Coriolis effect influences any object with mass—*as long as that object is moving*—it plays a large role in the movements of air and water on Earth. The Coriolis effect is most apparent in mid-latitude situations involving the almost frictionless flow of fluids: between layers of water in the ocean and in the circuits of winds. Does the Coriolis effect influence the directions of cars and airplanes? Yes, but in these cases friction (of tires on pavement, of wings on air) is much greater than the influence of the Coriolis effect, so the deflection is not observable.

> To observers on the surface, Earth's rotation causes moving air (or any moving mass) in the Northern Hemisphere to curve to the right of its initial path, and in the Southern Hemisphere to the left. The apparent curvature of path is known as the Coriolis effect.

The Coriolis Effect and Atmospheric Circulation Cells

We can now modify our first model of atmospheric circulation (Figure 7.5) to the more correct representation in **Figure 7.9.** Yes, air does warm, expand, and rise at the equator; air does cool, contract, and fall at the poles. But instead of continuing all the way from equator to pole in a continuous loop in each hemisphere, air rising from the equatorial

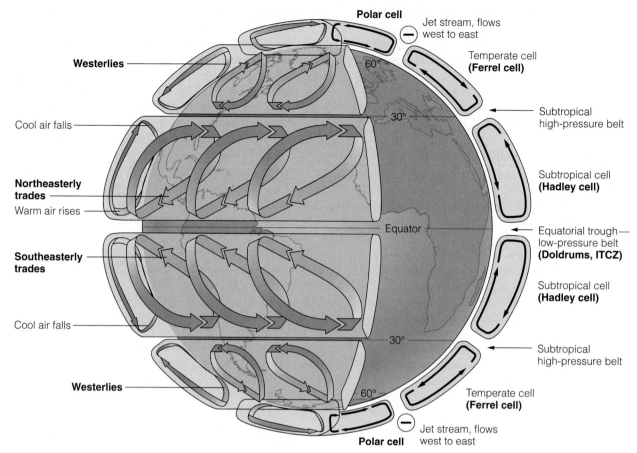

Figure 7.9 Global air circulation as described in the six-cell circulation model. As in Figure 7.5, air rises at the equator and falls at the poles. But instead of one great circuit in each hemisphere from equator to pole, there are three circuits in each hemisphere. Note the influence of the Coriolis effect on wind direction. The circulation shown here is ideal—that is, a *long-term average* of wind flow. Contrast this view with Figure 7.10, a snapshot of a moment in 1996.

region moves poleward and is gradually deflected eastward; that is, it turns to the *right* in the Northern Hemisphere and to the *left* in the Southern Hemisphere. This eastward deflection is caused by the Coriolis effect. (Note that the Coriolis effect does not *cause* the wind; it only *influences* the wind's direction.)

As air rises at the equator, it loses moisture by precipitation (rainfall) caused by expansion and cooling. This drier air now grows denser in the upper atmosphere as it radiates heat to space and cools. When it has traveled about a third of the way from the equator to the pole—that is, to about 30°N and 30°S latitude—the air becomes dense enough to fall back toward the surface. Most of the descending air turns back toward the equator when it reaches the surface. In the Northern Hemisphere the Coriolis effect again deflects this surface air to the right, and the air blows across the ocean or land from the northeast. (This air is represented by the arrows labeled "Northeasterly trades" in Figure 7.9.) Though it has been heated by compression during its descent, this air is generally still colder than the surface over which it flows. The air soon warms as it moves equatorward, however, evaporating surface water and becoming humid. The warm,

moist, less dense air then begins to rise as it approaches the equator and completes the circuit.

Such a large circuit of air is called an **atmospheric circulation cell.** A pair of these tropical cells exists, one on each side of the equator. They are known as **Hadley cells** in honor of George Hadley, the London lawyer and philosopher who worked out an overall scheme of wind circulation in 1735. Look for them in Figure 7.9.

A more complex pair of circulation cells operates at mid-latitudes in each hemisphere. Some of the air descending at 30° latitude turns poleward rather than equatorward. Before this air descends to the surface, it is joined at high altitude by air returning from the north. As can be seen in Figure 7.9, a loop of air forms between 30° and about 50°–60° latitude. As before, the air is driven by uneven heating and influenced by the Coriolis effect. Surface wind in this circuit is again deflected to the right, this time flowing from the west to complete the circuit. (This air is represented by the arrows labeled "Westerlies" in Figure 7.9.) The mid-latitude circulation cells of each hemisphere are named **Ferrel cells** after William Ferrel, the American who discovered their inner workings in the mid-nineteenth century. They, too, can be seen in Figure 7.9.

Meanwhile, air that has grown cold over the poles begins blowing toward the equator at the surface, turning to the west as it does so. At 50°–60° latitude in each hemisphere, this air has taken up enough heat and moisture to ascend. However, this polar air is denser than the air in the adjacent Ferrel cell and does not mix easily with it. The unstable zone between these two cells generates most mid-latitude weather. At high altitude the ascending air from 50°–60° latitude turns poleward to complete a third circuit. These are the **polar cells.**

Each hemisphere has three large atmospheric circulation cells: a Hadley cell, a Ferrel cell, and a polar cell. Air circulation within each cell is powered by uneven solar heating and influenced by the Coriolis effect.

Wind Patterns

The model of atmospheric circulation described above has many interesting features. At the bands between circulation cells, the air is moving *vertically* and surface winds are weak and erratic. Such conditions exist at the equator (where air rises and atmospheric pressure is generally low) or at 30° latitude in each hemisphere (where air falls and atmospheric pressure is generally high). Places within these circulation cells where air moves rapidly *horizontally* across the surface from zones of high pressure to zones of low pressure are characterized by strong, dependable winds.

Sailors have a special term for the calm equatorial areas where the surface winds of the two Hadley cells converge: the equatorial low called the **doldrums.** The word has come to be associated with a gloomy, listless mood, perhaps reflecting the sultry air and variable breezes found there. Scientists who study the atmosphere call this area the **intertropical convergence zone (ITCZ)** to reflect the influence of wind convergence on conditions near the equator. Strong heating in the ITCZ causes surface air to expand and rise. The humid, rising, expanding air loses moisture as rain, some of which contributes to the success of tropical rain forests.

Sinking air, in contrast, is generally arid. The great deserts of both hemispheres, dry bands centered around 30° latitude, mark the intersection of the Hadley and Ferrel cells. Because evaporation is higher than precipitation in these areas, ocean surface salinity tends to be highest at these latitudes. At sea, these areas of high atmospheric pressure and little surface wind are called the subtropical high, or **horse latitudes.** Spanish ships laden with supplies for the New World were often becalmed there, sometimes for weeks on end. When the mariners ran out of water and feed for their livestock, they were forced to throw the dead horses over the side.

Of much more interest to sailing masters were the bands of dependable surface winds *between* the zones of ascending and descending air. Most constant of these are the persistent **trade winds,** or easterlies, centered at about 15°N and 15°S latitude. The trade winds are the surface winds of the Hadley cells as they move from the horse latitudes to the doldrums. In the Northern Hemisphere they are the northeast trade winds; the southeast trade winds are the Southern Hemisphere counterpart.[1] The **westerlies,** surface winds of the Ferrel cells centered at about 45°N and 45°S latitude, flow between the horse latitudes and the boundaries of the polar cells in each hemisphere. The westerlies, then, approach from the southwest in the Northern Hemisphere and from the northwest in the Southern Hemisphere. Sailors outbound from Europe to the New World learned to drop south to catch the trade winds and to return home by a more northerly route to take advantage of the westerlies. Trade winds and westerlies are shown in Figure 7.9.

The six-cell model of atmospheric circulation (three cells in each hemisphere) discussed here represents an *average* of air flow through many years over the planet as a whole. Though the model is accurate in a general sense, local details of cell circulation vary because surface conditions are different at different longitudes. The ocean's thermostatic effect is the major factor reducing irregularities in cell circulation over water. **Figure 7.10** is a depiction of winds over the Pacific on two days in September 1996. As you can see, the patterns depart substantially from what we would expect from the six-cell model of Figure 7.9. Most of the difference is caused by the geographical distribution of landmasses, the different responses of land and ocean to solar heating, and chaotic flow. But, as noted above, over long periods of time (many years), *average* flow looks remarkably like what we would expect.

The atmosphere responds to uneven solar heating by flowing in three great circulating cells over each hemisphere. The flow of air within these cells is influenced by Earth's rotation.

Monsoons

A **monsoon** is a pattern of wind circulation that changes with the season. (The word *monsoon* is derived from *mausim*, the Arabic word for "season.") Areas subject to monsoons generally have wet summers and dry winters.

Monsoons are linked to the different specific heats of land and water and to the annual north-south movement of

---

[1] Winds are named by the direction *from* which they blow. A west wind blows from the west toward the east; a northeast wind blows from the northeast toward the southwest. The trade winds are named in honor of their ability to blow steadily—the early English word *trade* meant "steadily" or "constantly."

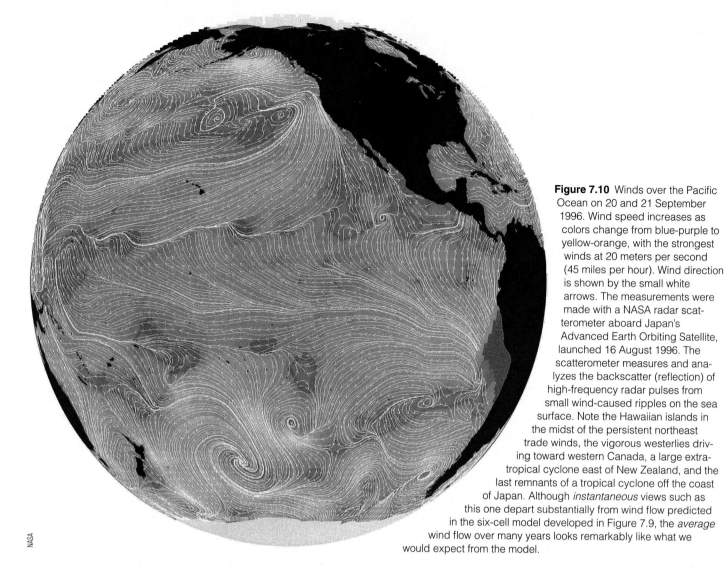

NASA

Figure 7.10 Winds over the Pacific Ocean on 20 and 21 September 1996. Wind speed increases as colors change from blue-purple to yellow-orange, with the strongest winds at 20 meters per second (45 miles per hour). Wind direction is shown by the small white arrows. The measurements were made with a NASA radar scatterometer aboard Japan's Advanced Earth Orbiting Satellite, launched 16 August 1996. The scatterometer measures and analyzes the backscatter (reflection) of high-frequency radar pulses from small wind-caused ripples on the sea surface. Note the Hawaiian islands in the midst of the persistent northeast trade winds, the vigorous westerlies driving toward western Canada, a large extratropical cyclone east of New Zealand, and the last remnants of a tropical cyclone off the coast of Japan. Although *instantaneous* views such as this one depart substantially from wind flow predicted in the six-cell model developed in Figure 7.9, the *average* wind flow over many years looks remarkably like what we would expect from the model.

the ITCZ. In the spring, land heats more rapidly than the adjacent ocean. Air above the land becomes warmer and so rises. Relatively cool air flows from over the ocean to the land to take its place. Continued heating causes this humid air to rise, condense, and form clouds and rain. In autumn, the land cools more rapidly than the adjacent ocean. Air cools and sinks over the land, and dry surface winds move seaward. The intensity and location of monsoon activity depend on the position of the ITCZ. Note in **Figure 7.11** that the monsoons follow the ITCZ south in the Northern Hemisphere's winter, and north in its summer.

In Africa and Asia, more than 2 billion people depend on summer monsoon rains for drinking water and agriculture. The most intense summer monsoons occur in Asia. The great landmass of Asia draws vast quantities of warm, moist air from the Indian Ocean (see again Figure 7.11). Southerly winds drive this moisture toward Asia, where it rises and condenses to produce a months-long deluge. Much smaller monsoons occur in North America as warming and rising air over the South and West draws humid air and thunderstorms from the Gulf of Mexico.

Sea Breezes and Land Breezes

Land breezes and sea breezes are small, daily mini-monsoons. Morning sunlight falls on land and adjacent sea, warming both. The temperature of the water doesn't rise as much as the temperature of the land, however. The warmer inland rocks transfer heat to the air, which expands and rises, creating a zone of low atmospheric pressure over the land. Cooler air from over the sea then moves toward land; this is the **sea breeze (Figure 7.12a)**. The situation reverses after sunset, with land losing heat to space and falling rapidly in temperature. After a while, the air over the still-warm ocean will be warmer than the air over the cooling land. This air will then rise, and the breeze direction will reverse, becoming a **land breeze (Figure 7.12b)**. Land breezes and sea breezes are common and welcome occurrences in coastal areas.

El Niño and La Niña

Sometimes cell circulation doesn't seem to play by the rules. In three- to eight-year cycles, atmospheric circulation changes

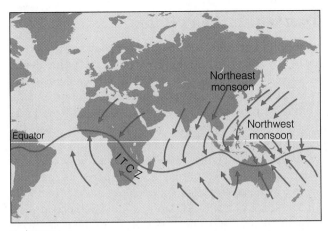

a January

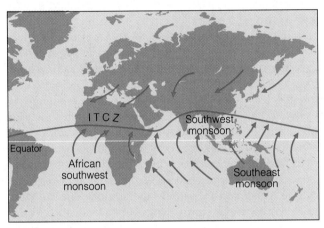

b July

Figure 7.11 Monsoon patterns. During the monsoon circulations of January and July, surface winds are deflected to the right in the Northern Hemisphere and to the left in the Southern Hemisphere. (Reprinted by permission of Alan D. Iselin.)

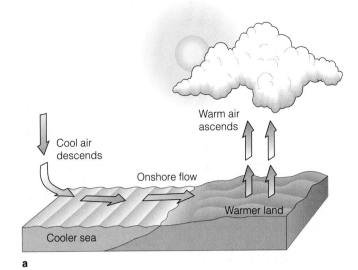

a

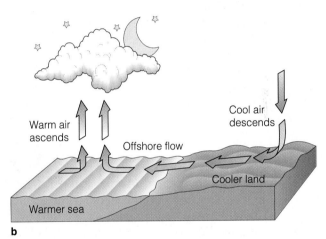

b

Figure 7.12 The flow of air in coastal regions during stable weather conditions. (**a**) In the afternoon, the land is warmer than the ocean surface, and warm air rising from the land is replaced by an onshore sea breeze. (**b**) At night, as the land cools, the air over the ocean is now warmer than the air over the land. The ocean air rises. Air flows offshore to replace it, generating an offshore flow (a land breeze).

significantly from the patterns shown in Figure 7.9. A reversal in the distribution of atmospheric pressure between the eastern and western Pacific causes the trade winds to weaken or reverse. The trade winds normally drag huge quantities of water westward along the ocean's surface near the equator, but without the winds these equatorial currents crawl to a stop. Warm water that has accumulated at the western side of the Pacific can build eastward along the equator toward the coast of Central and South America, greatly changing ocean conditions there.

El Niño and La Niña are primarily ocean current phenomena, so we will study them in Chapter 8's discussion of ocean currents.

Storms

Storms are regional atmospheric disturbances characterized by strong winds often accompanied by precipitation. Few natural events underscore human insignificance like a great

storm. When powered by stored sunlight, the combination of atmosphere and ocean can do fearful damage.

In Bangladesh, on 13 November 1970, a tropical cyclone (a hurricane) with wind speeds of more than 200 kilometers (125 miles) per hour roared up the mouths of the Ganges River, carrying with it masses of seawater up to 12 meters (40 feet) high. Water and wind clawed at the aggregation of small islands, most just above sea level, that makes up this impoverished country. In only 20 minutes at least 300,000 lives were lost, and estimates ranged up to a million dead! Property damage was absolute. Photographs taken soon after the storm showed a horizon-to-horizon morass of flooded, deep-gashed ground tortured by furious winds. There was almost no trace of human inhabitants, farms, domestic animals, or villages. Another great storm that struck in May 1991 killed another 200,000 people. The economy of the shattered country may not recover for decades.

A much different type of storm hammered the U.S. East Coast in March 1993. Mountainous snows from New York to North Carolina (1.3 meters, or 50 inches, at Mount Mitchell), 175-kilometer- (109-mile-) per-hour winds in Florida, and record cold in Alabama (−17°C, or 2°F, in Birmingham) were elements of a four-day storm that spread chaos from Canada to Cuba. At least 238 people died on land; another 48 were lost at sea. At one point more than 100,000 people were trapped in offices, factories, vehicles, and homes; 1.5 million were without electricity. Damage exceeded $1 billion.

These two great storms are examples of *tropical cyclones* and *extratropical cyclones* at their worst.[2] As the name implies, tropical cyclones like the hurricane that struck Bangladesh are primarily a tropical phenomenon. Extratropical cyclones—the winter weather disturbances with which residents of the U.S. Eastern Seaboard and other mid-latitude dwellers are most familiar—are found mainly in the Ferrel cells of each hemisphere.

Both kinds of storms are **cyclones,** huge rotating masses of low-pressure air in which winds converge and ascend. The word *cyclone,* derived from the Greek noun *kyklon* (meaning "an object moving in a circle"), underscores the spinning nature of these disturbances. (Don't confuse a cyclone with a **tornado,** a much smaller funnel of fast-spinning wind associated with severe thunderstorms.)

Air Masses

Cyclonic storms form between or within air masses. An **air mass** is a large body of air with nearly uniform temperature, humidity, and therefore density throughout. Air pausing over water or land will tend to take on the characteristics of the surface below. Cold, dry land causes the mass of air above to become chilly and dry. Air above a warm ocean surface will become hot and humid. Cold, dry air masses are dense and form zones of high atmospheric pressure. Warm, humid air masses are less dense and form zones of lower atmospheric pressure.

Air masses can move within or between circulation cells. Density differences, however, will prevent the air masses from mixing when they approach one another. Energy is required to mix air masses. Since that energy is not always available, a dense air mass may slide beneath a lighter air mass, lifting the lighter one and causing its air to expand and cool. Water vapor in the rising air may condense. All of these effects contribute to turbulence at the boundaries of the air masses.

The boundary between air masses of different density is called a **front.** The term was coined by a meteorologist who saw a similarity between the zone where air masses meet and the violent battle fronts of World War I.

Extratropical cyclones form at a front between *two* air masses. Tropical cyclones form from disturbances within *one* warm and humid air mass.

Extratropical Cyclones

Extratropical cyclones form at the boundary between each hemisphere's polar cell and its Ferrel cell—the **polar front.** These great storms occur mainly in the winter hemisphere when temperature and density differences across the polar front are most pronounced. Remember, the cold wind poleward of the front is generally moving from the east; the warmer air equatorward of the front is generally moving from the west (see again Figure 7.9). The smooth flow of winds past each other at the front may be interrupted by zones of alternating high and low atmospheric pressure that bend the front into a series of waves. Because of the difference in wind direction in the air masses north and south of the polar front, the wave shape will enlarge, and a twist will form along the front. The different densities of the air masses prevent easy mixing, so the cold, dense air mass will slide beneath the warmer, lighter one. Formation of this twist in the Northern Hemisphere, as seen from above, is shown in **Figure 7.13.** The twisting mass of air becomes an extratropical cyclone.

The twist that generates an extratropical cyclone circulates to the left in the Northern Hemisphere, seemingly in opposition to the Coriolis effect. The reasons for this paradox become clear, however, when we consider the wind directions and the nature of interruption of the air flow between the cells. (In fact, the leftward motion of the cyclone *is* Coriolis-driven because the large-scale air flow pattern at the *edges* of the cells is generated in part by the Coriolis effect.) Wind speed increases as the storm "wraps up" in much the same way that a spinning skater increases rotation speed by pulling in his or her arms close to the body. Air rushing toward the center of the spinning storm rises to form a low-pressure zone at the center. Extratropical cyclones are embedded in the westerly winds and thus move eastward. They are typically 1,000 to 2,500 kilometers (620 to 1,600 miles) in diameter and last from two to five days. **Figure 7.14** provides a beautiful example. The wind and precipitation associated with these fronts are sometimes referred to as **frontal storms.**

North America's most violent extratropical cyclones are the **nor'easters** (northeasters) that sweep the Eastern Seaboard in winter. The name indicates the direction from which the storm's most powerful winds approach. About 30 times a year, nor'easters moving along the mid-Atlantic and New England coasts generate wind and waves with enough force to erode beaches and offshore barrier islands, disrupt communication and shipping schedules, damage shore and harbor installations, and break power lines. About every hundred years a nor'easter devastates coastal settlements. In spite of a long history of destruction, people continue to build on unstable exposed coasts (**Figure 7.15**).

Large storms are spinning areas of unstable air that develop between or within air masses. Extratropical cyclones originate at the boundary between air masses.

---

[2] Note that the prefix *extra-* means "outside" or "beyond." *Extratropical* refers to the location of the storm, not its intensity.

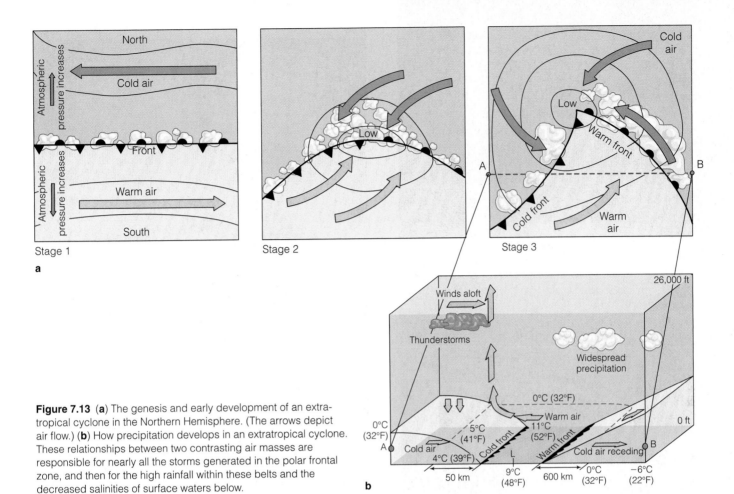

Stage 1

Stage 2

Stage 3

a

Figure 7.13 (**a**) The genesis and early development of an extratropical cyclone in the Northern Hemisphere. (The arrows depict air flow.) (**b**) How precipitation develops in an extratropical cyclone. These relationships between two contrasting air masses are responsible for nearly all the storms generated in the polar frontal zone, and then for the high rainfall within these belts and the decreased salinities of surface waters below.

b

Tropical Cyclones

7.17

Tropical cyclones are great masses of warm, humid, rotating air. They occur in all tropical oceans except the equatorial South Atlantic. Large tropical cyclones are called **hurricanes** (*Huracan* is the god of the wind of the Caribbean Taino people) in the North Atlantic and eastern Pacific,

Figure 7.14 A well-developed extratropical cyclone over the northeastern Pacific on 27 October 2000. Looking like a huge comma, the cloud-dense cold front extends southward and westward from the storm's center. Spotty cumulus clouds and thunderstorms have formed in the cold, unstable air behind the front. This picture was taken by *GOES-10* in visible light.

Figure 7.15 Storm damage from a particularly severe nor'easter, the Ash Wednesday (7 March) storm of 1962. Massive damage such as this, on Fire Island, New York, is most common for buildings near the shoreline.

Figure 7.16 Hurricane Florence over the North Atlantic, photographed from the space shuttle *Discovery* in November 1994. The cloud-free eye of the storm is clearly visible. Note the spiral bands of cloud extending in toward the eye.

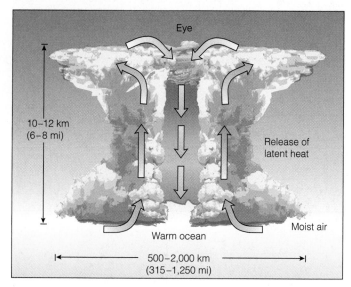

Figure 7.17 The internal structure of a hurricane. (The vertical dimension is greatly exaggerated in this drawing.)

typhoons (*Tai fung* is Chinese for "great wind") in the western Pacific, *tropical cyclones* in the Indian Ocean, and *williwillis* in the waters near Australia. To qualify formally as a hurricane or typhoon, the tropical cyclone must have sustained winds of at least 119 kilometers (74 miles) per hour. About a hundred tropical cyclones grow to hurricane status each year. A very few of these develop into superstorms, with winds near the core that exceed 250 kilometers (155 miles) per hour! (To imagine what winds in such a storm might feel like, picture yourself clinging to the wing of a twin-engine private airplane in flight!) Tropical cyclones containing winds less than hurricane force are called *tropical storms* and *tropical depressions.*

From above, tropical cyclones appear as circular spirals (**Figure 7.16**). They may be 1,000 kilometers (620 miles) in diameter and 15 kilometers (9.3 miles or 50,000 feet) high. The calm center, or *eye* of the storm—a zone some 13 to 16 kilometers (8 to 10 miles) in diameter—is sometimes surrounded by clouds so high and dense that the daytime sky above looks dark. Farther out, churned by furious winds, the rainband clouds condense huge amounts of water vapor into rain. A tropical cyclone is diagrammed in **Figure 7.17.**

Unlike extratropical cyclones, these greatest of storms form within *one* warm, humid air mass between 10° and 25° latitude in both hemispheres. (Though air conditions would be favorable, the Coriolis effect closer to the equator is too weak to initiate rotary motion.) The origins of tropical cyclones are not well understood. A tropical cyclone usually develops from a small tropical depression. Tropical depressions form in easterly waves, areas of lower pressure within the easterly trade winds that are thought to originate over a large, warm landmass. When air containing the disturbance is heated over tropical water with a temperature of about 26°C (79°F) or higher, circular winds begin to blow in the vicinity of the wave, and some of the warm, humid air is forced upward. Condensation begins, and the storm takes shape. Under ideal conditions the embryo storm reaches hurricane status—that is, with wind speeds in excess of 119 kilometers (74 miles) per hour—in two to three days.

The centers of most tropical cyclones move westward and poleward at 5 to 40 kilometers (3 to 25 miles) per hour. Typical tracks of these storms are shown in **Figure 7.18.**

Although its birth process is somewhat mysterious, the source of the storm's power is well understood. Its strength comes from the same seemingly innocuous process that warms a chilled soft-drink can when water from the atmosphere condenses on its surface. As you may recall from Chapter 6, it takes quite a bit of energy to break the bonds that hold water molecules together and evaporate water into the atmosphere—water's latent heat of vaporization is very high. That heat energy is released when the water vapor recondenses as liquid. It tends to warm your drink very quickly, and the more humid the air, the faster the condensing and warming. In tropical cyclones, the condensation energy generates air movement (wind), not more heat. Fortunately only 2% to 4% of this energy of condensation is converted into motional energy!

A tropical cyclone is an ideal machine for "cashing in" water vapor's latent heat of vaporization. Warm, humid air

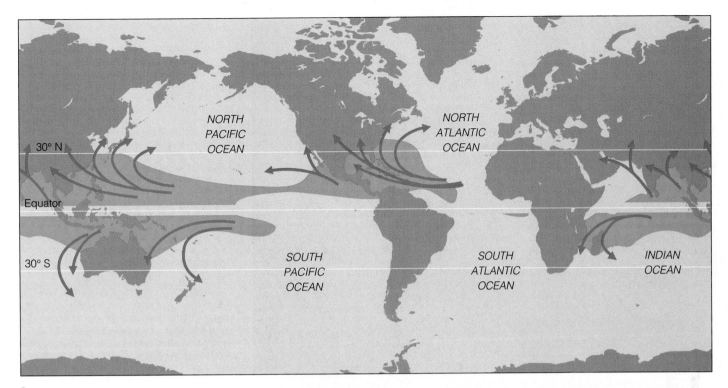

a

Figure 7.18 (**a**) The breeding grounds of tropical cyclones are shown as orange areas. The storms follow curving paths: First they move westward with the trade winds. Then they either die over land or turn eastward until they lose power over the cooler ocean of mid-latitudes. Cyclones are not spawned over the South Atlantic or the southeast Pacific because the waters are too chilly, nor in the still air—the doldrums—within 5° of the equator. (**b**) A composite of infrared satellite images of Hurricane Georges from 18 September to 28 September 1998. Its westward and poleward trek across the Caribbean and into the United States is clearly shown.

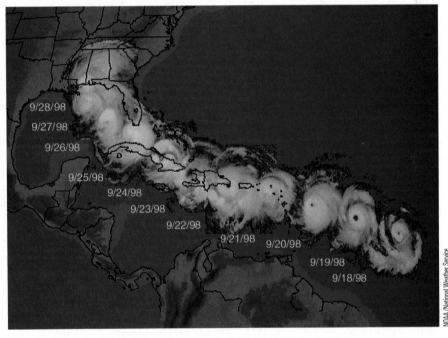

b

forms in great quantity only over a warm ocean. As hot, humid tropical air rises and expands, it cools and is unable to retain the moisture it held when warm. Rainfall begins. The rainfall rate in some parts of the storm routinely exceeds 2.5 centimeters (1 inch) per hour, and 20 billion metric tons of water can fall from a large tropical cyclone in a day! Tremendous energy is released as this moisture changes from water vapor to liquid. In one day, a large tropical cyclone generates about 2.4 trillion kilowatt-hours of power, equivalent to the electrical energy needs of the entire United States for a year! So, solar energy ultimately powers the

storm in a cycle of heat absorption, evaporation, condensation, and conversion of heat energy to kinetic energy. This energy is available as long as the storm stays over warm water and has a ready source of hot, humid air.

Three aspects of a tropical cyclone can cause property damage and loss of life: wind, rain, and storm surge. The destructive force of winds of 250 kilometers (155 miles) per hour—or more—is self-evident. Rapid rainfall can cause severe flooding when the storm moves onto land. But the most danger lies in a **storm surge,** a mass of water driven by the storm (**Figure 7.19**). The low atmospheric pressure

Figure 7.19 High winds and strong rain can inflict great damage when a large tropical storm comes ashore, but the greatest destruction is often caused by a storm surge, a dome of wind-driven water. Here Hurricane Kate slams into Key West, Florida, on 20 November 1985. This hurricane was responsible for seven deaths in the United States and more than $1 billion in damages.

at the storm's center produces a dome of seawater that can reach a height of 1 meter (3.3 feet) in the open sea. The water height increases when waves and strong hurricane winds ramp the water mass ashore. If a high tide coincides with the arrival of all this water at a coast, or if the coastline converges (as is the case at the mouths of the Ganges in Bangladesh), rapid and catastrophic flooding will occur. Storm surges of up to 12 meters (40 feet) were reported at Bangladesh in 1970. Much of the $3 billion in damage done by Hurricane Andrew to south Florida in August 1992 was caused by a 4-meter (13-foot) storm surge arriving near high tide.

You may have noticed that tropical cyclones turn leftward (or counterclockwise when viewed from above) in the Northern Hemisphere and rightward (clockwise) in the Southern Hemisphere. Does this mean that the Coriolis effect does not apply to tropical cyclones? No. Their spin is caused by the Coriolis deflection of winds approaching the center of a low-pressure area from great distances. In the Northern Hemisphere, there is rightward deflection of the *approaching air*. The edge spin given by this incoming air causes the storm to spin to the left in the Northern Hemisphere.

Tropical cyclones last from three hours to three weeks; most have lives of five to ten days. They eventually run down when they move over land or over water too cool to supply the humid air that sustains them. The friction of a land encounter rapidly drains a tropical cyclone of its energy, and a position above ocean water cooler than 24°C (75°F) is a sure harbinger of the storm's demise. When deprived of energy, the storm "unwinds" and becomes a mass of unstable, humid air pouring rain, lightning, and even tornadoes from its clouds. Tropical cyclones can be dangerous to the

end: Torrential rain streaming from the remnants of Hurricane Agnes in 1976 caused more than $2 billion in damage, mostly to Pennsylvania. Chesapeake and Delaware Bays were flooded with fresh water and sediments, destroying much of the shellfish industry there.

Tropical cyclones are nature's escape valves, flinging solar energy poleward from the tropics. They are beautiful, dangerous examples of the energy represented by water's latent heat of fusion.

Tropical cyclones, the most powerful of Earth's atmospheric storms, occur within a single humid air mass.

QUESTIONS FROM STUDENTS

1. Earth's orbit brings it closer to the sun in the Northern Hemisphere's winter than in its summer. Yet it's warmer in summer. Why?

Earth's orbit around the sun is elliptical, not circular. The whole Earth receives about 7% more solar energy through the half of the orbit during which we are closer to the sun than through the other half. The time of greater energy input comes during our winter, but the entire Northern Hemisphere is tilted toward the sun during the summer, which results in much more light reaching it in the summer. Three times more energy enters the Northern Hemisphere each day at midsummer than at midwinter.

2. Does the ocean affect weather at the centers of continents?

Absolutely. In a sense, *all* large-scale weather on Earth is oceanically controlled. The ocean acts as a solar collector and heat sink, storing and releasing heat. Most great storms (tropical and extratropical cyclones alike) form over the ocean and then sweep over land.

3. Are any of the results of large-scale atmospheric circulation apparent to the casual observer?

Yes. The view of atmospheric circulation developed in this chapter explains some phenomena you may have experienced. For instance, flying from Los Angeles to New York takes about 40 minutes less than flying from New York to Los Angeles because of westerly headwinds (indicated in Figure 7.9).

Because of these same prevailing westerlies, most storms travel over the United States from west to east. Weather prediction is based on observations and samples taken from the air masses as they move. Forecasting is often easier in the East and Midwest than in the West because more data are available from an air mass when it's over land.

Temperatures are milder (not as hot in summer, not as cold in winter) on the West Coast than on the East Coast, again because of the prevailing westerlies. The wind blows over water (which moderates temperatures) toward the West Coast, but over land toward the East Coast.

4. Does the Coriolis effect really make explorers wander to the left in the snows of Antarctica, or tree trunks grow in rightward spirals in Canadian forests, or water swirl counterclockwise down a washbasin in Las Vegas?

The Coriolis effect depends on the speed, mass, and latitude of the moving object. Small, lightweight objects moving slowly are subject to many forces and conditions (such as wind currents, natural variations in basin shape, and friction) that overwhelm Coriolis acceleration. For example, think of how *very* small the difference in eastward speed of the northern edge of a washbasin is in comparison to the southern edge. Any small irregularity in the washbasin's shape will be hundreds of times more important than the Coriolis effect in determining whether water will exit in a rightward spin or a leftward spin! Explorers and trees aren't massive enough and don't move quickly enough to be affected by the Coriolis effect.

But if the moving object is at mid-latitude, is heavy, and is moving quickly, it *will* be deflected to the right of its intended path. Artillery shells and weather systems are noticeably influenced by the Coriolis effect.

CHAPTER SUMMARY

The water, gases, and energy at Earth's surface are shared between the atmosphere and the ocean. The two bodies are in continuous contact, and conditions in one are certain to influence conditions in the other. The interaction of ocean and atmosphere moderates surface temperatures, shapes Earth's weather and climate, and creates most of the sea's waves and currents.

The atmosphere responds to uneven solar heating by flowing in three great circulating cells over each hemisphere. This circulation of air is responsible for about two-thirds of the heat transfer from tropical to polar regions. The flow of air within these cells is influenced by Earth's rotation. To observers on the surface, Earth's rotation causes moving air (or any moving mass) in the Northern Hemisphere to curve to the right of its initial path, and in the Southern Hemisphere to the left. The apparent curvature of path is known as the Coriolis effect.

Uneven flow of air within cells is one cause of the atmospheric changes we call *weather*. Large storms are spinning areas of unstable air that occur between or within air masses. Extratropical cyclones originate at the boundary between air masses; tropical cyclones, the most powerful of Earth's atmospheric storms, occur within a single humid air mass. The immense energy of tropical cyclones is derived from water's latent heat of vaporization.

ON-LINE STUDY RESOURCES

The Web site for this book contains a wealth of helpful study aids. Log on to:

http://info.brookscole.com/garrison

and select *Essentials of Oceanography,* 3rd edition. Select Chapter 7 from the drop-down menu or click on one of these resource areas:

- For study and review, **Chapter at a Glance** gives you an outline of the chapter, **Chapter Summary** allows you to review the chapter's main ideas, and **Glossary** lists concepts and terms for the chapter along with their definitions.

- To test your mastery of the Terms and Concepts to Remember for this chapter, you can use the electronic **Flash Cards,** play the **Concentration** game, or work the **Crossword Puzzle.**

- For practice quizzing, try the multiple-choice **Tutorial Quiz,** the ten-question **True/False Quiz,** or the **Image Analysis Quiz,** which poses questions based on art and photos from the chapter.

TERMS AND CONCEPTS TO REMEMBER

air mass	Hadley cell
atmospheric circulation cell	horse latitudes
climate	hurricane
convection current	intertropical convergence
Coriolis, Gaspard Gustave de	zone (ITCZ)
Coriolis effect	land breeze
cyclone	monsoon
doldrums	nor'easter (northeaster)
extratropical cyclone	polar cell
Ferrel cell	polar front
front	precipitation
frontal storm	sea breeze

storm

storm surge

tornado

trade winds

tropical cyclone

water vapor

weather

westerlies

STUDY QUESTIONS

1. What happens when air containing water vapor rises?

2. What factors contribute to the uneven heating of Earth by the sun?

3. How does the atmosphere respond to uneven solar heating? How does Earth's rotation affect the resultant circulation? How many atmospheric circulation cells exist in each hemisphere?

4. Describe the atmospheric circulation cells in the Northern Hemisphere. At what latitudes does air move vertically? Horizontally? What are the trade winds? The westerlies? Where are deserts located? Why? What do you think ocean surface salinity is like in these desert bands?

5. How do the two kinds of large storms differ? How are they similar? What causes an extratropical cyclone? What happens in one?

6. What triggers a tropical cyclone? From what is its great power derived? What causes the greatest loss of life and property when a tropical cyclone reaches land?

7. If the Coriolis effect causes the rightward deflection of moving objects in the Northern Hemisphere, why does air rotate to the left around zones of low pressure in that hemisphere?

 Earth Systems Today CD Questions

1. Go to "Weather and Climate," then to "Weather Forecasting." Note the many parameters from which to select. For our purposes, temperature, humidity, wind speed, and satellite pictures will be most useful. Review the general flow pattern of Earth's surface winds (see Figure 7.9). Now predict wind direction over the continental United States. Toggle the wind vector and jet stream overlays and see if your predictions hold up. How does the wind flow affect temperature and humidity?

2. Go to "Weather and Climate" and connect to the Internet through **Earth Systems Today** to see today's satellite images and forecasts for your zip code.

RESOURCES FOR FURTHER READING AND RESEARCH

The Web site for this book contains many ideas for further reading and research. Log on to:

http://info.brookscole.com/garrison

and select *Essentials of Oceanography*, 3rd edition. Select Chapter 7 from the drop-down menu or click on one of these resource areas:

- **Additional Readings** lists the major books and articles consulted in writing this chapter, along with comments from the author about their content and reading level.

- **Hypercontents** takes you to an extensive list of current links to Internet sites with news, research, and images related to individual subjects in the chapter. Just click on the icon that corresponds to a numbered section to see the links for that subject.

- **Internet Exercises** are critical thinking questions that involve research on the Internet with starter URLs provided.

- **InfoTrac College Edition Exercises** leads you to Critical Thinking Projects that use InfoTrac College Edition as a research tool.

- **Regional InfoTrac College Edition** articles are organized into East Coast, West Coast, and Gulf Coast regions, allowing you to study oceanography on a more local level.

 For more readings, go to InfoTrac College Edition, your on-line research library, at:

http://infotrac.thomsonlearning.com

Palmlike trees grow on the northwestern coast of Ireland. Wind blowing over the warm waters of the Gulf Stream and the North Atlantic Current brings heat to the coast, making it possible for these subtropical plants to survive.

Ocean Circulation

PALM TREES IN IRELAND?

Commercial airliners flying to Europe from West Coast cities of the United States fly over central Ontario. In winter and spring months, the ground is hidden beneath masses of ice and snow, but passengers can often make out the frozen surface of James Bay. When those passengers land in London or Dublin, they find a much milder climate than the barren whiteness of central Canada. Yet London's latitude of 51°N is the same as that of the southern tip of James Bay, and Dublin lies at 54°N, the same latitude as Ontario's Polar Bear Provincial Park, a sanctuary for migrating polar bears. Why the great difference in climate? The predominant direction of air flow is eastward at these latitudes. Air that is flowing toward James Bay loses heat as it passes over the frozen landmass of Canada, but air that is moving over the ocean toward the British Isles is heated by contact with the warm Gulf Stream and

139

North Atlantic Current. Ireland and England therefore have a mild maritime climate, and polar bears are found only in zoos. These cities are warmed by the energy of tropical sunlight transported to their northern latitudes by ocean currents.

Surface Currents

About 10% of the water in the world ocean is involved in **surface currents,** water flowing horizontally in the uppermost 400 meters (1,300 feet) of the ocean's surface, driven mainly by wind friction. Most surface currents move water above the pycnocline, the zone of rapid density change with depth.

Water within and below the pycnocline also circulates, but the power for this slower, deeper circulation comes from the action of gravity on adjacent water masses of different densities. Since density is largely a function of temperature and salinity, circulation due to density differences is called *thermohaline circulation*. We'll investigate thermohaline circulation after a discussion of surface currents.

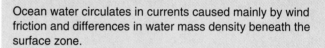

Ocean water circulates in currents caused mainly by wind friction and differences in water mass density beneath the surface zone.

The primary force responsible for surface currents is wind. As you read in Chapter 7, surface winds form global patterns within latitude bands (see Figure 7.9 and **Figure 8.1**). Most of Earth's surface wind energy is concentrated in each hemisphere's trade winds (easterlies) and westerlies. Waves on the sea surface transfer some of the energy from the moving air to the water by friction. This tug of wind on the ocean surface begins a mass flow of water. The water flowing beneath the wind forms a surface current.

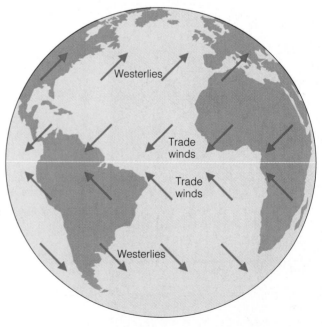

Figure 8.1 Winds, driven by uneven solar heating and Earth's spin, drive the movement of the ocean's surface currents. The prime movers are the powerful westerlies and the persistent trade winds (easterlies).

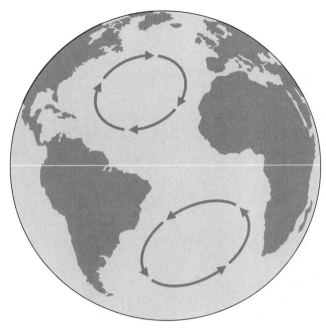

Figure 8.2 A combination of four forces—surface winds, the sun's heat, the Coriolis effect, and gravity—circulates the ocean surface clockwise in the Northern Hemisphere and counterclockwise in the Southern Hemisphere, forming gyres.

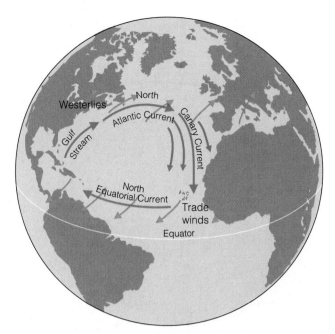

Figure 8.3 The North Atlantic gyre, a series of four interconnecting currents with different flow characteristics and temperatures.

The moving water will "pile up" in the direction the wind is blowing. Water pressure will be higher on the "piled-up" side, and the force of gravity will act to pull the water down the slope—against the *pressure gradient*—in the direction from which it came. But the Coriolis effect intervenes. Because of the Coriolis effect, Northern Hemisphere surface currents flow to the *right* of the wind direction. Southern Hemisphere currents flow to the *left*. Continents and basin topography often block continuous flow and help to deflect the moving water into a circular pattern. This flow around the periphery of an ocean basin is called a **gyre.** Two gyres are shown in **Figure 8.2.**

Flow Within a Gyre 8.3

Figure 8.3 shows the North Atlantic gyre in more detail. Though it flows continuously without obvious places where one current ceases and another begins, oceanographers subdivide the North Atlantic gyre into four interconnected currents because each has distinct flow characteristics and temperatures. (Gyres in other ocean basins are similarly divided.) Notice that the east-west currents in the North Atlantic gyre flow to the right of the driving winds; once initiated, water flow in these currents continues in a roughly east-west direction. Where their flow is blocked by continents, the currents turn clockwise to complete the circuit.

Why does water flow around the *periphery* of the ocean basin instead of spiraling to the center? After all, the Coriolis effect influences any moving mass *as long as it moves*, so water in a gyre might be expected to curve to the center of the North Atlantic and stop. To understand this aspect of current movement, imagine the forces acting on the surface water at 45°N latitude (point **A** in **Figure 8.4**). Here the west-

erlies blow from the southwest, so initially the water will move toward the northeast. The rightward Coriolis deflection then causes the water to flow almost due east. A particle at 15°N latitude (point **B**) responds to the push of the trade winds from the northeast, however, and with Coriolis deflection it will flow almost due west.

When driven by the wind, the topmost layer of ocean water in the Northern Hemisphere flows at about 45° to the right of the wind direction, a flow consistent with the arrows leading away from points **A** and **B** in Figure 8.4. But what

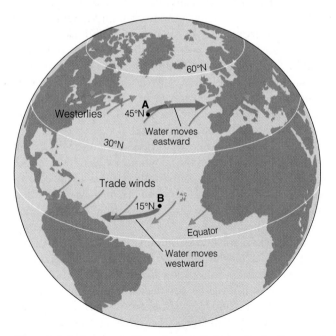

Figure 8.4 Surface water blown by the winds at point **A** will veer to the right of its initial path and continue eastward. Water at point **B** veers to the right and continues westward.

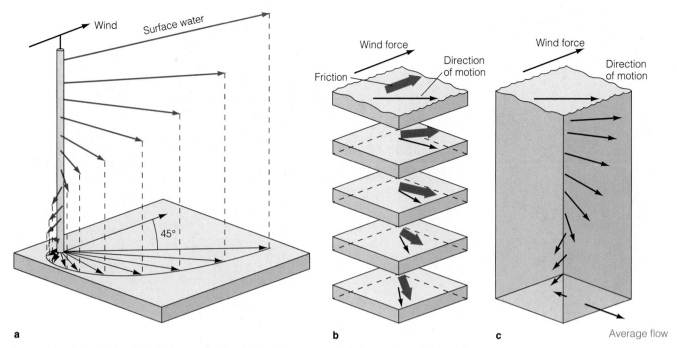

Figure 8.5 The Ekman spiral and the mechanism by which it operates. (**a**) The Ekman spiral model. (**b**) A body of water can be thought of as a set of layers. The top layer is driven forward by the wind, and each layer below is moved by friction. Each succeeding layer moves with a slower speed and at an angle to the layer immediately above it—to the right in the Northern Hemisphere, to the left in the Southern Hemisphere—until water motion becomes negligible. (**c**) Though the direction of movement varies for each layer in the stack, the theoretical net flow of water in the Northern Hemisphere is 90° to the right of the prevailing wind force. The length of the arrows is proportional to the speed of the current in each layer. (From *Laboratory Exercises in Oceanography*, 4/e by Pipkin, Gorslin, Casey, and Hammond. © 1987 by W. H. Freeman and Company. Reprinted by permission.)

about the water in the next layer down? It can't "feel" the wind at the surface; it "feels" only the movement of the water immediately above. This deeper layer of water moves *at an angle to the right* of the overlying water. The same thing happens in the layer below that, and the next layer, and so on, to a depth of about 100 meters (330 feet) at mid-latitudes. Each layer slides horizontally over the one beneath it like cards in a deck, with each lower card moving at an angle slightly to the right of the one above. Because of frictional losses, each lower layer also moves more slowly than the layer above. The resulting situation, portrayed in **Figure 8.5,** is known as an *Ekman spiral* after the Swedish oceanographer who worked out the mathematics involved.

The term *spiral* is somewhat misleading; the water itself does not spiral downward in a whirlpool-like motion like water going down a drain. Rather, the spiral is a way of conceptualizing the horizontal movements in a layered water column, each layer moving in a slightly different horizontal direction. An unexpected result of the Ekman spiral is that at some depth (known as the friction depth), water will be flowing in the opposite direction from the surface current!

The *net* motion of the water down to about 100 meters (330 feet), after allowance for the summed effects of the Ekman spiral (the sum of all the arrows indicating water direction in the affected layers), is known as **Ekman transport.** In theory, the direction of Ekman transport is 90° to the *right* of the wind direction in the Northern Hemisphere and 90° to the *left* in the Southern Hemisphere.

Water near the ocean surface moves to the right of the wind direction in the Northern Hemisphere and to the left in the Southern Hemisphere.

Armed with this information, we can look in more detail at the area around point **B** in Figure 8.4, which is enlarged in **Figure 8.6.** In nature, Ekman transport in gyres is less than 90°; in most cases the deflection barely reaches 45°. This deviation from theory occurs because of an interaction between the Coriolis effect and the pressure gradient. Some flowing Atlantic water has turned to the right to form a hill of water; it followed the rightward dotted-line arrow in Figure 8.6. Why does the water now go straight west from point **B** without turning? Because, as **Figure 8.7a** shows, to turn farther *right* the water would have to move uphill against the pressure gradient (and in defiance of gravity), but to turn *left* in response to the pressure gradient would defy the Coriolis effect. So the water continues westward and then clockwise around the whole North Atlantic gyre, dynamically balanced between the downhill urge of the pressure gradient and the uphill tendency of Coriolis deflection.

Yes, there really *is* a hill near the middle of the North Atlantic, centered in the area of the Sargasso Sea (**Figure 8.7b**). This hill is formed of surface water gathered at the

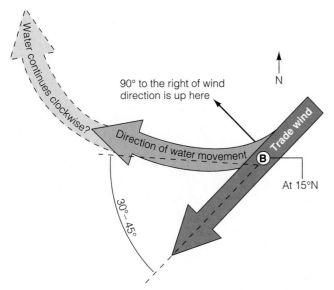

Figure 8.6 The movement of water away from point **B** in Figure 8.4 is influenced by the rightward tendency of the Coriolis effect and the gravity-powered movement of water down the pressure gradient.

ocean's center of circulation. It is not a steep mountain of water—its maximum height is an unspectacular 2 meters (6.5 feet)—but rather a gradual rise and fall from coastline to open ocean and back to opposite coastline. Its slope is so gradual you wouldn't notice it on a transatlantic crossing.

The hill is maintained by wind energy. If the winds did not continuously inject new energy into currents, then friction within the fluid mass and with the surrounding ocean basins would slow the flowing water, gradually converting its motion into heat. The balance of wind energy and friction, and of the Coriolis effect and the pressure gradient (through the effect of gravity), propels the currents of the gyre and holds them along the outside edges of the ocean basin.

Geostrophic Gyres

Gyres in balance between the pressure gradient and the Coriolis effect are called **geostrophic gyres,** and their currents are *geostrophic currents.* Because of the patterns of driving winds and the present positions of continents, the geostrophic gyres are largely independent of one another in each hemisphere.

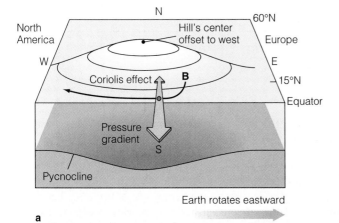

Figure 8.7 (**a**) The surface of the North Atlantic is raised through wind motion and Ekman transport to form a low hill. The eastward rotation of Earth offsets the center of the hill to the west. Water from point **B** (see also Figures 8.4 and 8.6) turns westward and flows along the side of this hill. The westward-moving water is balanced between the Coriolis effect (which would turn the water to the right) and flow down the pressure gradient, driven by gravity (which would turn it to the left). Thus, water in a gyre moves along the outside edge of an ocean basin. Note that the hill has a submerged component: The pycnocline is depressed by the mass of water balanced above. (**b**) The average height of the surface of the North Atlantic is shown in color in this image derived from data taken in 1992 by the *TOPEX/Poseidon* satellite. Red indicates the highest surface; green and blue the lowest. Note that the measured position of the hill is offset to the west as seen in (**a**). The gradually sloping hill is only 2 meters (6.5 feet) high and would not be apparent to anyone traveling from coast to coast.

Center of hill

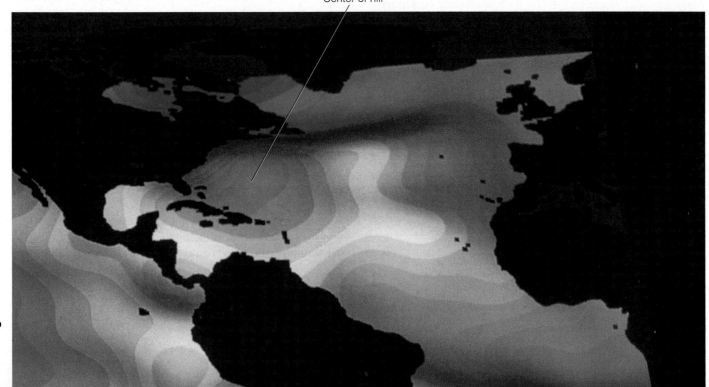

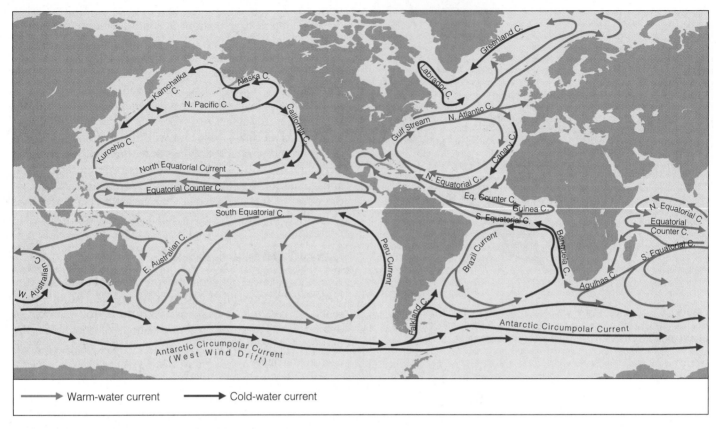

Figure 8.8 The names and usual directions of the world ocean's major surface currents. The powerful western boundary currents flow along the western boundaries of ocean basins in *both* hemispheres.

There are six great current circuits in the world ocean: two in the Northern Hemisphere and four in the Southern Hemisphere (**Figure 8.8**). Five are geostrophic gyres: the North Atlantic gyre, the South Atlantic gyre, the North Pacific gyre, the South Pacific gyre, and the Indian Ocean gyre. Though it is a closed circuit, the sixth and largest current is technically not a gyre because it does not flow around the periphery of an ocean basin. The West Wind Drift, or **Antarctic Circumpolar Current,** as this exception is called, flows endlessly eastward around Antarctica, driven by powerful, nearly ceaseless westerly winds. This greatest of all the surface ocean currents is never deflected by a continent.

> Surface currents affect the uppermost 10% of the world ocean. Some surface currents are rapid and riverlike, with well-defined boundaries; others are slow and diffuse. The largest surface currents are organized into huge circuits known as gyres.

Currents Within Gyres   8.5

Because of the different factors that drive and shape them, the currents that form geostrophic gyres have different characteristics. Geostrophic currents may be classified by their position within the gyre as western boundary currents, eastern boundary currents, or transverse currents.

WESTERN BOUNDARY CURRENTS   The fastest and deepest geostrophic currents are found at the western boundaries of ocean basins (that is, off the *east* coast of continents). These narrow, fast, deep currents move warm water poleward in each of the gyres. There are five large **western boundary currents:** the Gulf Stream (in the North Atlantic), the Japan or Kuroshio Current (in the North Pacific), the Brazil Current (in the South Atlantic), the Agulhas Current (in the Indian Ocean), and the East Australian Current (in the South Pacific).

The **Gulf Stream** is the largest western boundary current. Studies of the Gulf Stream have revealed that off Miami, the Gulf Stream moves at an average speed of 2 meters per second (5 miles per hour) to a depth of more than 450 meters (1,500 feet). Water in the Gulf Stream can move more than 160 kilometers (100 miles) in a day. Its average width is about 70 kilometers (43 miles).

The volume of water transported in western boundary currents is extraordinary. The unit used to express volume transport in ocean currents is the **sverdrup (sv)**, named in honor of Harald Sverdrup, one of this century's pioneering oceanographers. A sverdrup equals 1 million cubic meters per second.[1] The Gulf Stream flow is at least 55 sv (55 mil-

---

[1] One million cubic meters is about half the volume of the Louisiana Superdome.

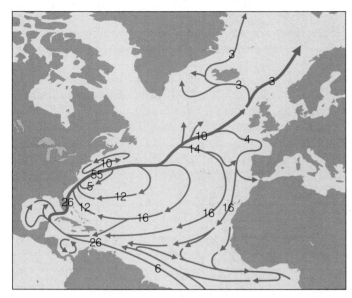

Figure 8.9 The general surface circulation of the North Atlantic. The numbers indicate flow rates in sverdrups (1 sv = 1 million cubic meters of water per second).

lion cubic meters per second), about 300 times the usual flow of the Amazon, the greatest of rivers. In **Figure 8.9** the surface currents of the North Atlantic gyre are shown with their volume transport (in sverdrups) indicated.

Water in a current, especially a western boundary current, can move for surprisingly long distances within well-defined boundaries. In the Gulf Stream, the current-as-river analogy can be startlingly apt: The western edge of the current is often clearly visible. Water within the current is usually warm, clear, and blue, often depleted of nutrients and incapable of supporting much life. By contrast, water over the continental slope adjacent to the current is often cold, green, and teeming with life.

Long, straight edges are the exception rather than the rule in western boundary currents, however. Unlike rivers, ocean currents lack well-defined banks, and friction with adjacent water can cause a current to form waves along its edges. Western boundary currents meander as they flow poleward. The looping meanders sometimes connect to form turbulent rings, or **eddies,** that trap cold or warm water in their centers and then separate from the main flow. For example, *cold-core eddies* form in the Gulf Stream as it meanders eastward upon leaving the coast of North America off Cape Hatteras (**Figure 8.10**). *Warm-core eddies* can form north of the Gulf Stream when the warm current loops into the cold water lying to the north. When these loops are cut off, they become freestanding spinning masses of water. Warm-core eddies rotate clockwise, and cold-core eddies rotate counterclockwise.

The slowly rotating eddies move away from the current and are distributed across the North Atlantic. Some may be 1,000 kilometers (620 miles) in diameter and retain their identity for more than three years. In mid-latitudes as much as one-fourth of the surface of the North Atlantic may consist of old, slow-moving, cold-core eddy remnants! Both cold and warm eddies are visible in the satellite image of **Figure 8.11.**

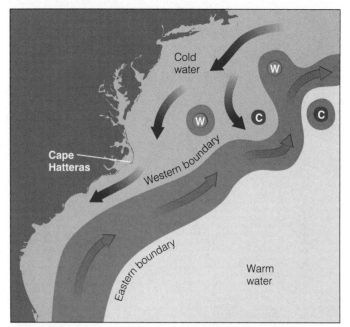

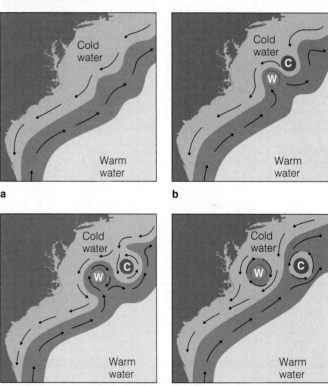

Figure 8.10 Eddy formation. The western boundary of the Gulf Stream is usually distinct, marked by abrupt changes in water temperature, speed, and direction. Meanders (eddies) form at this boundary as the Gulf Stream leaves the U.S. coast at Cape Hatteras (**a**). The meanders can pinch off (**b**) and eventually become isolated cells of warm water between the Gulf Stream and the coast (**c**). Likewise, cold cells can pinch off and become entrained in the Gulf Stream itself (**d**). (C = cold water, W = warm water; blue = cold, red = warm). Figure 8.11 shows the Gulf Stream from space with meanders and eddies clearly visible.

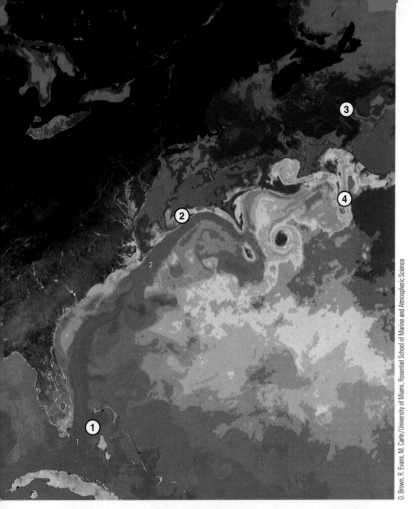

Figure 8.11 The Gulf Stream viewed from space. This image is a composite of temperature data returned from NOAA polar-orbiting meteorological satellites during the first week of April 1984. The composite image is printed with an artificial color scale: Reds and oranges are a warm 24°–28°C (76°–84°F), yellows and greens are 17°–23°C (63°–74°F), blues are 10°–16°C (50°–61°F), and purples are a cold 2°–9°C (36°–48°F). The Gulf Stream appears like a red (warm) river as it moves from the southern tip of Florida ① north along the East Coast. Moving offshore at Cape Hatteras ②, it begins to meander, with some meanders pinching off to form warm-core ③ and cold-core ④ eddies. As it moves northeastward, the water cools dramatically, releasing heat to the atmosphere and mixing with the cooler surrounding waters. By the time it reaches the middle of the North Atlantic, it has cooled so much that its surface temperature can no longer be distinguished from that of the surrounding waters.

Their influence reaches to the seafloor. Warm- and cold-core eddies are probably responsible for the slowly moving *abyssal storms* that leave often-observed ripple marks in deep sediments.

EASTERN BOUNDARY CURRENTS   Five **eastern boundary currents** are found at the eastern edge of ocean basins (that is, off the *west* coast of continents): the Canary Current (in the North Atlantic), the Benguela Current (in the South Atlantic), the California Current (in the North Pacific), the West Australian Current (in the Indian Ocean), and the Peru or Humboldt Current (in the South Pacific).

Eastern boundary currents are the opposite of their western boundary counterparts in nearly every way: They carry cold water equatorward; they are shallow and broad, sometimes more than 1,000 kilometers (620 miles) across; their boundaries are not well defined; and eddies tend not to form. Their total flow is less than that of their western counterparts. The Canary Current in the North Atlantic carries only 16 sv of water at about 2 kilometers (1.2 miles) per hour. The current is so shallow and broad that sailors may not notice it. Contrast the flow rates of the North Atlantic's western and eastern boundary currents in Figure 8.9. **Table 8.1** summarizes the major differences between boundary currents in the Northern Hemisphere.

TRANSVERSE CURRENTS   As we have seen, most of the power for ocean currents is derived from the trade winds at the fringes of the tropics, and from the mid-latitude westerlies. The stress of winds on the ocean in these bands gives rise to the **transverse currents**—currents that flow from east to west and west to east, linking the eastern and western boundary currents.

The trade wind–driven North and South Equatorial Currents in the Atlantic and Pacific are moderately shallow and broad, but each transports about 30 sv westward. Because of the thrust of the trades, Atlantic water at Panama is usually 20 centimeters (8 inches) higher, on average, than water across the isthmus in the Pacific. The Pacific's greater expanse of water at the equator and stronger trade winds develop more powerful westward-flowing equatorial currents, and the height

Table 8.1 Boundary Currents in the Northern Hemisphere				
Type of Current (example)	**General Features**	**Speed**	**Transport (millions of cubic meters per second)**	**Special Features**
Western Boundary Currents Gulf Stream, Kuroshio (Japan) Current	**Warm** Narrow, <100 km. Deep—substantial transport to depths of 2 km.	Swift, hundreds of kilometers per day.	Large, usually 50 sv or greater.	Sharp boundary with coastal circulation system; little or no coastal upwelling; waters tend to be depleted in nutrients, unproductive; waters derived from trade wind belts.
Eastern Boundary Currents California Current, Canary Current	**Cold** Broad, ~1000 km. Shallow, <500 m.	Slow, tens of kilometers per day.	Small, typically 10–15 sv.	Diffuse boundaries separating from coastal currents; coastal upwelling common; waters derived from mid-latitudes.

Source: From M. Grant Gross, *Oceanography: A View of the Earth,* 5/e, © 1990, p. 173. Prentice-Hall. Reprinted by permission.

differential between the western and eastern Pacific is thought to approach 1 meter (3.3 feet)!

As can be seen in Figure 8.8, the westward flow of the transverse currents near the equator proceeds unimpeded for great distances, but the eastward flow of transverse currents at middle and high latitudes in the northern ocean basins is interrupted by continents and island arcs. In the far south, however, eastward flow is almost completely free. Intense westerly winds over the southern ocean drive the greatest of all ocean currents, the unobstructed Antarctic Circumpolar Current (or West Wind Drift). This current carries more water than any other—at least 100 sv west to east in the Drake Passage between the tip of South America and the adjacent Palmer Peninsula of Antarctica.

WESTWARD INTENSIFICATION   Why should western boundary currents be concentrated and eastern boundary currents be diffuse? Due to the Coriolis effect—which increases as water moves farther from the equator—eastward-moving water on the north side of the North Atlantic gyre is turned sooner toward the equator than westward-flowing water at the equator is turned toward the pole. So the peak of the hill described in Figure 8.7 is not in the center of the ocean basin, but closer to its western edge. If an equal volume of water flows around the gyre, this means the current on the eastern boundary (off the coast of Europe) is spread out and slow, and the current on the western boundary (off the American coast) is concentrated and rapid. The effect on current flow is known as **westward intensification (Figure 8.12)**, a phenomenon clearly visible in Figures 8.7 and 8.9.

Westward intensification doesn't just happen in the North Atlantic. The western boundary currents in the gyres of both hemispheres are more intense than their eastern counterparts.

> The Coriolis effect modifies the courses of currents, with currents turning clockwise in the Northern Hemisphere and counterclockwise in the Southern Hemisphere. The Coriolis effect is largely responsible for the phenomenon of westward intensification in both hemispheres.

Gyres: A Final Word

Although we have stressed individual currents in our discussion, remember that gyres consist of currents that blend into one another. Flow is continuous without obvious places where one current ceases and another begins. The balance of wind energy, friction, the Coriolis effect, and the pressure gradient propels gyres and holds them along the outside of ocean basins.

Effects of Surface Currents on Climate

Along with the winds, surface currents distribute tropical heat worldwide. Warm water flows to higher latitudes, transfers heat to the air and cools, moves back to low latitudes, absorbs heat again, and the cycle repeats. The greatest amount of heat transfer occurs at mid-latitudes, where about 10 million billion calories of heat are transferred *each second*—more than a million times the power consumed by all the world's human population in the same length of time! This combination of water flow and heat transfer from and to water influences climate and weather in several ways.

In winter, for example, Edinburgh, Dublin, and London are bathed in eastward-moving air only recently in contact with the relatively warm North Atlantic Current. Scotland, Ireland, and England have a maritime climate. As you read in this chapter's opener, they are warmed in part by the energy of tropical sunlight transported to their high latitudes by the Gulf Stream (see once again Figure 8.8).

At lower latitudes on an ocean's eastern boundary the situation is often reversed. Mark Twain is supposed to have said that the coldest winter he ever spent was a summer in San Francisco. Summer months in that West Coast city are cool, foggy, and mild, while Washington, D.C., on nearly the same line of latitude (but on the western boundary of an ocean basin), is known for its August heat and humidity. Why the difference? Look at Figure 8.8 and follow the currents responsible. The California Current, carrying cold water from

Figure 8.12 Geostrophic flow in the North Atlantic. The Gulf Stream, a western boundary current, is narrow and deep and carries warm water northward. The Canary Current, an eastern boundary current, carries cold water at a much more leisurely pace. This happens because eastward-moving water on the north side of the North Atlantic gyre is turned sooner toward the equator than westward-flowing water at the equator is turned toward the pole. The peak of the geostrophic hill is forced toward the west. Although the gradually sloping hill is only 2 meters (6.5 feet) high and would not be apparent to anyone traveling from coast to coast, it is large enough to steer currents in the North Atlantic gyre. (This figure is vertically exaggerated.)

Gulf Stream · Steep slope · Top of hill · Gentle slope · Canary Current · Sargasso Sea · Water surface · Narrow, deep, warm, strong currents · Broad, shallow, cold, weak currents

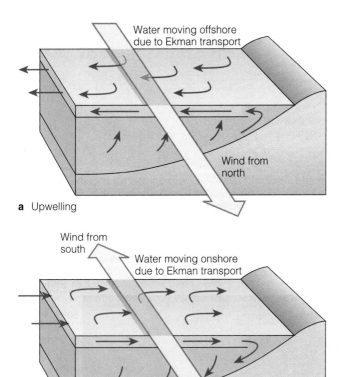

a Upwelling

b Downwelling

Figure 8.15 Coastal upwelling and downwelling. (**a**) In the Northern Hemisphere, coastal upwelling can be caused by winds from the north blowing along the west coast of a continent. Water moved offshore by Ekman transport is replaced by cold, deep, nutrient-laden water. (**b**) A prolonged southerly wind along a Northern Hemisphere west coast can result in downwelling.

falter during an ENSO event, warm equatorial water that would normally flow westward in the equatorial Pacific backs up to flow east (**Figure 8.16c, d**). The normal northward flow of the cold Peru Current is interrupted or overridden by the warm water. Upwelling within the nutrient-laden Peru Current is responsible for the great biological productivity of the ocean off the coasts of Peru and Chile. Although upwelling may continue during an ENSO event, the source of the upwelled water is nutrient-depleted water in the thickened surface layer approaching from the west. When the Peru Current slows and its upwelled water lacks nutrients, fish and seabirds dependent on the abundant life it contains die or migrate elsewhere. Peruvian fishermen are never cheered by this Christmas gift!

During major ENSO events, sea level rises in the eastern Pacific, sometimes by as much as 20 centimeters (8 inches) in the Galápagos. Water temperature also increases by up to 7°C (13°F). The warmer water causes more evaporation, and the area of low atmospheric pressure over the eastern Pacific intensifies. Humid air rising in this zone, centered some 2,000 kilometers (1,200 miles) west of Peru, causes high precipitation in normally dry areas. The increased evaporation intensifies coastal storms, and rainfall inland may be much higher than normal. Marine and terrestrial habitats and organisms can be affected by these changes.

The two most severe ENSO events of this century occurred in 1982–83 and 1997–98 (**Figure 8.17**). In both cases, effects associated with El Niño were spectacular over much of the Pacific and some parts of the Atlantic and Indian Oceans. In February of 1998, 40 people were killed and 10,000 buildings damaged by a "wall" of tornadoes advancing over the southeastern United States. This record-breaking tornado event was spawned by the collision of warm, moist air that had lingered over the warm Pacific and a polar front that dropped from the north. In the eastern Pacific, heavy rains throughout the 1997–98 winter in Peru left at least 250,000 people homeless, destroyed 16,000 dwellings, and closed every port in the country for at least one month. Hawaii, however, was left with a record drought, and some parts of southwestern Africa and Papua New Guinea received so little rain that crops failed completely and whole villages were abandoned due to starvation. Most of the United States escaped serious consequences—indeed, the midwestern states, Pacific Northwest, and Eastern Seaboard enjoyed a relatively mild fall, winter, and spring. But California's trials were widely reported: Rainfall in most of the state exceeded twice normal amounts, and landslides, avalanches, and other weather-related disasters crowded the evening news. Conditions did not return to near-normal until the late spring of 1998. Estimates of worldwide 1997–98 ENSO-related damage exceed 23,000 deaths and $33 billion.

Normal circulation sometimes returns with surprising vigor, producing strong currents, powerful upwelling, and chilly and stormy conditions along the South American coast. These contrasting colder-than-normal events are given a contrasting name: **La Niña** ("the girl"). As conditions to the east cool off, the ocean to the west (north of Australia) warms rapidly. The renewed thrust of the trade winds piles this water upon itself, depressing the upper curve of the thermocline to more than 100 meters (328 feet). In contrast, the thermocline during a La Niña event in the eastern equatorial Pacific rests at about 25 meters (82 feet). A vigorous La Niña followed the 1997–98 El Niño and persisted for nearly a year (**Figure 8.17e**).

Studies of the ocean and atmosphere in 1982–83 and 1997–98 have given researchers new insight into the behavior and effects of the Southern Oscillation. Some researchers believe that the 1982–83 event was triggered by the violent 1982 eruption of El Chichón, a Mexican volcano, which injected huge quantities of obscuring dust and sulfur-rich gases into the atmosphere. No similar trigger occurred before the 1997–98 ENSO, however, and the exact cause or causes of the Southern Oscillation are not yet understood. Subtle changes in the atmosphere permit meteorologists to predict a severe El Niño nearly a year in advance of its most serious effects.

El Niño, an anomaly in surface circulation, occurs when the trade winds falter, allowing warm water to build eastward across the Pacific at the equator.

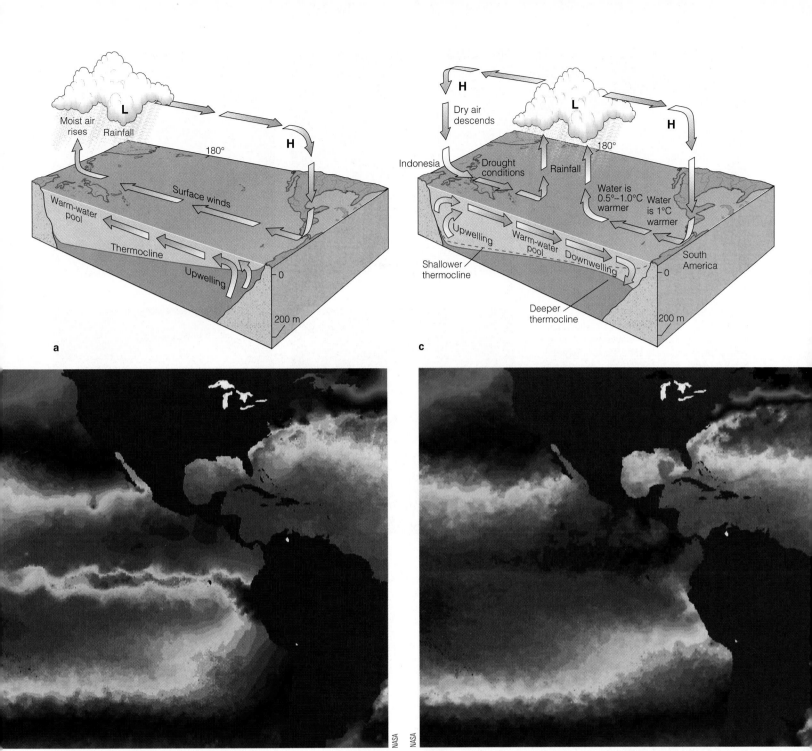

a

b

c

d

Figure 8.16 A non–El Niño year: (**a**) Normally the air and surface water flow westward, the thermocline rises, and upwelling of cold water occurs along the west coast of Central and South America. (**b**) This map from satellite data shows the temperature of the equatorial Pacific on 31 May 1988. The warmest water is indicated by dark red, and progressively cooler water by yellow and green. Note the coastal upwelling along the coast at the lower right of the map, and the tongue of recently upwelled water extending westward along the equator from the South American coast.

An El Niño year: (**c**) When the Southern Oscillation develops, the trade winds diminish and then reverse, leading to an eastward movement of warm water along the equator. The surface waters of the central and eastern Pacific become warmer, and storms over land may increase. (**d**) Sea-surface temperatures on 13 May 1992, a time of El Niño conditions. The thermocline was deeper than normal, and equatorial upwelling was suppressed. Note the absence of coastal upwelling along the coast, and the lack of the tongue of recently upwelled water extending westward along the equator.

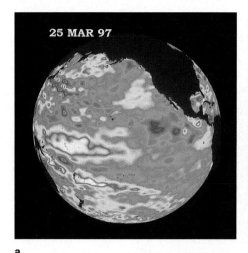

25 MAR 97

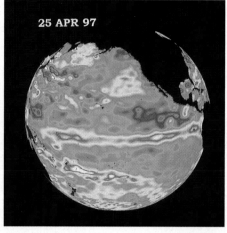

25 APR 97

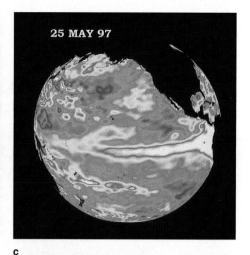

25 MAY 97

a

b

c

Figure 8.17 Development of the 1997–98 El Niño, observed by the *TOPEX/Poseidon* satellite. (**a**) March 1997. The slackening of the trade winds and westerly wind bursts allow warm water to move away from its usual location in the western Pacific Ocean. Red and white indicate sea level above average height. (**b**) April 1997. About a month after it began to move, the leading edge of the warm water reaches South America. (**c**) May 1997. Warm water piles up against the South American continent. The white area of sea level is 13 to 30 centimeters (5 to 12 inches) above normal height, and 1.6°–3°C (3°–5°F) warmer. (**d**) October 1997. By October, sea level is as much as 30 centimeters (12 inches) lower than normal near Australia. The bulge of warm water has spread northward along the coast of North America from the equator to Alaska. Fisheries in Peru are severely affected: The warm water prevents upwelling of cold, nutrient-rich water necessary for the support of large fish populations. (**e**) Normal circulation sometimes returns with surprising vigor after an El Niño event, producing strong currents, powerful upwelling, and chilly and stormy conditions along the South American coast. This image was prepared from data for 17 January 1999. Note the mass of cold surface water and relatively low sea level (purple). Such cold water tends to deflect winds around it, changing the course of weather systems locally and the nature of weather patterns globally.

23 OCT 97

d

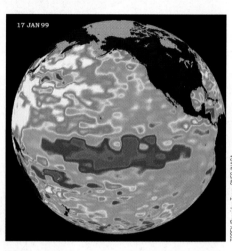

17 JAN 99

e

TOPEX/Poseidon Team, CNES/NASA

Thermohaline Circulation

8.12

The surface currents we have discussed affect the uppermost layer of the world ocean (about 10% of its volume), but horizontal and vertical currents also exist below the pycnocline in the ocean's deeper waters. The slow circulation of water at great depths is driven by density differences rather than by wind energy. Because density is largely a function of water temperature and salinity, the movement of water due to differences in density is called **thermohaline circulation.** Virtually the entire ocean is involved in slow thermohaline circulation, a process responsible for most of the vertical movement of ocean water and the circulation of the global ocean as a whole.

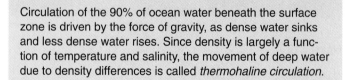

Circulation of the 90% of ocean water beneath the surface zone is driven by the force of gravity, as dense water sinks and less dense water rises. Since density is largely a function of temperature and salinity, the movement of deep water due to density differences is called *thermohaline circulation.*

Water Masses

8.13

As you may recall from Chapter 6, the ocean is density stratified, with the most dense water near the seafloor and the least dense near the surface. Each water mass has specific temperature and salinity characteristics. Density stratification is most pronounced at temperate and tropical latitudes because the temperature difference between surface water and deep water is greater there than near the poles.

The water masses possess distinct, identifiable properties. Like air masses, water masses don't often mix easily when they meet due to their differing densities; instead, they usually flow above or beneath each other. Water masses can be remarkably persistent and will retain their identity for great distances and long periods of time. Oceanographers name water masses according to their relative position.

In temperate and tropical latitudes, there are five common water masses:

- *Surface water*, to a depth of about 200 meters (660 feet)
- *Central water*, to the bottom of the main thermocline (which varies with latitude)
- *Intermediate water*, to about 1,500 meters (5,000 feet)
- *Deep water*, water below intermediate water but not in contact with the bottom, to a depth of about 4,000 meters (13,000 feet)
- *Bottom water*, water in contact with the seafloor

Surface currents move in the relatively warm upper environment of surface and central water. The boundary between central water and intermediate water is the most abrupt and pronounced.

No matter at what depth they are located, the characteristics of each water mass are usually determined by the conditions of heating, cooling, evaporation, and dilution that occurred at the ocean surface when the mass was formed. The densest (and deepest) masses were formed by surface conditions that caused the water to become very cold and salty. Water masses near the surface can be warmer and less saline; they may have formed in warm areas where precipitation exceeded evaporation. Water masses at intermediate depths are intermediate in density.

In spite of this differentiation, the relatively cold water masses lying beneath the thermocline exhibit smaller variations in salinity and temperature than the water in the currents that move across the ocean's surface.

Water masses almost always form at the ocean surface. The densest (and deepest) masses were formed by surface conditions that caused water to become very cold and salty.

Formation and Downwelling of Deep Water

Antarctic Bottom Water, the most distinctive of all deep-water masses, is characterized by a salinity of 34.65‰, a temperature of −0.5°C (30°F), and a density of 1.0279 grams per cubic centimeter. This water is noted for its extreme density (the densest in the world ocean), for the great amount of it produced near Antarctic coasts, and for its ability to migrate north along the seafloor.

Most Antarctic Bottom Water forms in the Weddell Sea during winter. Salt is concentrated in pockets between crystals of pure water and then squeezed out of the freezing mass to form a frigid brine. Between 20 and 50 million cubic meters of this brine form every second! The salty water's great density causes it to sink rapidly toward the continental shelf, where it mixes with nearly equal parts of water from the southern Antarctic Circumpolar Current.

The mixture settles along the edge of Antarctica's continental shelf, descends along the slope, and spreads along the deep-sea bed, creeping north in slow sheets. Antarctic Bottom Water flows many times more slowly than the water in surface currents: In the Pacific it may take a thousand years to reach the equator. Six hundred years later it may be as far away as the Aleutian Islands at 50°N! Antarctic Bottom Water also flows into the Atlantic Ocean basin, where it flows north at a faster rate than in the Pacific. Antarctic Bottom Water has been identified as high as 40° *north* latitude on the Atlantic floor, a journey that has taken some 750 years.

Some dense bottom water also forms in the northern polar ocean, but the topography of the Arctic Ocean basin prevents most of it from escaping, except in the deep channels formed in the submarine ridges separating Scotland, Iceland, and Greenland. These channels allow the cold, dense water formed in the Arctic to flow into the North Atlantic to form **North Atlantic Deep Water.** Very little deep Arctic Ocean water enters the Pacific.

Other distinct deep-water masses exist. Their positions in relation to each other are always determined by their relative densities. North Atlantic Deep Water forms when the relatively warm and salty North Atlantic Ocean cools as cold winds from northern Canada sweep over it. Exposed to the chilled air, water at the latitude of Iceland releases heat, cools from 10° to 2°C (50° to 36°F), and sinks. (Transferred to the air, this bonus heat helps to moderate European winters.)

A deep-water mass also forms in the enclosed Mediterranean Sea, where surface water is made more saline by the excess of evaporation over freshwater input. About 300,000 cubic kilometers (72,000 cubic miles) more water evaporates annually from the Mediterranean than is replaced by river runoff or precipitation. In the cool winter months, Mediterranean water with a salinity of about 38‰ flows past the lip of Gibraltar and spreads into the Atlantic as Mediterranean Deep Water. Mediterranean Deep Water underlies much of the central water mass in the Atlantic, and some of this water can be traced as far south as the basins of the Antarctic.

Thermohaline Circulation Patterns

The great quantities of dense water sinking at ocean basin edges must be offset by equal quantities of water rising elsewhere. **Figure 8.18** shows an idealized model of thermohaline flow. Note that water sinks relatively rapidly in a small area where the ocean is very cold, but it rises much more gradually across a very large area in the warmer temperate

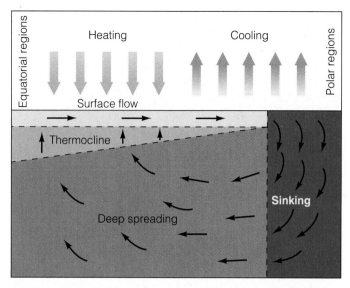

Figure 8.18 The classic model of a pure thermohaline circulation, caused by heating in lower latitudes and cooling in higher latitudes.

The water masses, each of distinct density and sandwiched in layers, are slowly propelled by gravity. Masses butt against one another in **convergence zones,** and the heavier water can slide beneath the lighter water. Hundreds of years may pass before water masses complete a circuit or blend to lose their identities. Remember, Antarctic Bottom Water in the Pacific retains its character for up to 1,600 years! The residence time of most deep water is less, however; it takes about 200 to 300 years to rise to the surface. (By contrast, a bit of surface water in the North Atlantic gyre may take only a little more than a year to complete a circuit.)

Not all thermohaline circulation is so sedate. Ripple marks in sediments, scour lines, and the erosion of rocky outcrops on deep-ocean floors are evidence that relatively strong, localized bottom currents exist (see Figure 5.3). Some of these currents may move as rapidly as 60 centimeters (24 inches) per second. These relatively fast currents are strongly influenced by bottom topography, and they are sometimes called *contour currents* because their dense water flows around (rather than over) seafloor projections.

and tropical zones. It then slowly returns poleward near the surface to repeat the cycle. The continual diffuse upwelling of deep water maintains the existence of the permanent thermocline found everywhere at low and mid-latitudes. This slow upward movement is estimated to be about 1 centimeter (½ inch) per day over most of the ocean. If this rise were to stop, downward movement of heat would cause the thermocline to descend and would reduce its steepness. In a sense, the thermocline is "held up" by the continual slow upward movement of cold water.

Most features of this ideal circulation pattern exist in nature. **Figure 8.19** shows deep circulation in the Atlantic.

Thermohaline Flow and Surface Flow: The Global Heat Connection

As we have seen, swift and narrow currents along the western margins of ocean basins carry warm tropical surface waters toward the poles. In a few places the water loses heat to the atmosphere and sinks to become deep water and bottom water. This sinking is most pronounced in the North Atlantic. The cold, dense water moves at great depths toward the Southern Hemisphere and eventually wells up into the surface lay-

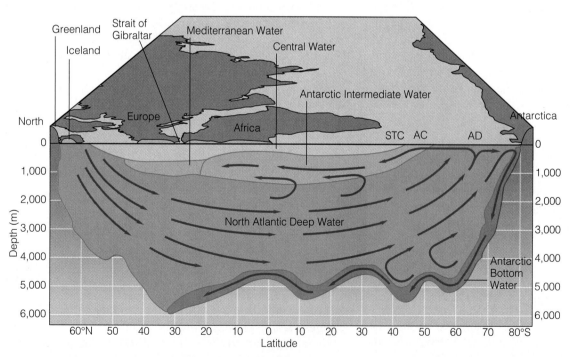

Figure 8.19 The water layers and deep circulation of the Atlantic Ocean. Arrows indicate the direction of water movement. STC indicates the position of the Subtropical Convergence; AC, the Antarctic Convergence; and AD, the Antarctic Divergence. The surface layer is too thin to show clearly in this scale. (The vertical scale is greatly exaggerated.)

ers of the Indian and Pacific Oceans. Almost a thousand years are required for this water to make a complete circuit.

The transport of tropical water to the polar regions is part of a global conveyor belt for heat. A simplified outline of the global circuit, the result of three decades of concentrated effort to understand deep circulation, is shown in **Figure 8.20.** This slow circulation straddles the hemispheres and is superimposed upon the more rapid flow of water in surface gyres. Recent analysis of this global circuit suggests that some of the heat warming the coast of Europe enters the ocean in the vicinity of Indonesia and Australia, travels to the Indian Ocean, and enters the Gulf Stream by way of the Agulhas Current rounding the southern tip of Africa. The surface water that leaves the Pacific is driven in part by excess rainfall and river runoff throughout the Pacific basin. The slow, steady, three-dimensional flow of water in the conveyor belt distributes dissolved gases and solids, mixes nutrients, and transports the juvenile stages of organisms among ocean basins.

Because they transfer huge quantities of heat, ocean currents greatly affect world weather and climate.

Studying Currents

Surface currents can be traced with drift bottles or drift cards. These tools are especially useful in determining coastal circulation, but they provide no information on the path the drift bottle or card may have taken between its release and collection points. Researchers who want to know the precise track taken by a drifting object can deploy more elaborate drift devices, such as drogues (**Figure 8.21a**), which can be tracked continuously by radio direction finders or radar. Surface currents can also be tracked by noting the difference between the daily expected and observed positions of ships at sea.

Bottle, card, or drogue studies are almost always carefully planned and executed, but not all surface drift releases have been intentional. In May 1990 a violent storm struck the containership *Hansa Carrier* enroute between Korea and Seattle. Twenty-one boxcar-sized cargo containers were lost overboard, among them, containers holding 30,910 pairs of Nike athletic shoes. About six months later, shoes from the broken containers began to wash up on beaches from British Columbia to Oregon. Because the shoes were not tied together in pairs, beachcombers placed advertisements in local newspapers and held swap meets to exchange the shoes (which were in excellent condition despite having been exposed to the ocean). Oceanographers noticed the ads and asked the media to request that individuals let them know where and when they found shoes. By knowing the place where the shoes were lost and the places where they were

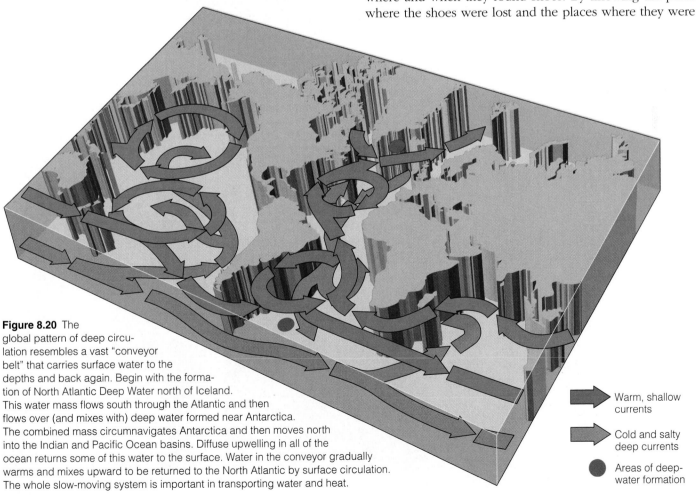

Figure 8.20 The global pattern of deep circulation resembles a vast "conveyor belt" that carries surface water to the depths and back again. Begin with the formation of North Atlantic Deep Water north of Iceland. This water mass flows south through the Atlantic and then flows over (and mixes with) deep water formed near Antarctica. The combined mass circumnavigates Antarctica and then moves north into the Indian and Pacific Ocean basins. Diffuse upwelling in all of the ocean returns some of this water to the surface. Water in the conveyor gradually warms and mixes upward to be returned to the North Atlantic by surface circulation. The whole slow-moving system is important in transporting water and heat.

Warm, shallow currents

Cold and salty deep currents

Areas of deep-water formation

found, researchers have been able to refine their computer models of the North Pacific gyre. Some of the shoes have completed a full circuit of the North Pacific. A similar spill occurred on 10 January 1992, when a storm-beset freighter lost a container filled with 29,000 rubber ducks, turtles, and other bathtub toys in the North Pacific (**Figure 8.21b**). At last report, 400 of the toys have been recovered from 500 miles of Alaskan shoreline.

Deeper currents can also be surveyed by free-floating devices. Sophisticated devices developed in the 1970s and 1980s depend on sofar channels to transmit sound (see Figure 6.14). Low-frequency tones are broadcast from autonomous submerged probes to moored listening stations (**Figure 8.21c**). These sofar probes have accumulated more than 240 float-years of data in the North Atlantic, at depths from 700 to 2,000 meters (2,300 to 6,600 feet). One of these drifting probes has been sending data for nine years!

Yet another method, developed for studying thermohaline circulation, senses the presence in seawater of tritium, a radioactive isotope of hydrogen. A small amount of tritium is produced in the upper atmosphere when hydrogen is bombarded by cosmic rays, but most of it was produced by hydrogen bomb testing in the 1960s. Most tritium combines with oxygen to become radioactive water, and some of this water enters the ocean by precipitation or river runoff. At high latitudes this water sinks and, labeled by its tritium content, can be traced by sampling. The speed of deep currents has been measured by analysis of their tritium content.

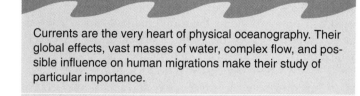

Currents are the very heart of physical oceanography. Their global effects, vast masses of water, complex flow, and possible influence on human migrations make their study of particular importance.

QUESTIONS FROM STUDENTS

1. If the Gulf Stream warms Britain during the winter and keeps Baltic ports free of ice, why doesn't it moderate New England winters? After all, Boston is closer to the warm core of the Gulf Stream than London is.

Yes, but remember the direction of prevailing winds in winter. Winter winds at Boston's latitude are generally from the west, so any warmth is simply blown out to sea. On the other side of the Atlantic the same winds blow toward Lon-

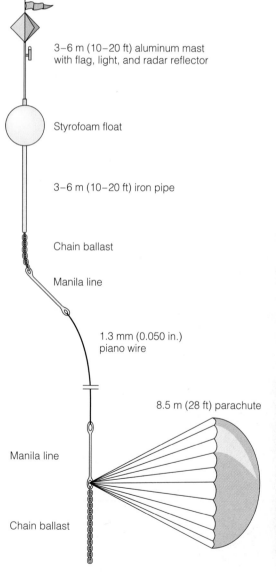

3–6 m (10–20 ft) aluminum mast with flag, light, and radar reflector

Styrofoam float

3–6 m (10–20 ft) iron pipe

Chain ballast

Manila line

1.3 mm (0.050 in.) piano wire

8.5 m (28 ft) parachute

Manila line

Chain ballast

a

Figure 8.21 Methods for measuring currents. (**a**) A drogue, which drifts with the current and is continuously tracked. (**b**) A rubber duck of the kind lost overboard in large numbers in 1992 in an unintentional drift experiment (see text). (**c**) A sofar float launched from the Woods Hole Oceanographic Institution's research ship *Oceanus*. The probe will drop to a depth of 3,500 meters (11,500 feet) and produce a low-frequency tone once each day for tracking.

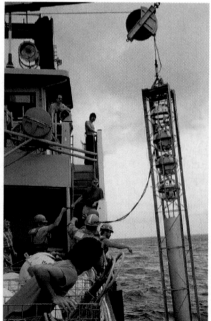

b

c

don. It does get cold in London, but generally winters in London are much milder than those in Boston.

2. Could currents be used as a source of electrical power? With the Gulf Stream so close to Florida, it seems that some way could be devised to take advantage of all that water flow to turn a turbine.

It's been considered. The total energy of the Gulf Stream flowing off Miami has been estimated at 25,000 megawatts! A Woods Hole Oceanographic Institution team has proposed a honeycomblike array of turbines for the layer between 30 and 130 meters (100 to 430 feet) across 20 kilometers (13 miles) of the current. They estimate a power output of around 1,000 megawatts, equal to the generation potential of two large nuclear power plants. Engineering difficulties would be considerable, however.

3. Are there any nongeostrophic currents? Are any currents not noticeably influenced by gravity, the Coriolis effect, uneven solar heating, planetary winds, and so forth?

Yes. Besides the high-latitude currents mentioned in the text, there are small-scale currents that are not noticeably affected. Currents of fresh water from river mouths, rip currents in surf, and tidal currents in small harbors are much more affected by basin and bottom topography than by the Coriolis effect and gravity.

4. Why are western boundary currents strong in both hemispheres? I thought things went the other way (counterclockwise) in the Southern Hemisphere. Shouldn't the eastern boundary currents be stronger down there?

Western boundary currents are strong in part because the "Coriolis hill" is offset to the west. Water moves in a relatively narrow path along the ocean's western boundary. Remember, the Coriolis effect becomes more pronounced as an object moves away from the equator. The Coriolis effect deflects and weakens the eastward-flowing currents in the gyres. Western boundary currents are therefore strong in both hemispheres.

5. How do oceanographers know how old a water mass is—that is, when it was last at the surface?

Researchers can determine the age of deep water by analyzing its dissolved oxygen content. Ocean water picks up free oxygen only at or near the surface by contact with the atmosphere or through the action of photosynthetic plants. After leaving the surface, the water gradually loses its oxygen through the respiration of organisms or by chemical reactions with sediments, rocks, or dissolved components. The dissolved oxygen content of a water mass is therefore a rough index of its age—the length of time since the water left the surface. Researchers have also found that water masses slowly mix with the surrounding water as they flow away from their sources, losing their individual identity as they grow older.

CHAPTER SUMMARY

Ocean water circulates in currents. Surface currents affect the uppermost 10% of the world ocean. The movement of surface currents is powered by the warmth of the sun and by winds. Water in surface currents tends to flow horizontally, but it can also flow vertically in response to wind blowing near coasts or along the equator. Surface currents transfer heat from tropical to polar regions, influence weather and climate, distribute nutrients, and scatter organisms. They have contributed to the spread of humanity to remote islands, and they are important factors in maritime commerce.

Circulation of the 90% of ocean water beneath the surface zone is driven by the force of gravity, as denser water sinks and less dense water rises. Since density is largely a function of temperature and salinity, the movement of deep water due to density differences is called *thermohaline circulation*. Currents near the seafloor flow as slow, riverlike masses in a few places, but the greatest volumes of deep water creep through the ocean at an almost imperceptible pace.

The Coriolis effect, gravity, and friction shape the direction and volume of surface currents and thermohaline circulation.

ON-LINE STUDY RESOURCES

The Web site for this book contains a wealth of helpful study aids. Log on to:

http://info.brookscole.com/garrison

and select *Essentials of Oceanography*, 3rd edition. Select Chapter 8 from the drop-down menu or click on one of these resource areas:

- For study and review, **Chapter at a Glance** gives you an outline of the chapter, **Chapter Summary** allows you to review the chapter's main ideas, and **Glossary** lists concepts and terms for the chapter along with their definitions.

- To test your mastery of the Terms and Concepts to Remember for this chapter, you can use the electronic **Flash Cards,** play the **Concentration** game, or work the **Crossword Puzzle.**

- For practice quizzing, try the multiple-choice **Tutorial Quiz,** the ten-question **True/False Quiz,** or the **Image Analysis Quiz,** which poses questions based on art and photos from the chapter.

TERMS AND CONCEPTS TO REMEMBER

Antarctic Bottom Water	ENSO
Antarctic Circumpolar Current	equatorial upwelling
	geostrophic gyre
coastal upwelling	Gulf Stream
convergence zone	gyre
current	La Niña
downwelling	North Atlantic Deep Water
eastern boundary current	Southern Oscillation
eddy	surface current
Ekman transport	sverdrup (sv)
El Niño	thermohaline circulation

transverse current
upwelling

western boundary current
westward intensification

STUDY QUESTIONS

1. What forces are responsible for the *movement* of ocean water in currents? What forces and factors influence the *direction* and *nature* of ocean currents?

2. What is a gyre? How many large gyres exist in the world ocean? Where are they located?

3. Why does water tend to flow around the periphery of an ocean basin? Why are western boundary currents the fastest ocean currents? How do they differ from eastern boundary currents?

4. What is El Niño? How does an El Niño situation differ from normal current flow? What are the usual consequences of El Niño?

5. What is the role of ocean currents in the transport of heat? How can ocean currents affect climate?

6. Contrast the climate of a mid-latitude coastal city at a western ocean boundary with a mid-latitude coastal city at an eastern ocean boundary.

7. What are water masses? Where are distinct water masses formed? What determines their relative position in the ocean?

8. What drives the vertical movement of ocean water? What is the general pattern of thermohaline circulation?

9. What methods are used to study ocean currents?

10. Can you think of ways ocean currents have (or *might* have) influenced history?

 Earth Systems Today CD Questions

1. Go to "Waves, Tides, and Currents," then to "World Currents." Toggle on the prevailing winds. Now predict the direction of the resulting surface currents. (Don't forget to include the Coriolis effect.) Now toggle the currents on, animate them, and check your predictions. Did you predict which currents were hot and which were cold?

2. Go to "Waves, Tides, and Currents," then to "Dissect a Current." Before beginning the animation, review the information on the Ekman spiral in the text. Now predict the resultant direction of flow of all the water in the surface current based only on the red wind arrow at the surface. Run the animation and see if you were successful.

3. Go to "Weather and Climate," then to "El Niño." Watch the sea surface temperatures from 1982 to 1990 in the animation. Note the back-and-forth oscillation of warm and cold water in the Pacific. Noting the pattern, could you predict ocean temperatures in the central eastern Pacific through 1992? How do you think warm water affects upwelling along the western coast of South America?

RESOURCES FOR FURTHER READING AND RESEARCH

The Web site for this book contains many ideas for further reading and research. Log on to:

http://info.brookscole.com/garrison

and select *Essentials of Oceanography*, 3rd edition. Select Chapter 8 from the drop-down menu or click on one of these resource areas:

- **Additional Readings** lists the major books and articles consulted in writing this chapter, along with comments from the author about their content and reading level.

- **Hypercontents** takes you to an extensive list of current links to Internet sites with news, research, and images related to individual subjects in the chapter. Just click on the icon that corresponds to a numbered section to see the links for that subject.

- **Internet Exercises** are critical thinking questions that involve research on the Internet with starter URLs provided.

- **InfoTrac College Edition Exercises** leads you to Critical Thinking Projects that use InfoTrac College Edition as a research tool.

- **Regional InfoTrac College Edition** articles are organized into East Coast, West Coast, and Gulf Coast regions, allowing you to study oceanography on a more local level.

 For more readings, go to InfoTrac College Edition, your on-line research library, at:

http://infotrac.thomsonlearning.com

CHAPTER 9

NOAA

The NOAA research ship *Surveyor* struggles through heavy Pacific swell west of Alaska.

SEA MONSTER

Wind waves like these can be huge. As you'll soon discover, the largest wave measured by sailors was 34 meters (112 feet) high. Astonishingly, even larger wind waves exist. Called "rogue waves," these monsters form when three or four large waves combine their energies. The combination appears momentarily as a single huge wave that can rise suddenly from a moderate sea to engulf and break a ship. They disappear as quickly as they come. A pair of rogue waves struck the Cunard

Waves

Liner *Queen Elizabeth* in 1943, sweeping over the bridge and shattering glass 28 meters (92 feet) above the water line.

Rogue waves kill people. Mobil Oil's drilling platform *Ocean Ranger* was collapsed by a giant wave off Newfoundland in February 1982. All 84 people aboard perished.

9.1

CHAPTER AT A GLANCE

To most people, an ocean wave in deep water appears to be a massive moving object—a ridge of water traveling across the sea surface. An ocean wave is one of several kinds of **waves,** all of which are disturbances caused by the movement of energy from a source through some medium (solid, liquid, or gas). As the energy of the disturbance travels, the medium through which it passes moves in specific ways. Sometimes this movement is visible to us as crests in the medium. The traveling crests produce the appearance of movement we see in a wave. In an ocean wave, a ribbon of *energy* is moving at the speed of the wave, but *water* is not.

Picture a resting sea gull as it bobs on the wavy ocean surface far from shore. The gull moves in *circles*—up and forward as the tops of the waves move to its position, down and backward as the tops move past. Each circle is equal in diameter to the wave's height. As may be seen in **Figure 9.1,** energy in waves flows past the resting bird, but the gull and its patch of water move only a very short distance forward in each up-and-forward, down-and-back wave cycle. The water on which the bird rests does not move continuously across the sea surface as the wave illusion suggests.[1]

Waves transmit energy, not water mass, across the ocean's surface.

The transfer of energy from water particle to water particle in these circular paths, or **orbits,** transmits wave energy across the ocean surface and causes the wave form to move. This kind of wave is known as an **orbital wave**—a wave in which particles of the medium (water) move in closed circles as the wave passes. Orbital ocean waves occur at the boundary between two fluid media (between air and water) and between layers of water of different densities. Because the wave form moves forward, these waves are a type of **progressive wave.**

The progressive wave that moved the gull was probably caused by wind. Other forces can generate much greater progressive waves in which water molecules move through much larger circular or elliptical orbits. Some of these waves are so large that they do not appear to us as waves at all, but rather as the slow sloshing of water in a harbor or bay, as dangerous flooding surges of water, or as rhythmic and predictable ocean tides.

Ocean waves have distinct parts. The **wave crest** is the highest part of the wave above average water level; the **wave**

---

[1] To clarify the important idea of wave-as-illusion, imagine yourself at a sports stadium where spectators are doing "the wave." Your role in wave propagation is simple: You stand up and sit down in precise synchronization with your neighbors. Though you move only a few feet vertically, the wave of which you were a part circles the arena at high speed. You and all the other participants stay in place, but the wave moves faster than anyone can run.

Figure 9.1 A floating sea gull demonstrates that wave forms travel but the water itself does not. In this sequence, a wave moves from left to right as the gull (and the water in which it is resting) revolves in a circle, moving slightly to the left up the front of an approaching wave, then to the crest, then sliding to the right down the back of the wave.

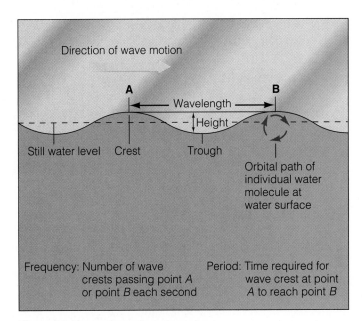

Figure 9.2 The anatomy of a progressive wave.

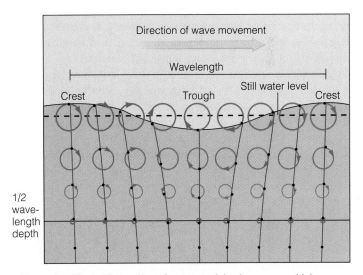

Figure 9.3 The orbital motion of water particles in a wave, which extends to a depth of about one-half of the wavelength.

trough is the valley between wave crests below average water level. **Wave height** is the vertical distance between a wave crest and the adjacent trough, while **wavelength** is the horizontal distance between two successive crests (or troughs). The relationship between these parts is shown in **Figure 9.2.** The time it takes for a wave to move a distance of one wave-length is known as the **wave period. Wave frequency** is the number of waves passing a fixed point per second.

The circular motion of water particles at the surface of a wave continues underwater. As **Figure 9.3** shows, the diameter of the orbits through which water particles move diminishes rapidly with depth. For all practical purposes wave motion is negligible below a depth of one-half the wavelength, where the circles are only $\frac{1}{23}$ the diameter of those at the surface. This means divers in 20 meters of water would not notice the passage of a wind wave of 30-meter wavelength and might barely notice the wave if they were at a depth of 15 meters. Since most ocean waves have moderate wavelengths, the circular disturbance of the ocean that propagates these waves affects only the uppermost layer of water. Note that the movement of water in circles doesn't resemble interlocking mechanical gears. Instead, there is a coordinated, uniform circular movement of water molecules in one direction as the waves pass.

Most waves affect only the ocean's surface layer. Movement ceases at a depth equal to about half the wave's wavelength.

Classifying Waves

Ocean waves are classified by the *disturbing force* that creates them, the *restoring force* that tries to flatten them, and their *wavelength*. (Wave height is not often used for classification because it varies greatly depending on water depth, interference between waves, and other factors.)

Disturbing Force

Energy that causes ocean waves to form is called a **disturbing force.** Wind blowing across the ocean surface provides the disturbing force for *wind waves*. The arrival of a storm surge or seismic sea wave in an enclosed harbor or bay, and a sudden change in atmospheric pressure, are the disturbing forces for a resonant rocking of water known as a *seiche*. Landslides, volcanic eruptions, and faulting of the seafloor associated with earthquakes are the disturbing forces for seismic sea waves (also known as *tsunami*). The disturbing forces for *tides* are changes in the magnitude and direction of gravitational forces among Earth, moon, and sun, combined with Earth's rotation.

Restoring Force

Restoring force is the dominant force that returns the water surface to flatness after a wave has formed in it. If the restoring force of a wave were quickly and fully successful, a disturbed sea surface would immediately become smooth and the energy of the embryo wave would be dissipated as heat. But that isn't what happens. Waves continue after they form because the restoring force overcompensates and causes oscillation. The situation is analogous to a weight bobbing at the bottom of a very flexible spring, constantly moving up and down past its normal resting point.

The dominant restoring force for very small water waves—those with wavelengths less than 1.73 centimeters (0.68 inch)—is surface tension (cohesion), the property that enables individual water molecules to stick to each other by means of hydrogen bonds (see Figure 6.4). These **capillary waves** are transmitted across a puddle because cohesion tugs the tiny wave troughs and crests toward flatness.

Capillary waves are the first waves to form when the wind blows. These small ripples are important in transferring energy from air to water to drive ocean currents but are of little consequence in the overall picture of ocean waves because they are tiny and carry very little energy.

All waves with wavelengths greater than 1.73 centimeters depend mostly on gravity to provide the restoring force. Gravity pulls the crests downward, but the momentum of the water causes the crests to overshoot and become troughs. The repetitive nature of this movement, like the spring weight moving up and down, gives rise to the circular orbits of individual water molecules in an ocean wave. Since the circular motion of water molecules in a wave is nearly friction-free, gravity waves can travel across thousands of miles of ocean surface without dissipating, eventually to break on a distant shore.

Wavelength

Wavelength is an important measure of wave size. **Table 9.1** lists the causes and typical wavelengths of the four types of ocean waves: wind waves, seiches, seismic sea waves, and tides. **Figure 9.4** shows the relationships between disturbing and restoring forces, period, and relative amount of energy present in the ocean's surface for each wave type. Note that more energy is stored in wind waves than in any of the other wave types.

Arranged from short to long wavelengths, ocean waves are generated by very small disturbances (capillary waves), wind (wind waves), rocking of water in enclosed spaces (seiches), seismic and volcanic activity or other sudden displacements (tsunami), and gravitational attraction (tides).

Table 9.1	Wavelengths and Disturbing Forces of the Four Types of Ocean Waves	
Wave Type	**Typical Wavelength**	**Disturbing Force**
Wind wave	60–150 m (200–500 ft)	Wind over ocean
Seiche	Large, variable; a fraction of basin size	Changes in atmospheric pressure, storm surge, tsunami
Tsunami	200 km (125 mi)	Faulting of sea floor, volcanic eruption, landside
Tide	½ circumference of Earth	Gravitational attraction, rotation of Earth

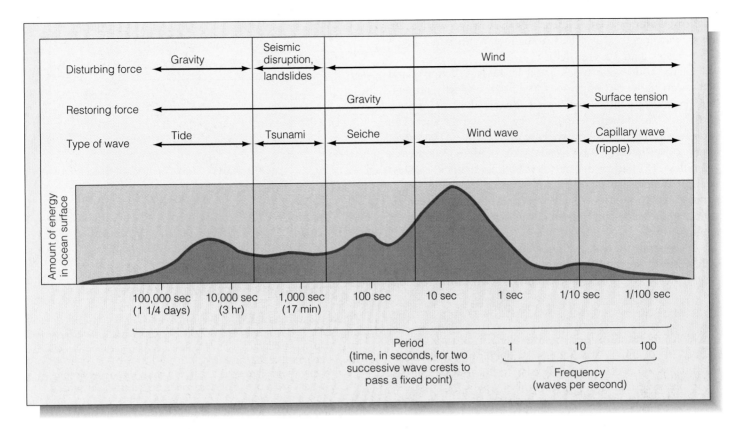

Disturbing force	Gravity		Seismic disruption, landslides	Wind						
Restoring force					Gravity				Surface tension	
Type of wave	Tide		Tsunami	Seiche		Wind wave			Capillary wave (ripple)	

Amount of energy in ocean surface

| 100,000 sec (1 1/4 days) | 10,000 sec (3 hr) | 1,000 sec (17 min) | 100 sec | 10 sec | 1 sec | 1/10 sec | 1/100 sec |

Period
(time, in seconds, for two
successive wave crests to
pass a fixed point)

1   10   100

Frequency
(waves per second)

Figure 9.4 Wave energy in the ocean as a function of the wave period. As the graph shows, most wave energy is typically concentrated in wind waves. However, large tsunami, rare events in the ocean, can transmit more energy than all wind waves for a brief time.

Deep-Water Waves and Shallow-Water Waves

 9.6

Most characteristics of ocean waves depend on the relationship between their wavelength and water depth. Wavelength determines the *size* of the orbits of water molecules within a wave, but water depth determines the *shape* of the orbits. The paths of water molecules in a wind wave are circular only when the wave is traveling in deep water. A wave cannot "feel" the bottom when it moves through water deeper than one-half its wavelength because too little wave energy is contained in the small circles below that depth. Waves moving through water deeper than one-half their wavelength are known as **deep-water waves.** A wave has no way of "knowing" how deep the water is, only that it is in water deeper than about half its wavelength. An example may help: A wind wave of 20 meter wavelength will act as a deep-water wave if it is passing through water more than 10 meters deep (**Figure 9.5a**).

The situation is different for wind-generated waves close to shore. The orbits of water molecules in waves moving through shallow water are flattened by the proximity of the bottom. Water just above the seafloor cannot move in a circular path, only forward and backward. Waves in water shallower than ¹⁄₂₀ their original wavelength are known as

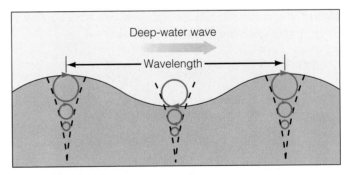

a Depth ≥ $\frac{1}{2}$ wavelength

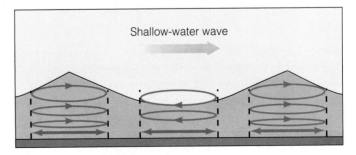

b Depth ≤ $\frac{1}{20}$ wavelength

Figure 9.5 Progressive waves: (**a**) a deep-water wave; (**b**) a shallow-water wave. (These diagrams are not to scale.)

shallow-water waves (Figure 9.5b). A wave with a 20-meter wavelength will act as a shallow water wave if the water is less than 1 meter deep.

Of the four wave types listed in Table 9.1, only wind waves can be deep-water waves. To understand why, remember that most of the ocean floor is deeper than 125 meters (400 feet), half the wavelength of very large wind waves. The wavelengths of the larger waves are *much* longer: The wavelength of seismic sea waves usually exceeds 100 kilometers (62 miles). No part of the ocean is 50 kilometers (31 miles) deep, so seiches, seismic sea waves, and tides are always in water that—to them—is shallow. Their huge orbital circles flatten against a distant bottom always less than half a wavelength away.

In general, the longer the wavelength of a wave, the faster the wave energy will move through the water. For *deep-water* waves this relationship is shown in the formula

$$S = L/T$$

in which S represents speed, L is wavelength, and T is time or period (in seconds). Wavelength is difficult to determine at sea, but period is comparatively easy to find—for example, by an observer timing the movement of waves past the bow of a stopped ship. If period (T) is known, then speed (S) can be calculated from the relationship

$$S(\text{in meters/sec}) = 1.56(T)$$

or, if you prefer,

$$S(\text{in feet/sec}) = 5(T).$$

Figure 9.6 shows the relationship between wavelengths of deep water waves and their speed and period.

The speed of ocean waves usually depends on their wavelength, with long waves moving fastest.

The action of shallow-water waves is described by a different equation that may be written as

$$S = \sqrt{gd} \qquad S = 3.1\sqrt{d}$$

where S is speed (in meters per second), g is acceleration due to gravity (an average of 9.8 meters per second per second), and d is the depth of water (in meters). The *period* of a wave remains unchanged regardless of the depth of water through which it is moving, but as deep-water waves enter the shallows and feel bottom, their velocity is reduced and their crests "bunch up," so their wavelength shortens.

Listed next are two very different kinds of ocean waves (a comparison of apples and oranges), but notice the general relationship between wavelength and wave velocity: The longer the wavelength, the greater the velocity.

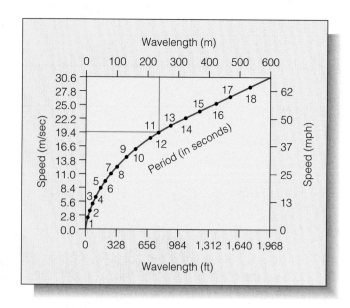

Figure 9.6 The theoretical relationship among speed, wavelength, and period in deep-water waves. Speed is equal to wavelength divided by period. If one characteristic of a wave can be measured, the other two can be calculated. The easiest to measure exactly is period. In the example shown in red, the speed of a wave with a wavelength of 228 meters and a period of 12 seconds is 19 meters per second.

WIND WAVES (DEEP-WATER WAVES):

- Period to about 20 *seconds*
- Wavelength to perhaps 600 meters (2,000 feet) in extreme cases
- Speed to perhaps 112 kilometers (70 miles) per hour in extreme cases

SEISMIC SEA WAVES (SHALLOW-WATER WAVES):

- Period to perhaps 20 *minutes*
- Wavelength typically 200 kilometers (125 miles)
- Speeds of 760 kilometers (470 miles) per hour

Remember that *energy*—not the water mass itself—is moving through the water at the astonishing speed of 760 kilometers (470 miles) per hour (the speed of a jet airliner!) in seismic sea waves.

The behavior of a wave depends largely on the relationship between the wave's size and the depth of water through which it is moving.

Wind Waves

Wind waves are gravity waves formed by the transfer of wind energy into water. Most wind waves are less than 3 meters (10 feet) high. Wavelengths from 60 to 150 meters (200 to 500 feet) are most common in the open ocean.

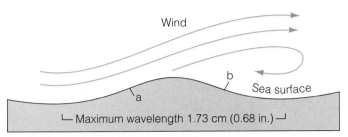

Figure 9.7 Wind forces acting on a capillary wave. Capillary waves interrupt the smooth sea surface, deflect surface wind upward, slow it, and cause some of the wind's energy to be transferred into the water to drive the capillary wave crest forward (point a). The wind may eddy briefly downwind of the tiny crest, creating a slight partial vacuum there. Atmospheric pressure pushes the trailing crest forward (downwind) toward the trough (point b), adding still more energy to the water surface. The increasing energy in the water surface expands the circular orbits of water particles in the direction of the wind, enlarging the small wave's size. The capillary wave becomes a wind wave when its wavelength exceeds 1.73 centimeters (0.68 inch), the wavelength at which gravity supersedes capillary action as the dominant restoring force.

Wind waves grow from capillary waves—tiny waves nearly always present on the ocean. Capillary waves interrupt the smooth sea surface and cause some of the wind's energy to be transferred into the water to drive the capillary wave crest forward. The wind may eddy briefly behind the tiny crest, creating a slight partial vacuum there. Atmospheric pressure pushes the trailing crest forward (downwind) toward the trough, adding still more energy to the water surface. The process is shown in **Figure 9.7.** The increasing energy in the water surface expands the circular orbits of water particles in the direction of the wind, enlarging the small wave's size. The capillary wave becomes a wind wave when its wavelength exceeds 1.73 centimeters

(0.68 inch), the wavelength at which gravity supersedes capillary action as the dominant restoring force.

If the wind wave remains in water deeper than half its wavelength, and the wind continues to blow, the wave becomes larger. Its crest is thrust higher into faster wind, extracting even more energy from the moving air. The circular orbits of water particles within the wave grow larger with more energy input; height, wavelength, and period increase proportionally. The irregular peaked waves in the area of wind wave formation are called **sea;** the chaotic surface is formed by simultaneous wind waves of many wavelengths, periods, and heights.

When the wind slows or ceases, as it does away from a storm, the wave crests become rounded and regular. These mature wind waves of uniform wavelength outside their original area of generation are called **swell.** Swell form smooth undulations in the ocean surface (**Figure 9.8**). Because waves with the longest wavelengths move fastest, swell are sorted by wavelength, with longer waves moving away from the storm most rapidly. Observers at a distance would first encounter large, quick-moving waves of long wavelength, then middle-sized waves, and then slow, small ones.

Factors That Affect Wind Wave Development

Three factors affect the growth of wind waves. First, the wind must be moving faster than the wave crests for energy transfer from air to sea to continue, so the mean speed of the wind, or **wind strength,** is clearly important to wind wave development. A second factor is the length of time the wind blows, or **wind duration;** high winds that blow only a short time will not generate large waves. The third is the uninterrupted

Figure 9.8 Swell—mature, regular wind waves—off the Oregon coast. The small waves superimposed on the swell are the result of local wind conditions.

Figure 9.9 The fetch, the uninterrupted distance over which the wind blows without significant change in direction. Wave size increases with increased wind speed, duration, and fetch. A strong wind must blow continuously in one direction for nearly three days for the largest waves to develop fully.

Ripples to chop to wind waves

Fully developed seas

Changing to swell

Wind

Direction of wave advance

Length of fetch

distance over which the wind blows without significant change in direction—the **fetch** (**Figure 9.9**).

A strong wind must blow continuously for nearly three days for the largest waves to develop fully. A **fully developed sea** is the maximum wave size theoretically possible for a wind of a specific strength, duration, and fetch. Longer exposure to wind at that speed will not increase the size of the waves.

The greatest potential for large waves occurs beneath the strong and nearly continuous winds of the Antarctic Circumpolar Current. The early nineteenth-century French explorer of the South Seas Jules Dumont d'Urville encountered waves with heights estimated "in excess" of 30 meters (100 feet) in Antarctic waters. Satellite observations (like those of **Figure 9.10**) have shown that wave heights to 11 meters (36 feet) are fairly common in this area.

In zones of high winds, a less than fully developed sea can also attract attention. Though wind speed within cyclonic storms is often very strong, the circular motion of air doesn't allow long fetches, and fully developed seas rarely occur beneath them. Officers standing deck watches during storms rarely quibble with theoretical maximum height versus observed height, however. Wind waves can be overwhelming even if they are not "fully developed."

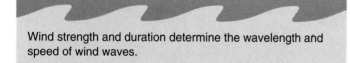

Wind strength and duration determine the wavelength and speed of wind waves.

Wind Wave Height and Wavelength

9.9

Wave height is not directly represented in the deep-water wave formula $S = L/T$. During their formation, moderate-

sized wind waves in the open ocean exhibit a maximum 1:7 ratio of wave height to wavelength (**Figure 9.11**); this ratio is the **wave steepness.** Waves 7 meters long will not be more than 1 meter high, and waves with a 70-meter wavelength will not exceed 10 meters of height. The angle at their crest will not exceed 120°. A peaked appearance usually indicates the continuing injection of wind energy. If a wave gets any higher than the 1:7 ratio for its wavelength, it will break and excess energy from the wind will be dissipated as turbulence—hence, the *whitecaps* or *combers* associated with a fully developed sea.

The highest wave ever measured was sighted on the night of 7 February 1933 by Lieutenant Frederick Marggraff, a watch officer aboard the U.S. Navy tanker *Ramapo*. USS *Ramapo* was steaming from Manila to San Diego through a furious storm—a storm made more intense by the coalescence of three low-pressure centers. For days a steady wind had blown at 107 kilometers (67 miles) per hour, and gusts to 126 kilometers (78 miles) per hour often lashed the decks. But the wind blew persistently from one direction, and though the monstrous waves it generated dwarfed the tanker, they were surprisingly orderly in form.

"The conditions for observing the seas from the ship were ideal," wrote *Ramapo*'s executive officer. "We were running directly down the wind with the sea. There were no cross seas and therefore no peaks along wave crests. There was practically no rolling, and the pitching motion was easy because of the fact that the sides of the waves were much longer than the ship. The moon was out astern and facilitated observations during the night. The sky was partly cloudy." At about three in the morning, Mr. Marggraff observed a train of tremendous waves looming in the moonlight. As the trough of the first wave approached the ship, he noted that its distant crest was on a level with the crow's nest on the mainmast. At that instant the ship's stern sank into the bottom of the onrushing trough. The next two waves were about the same size. Not surprisingly, such immense

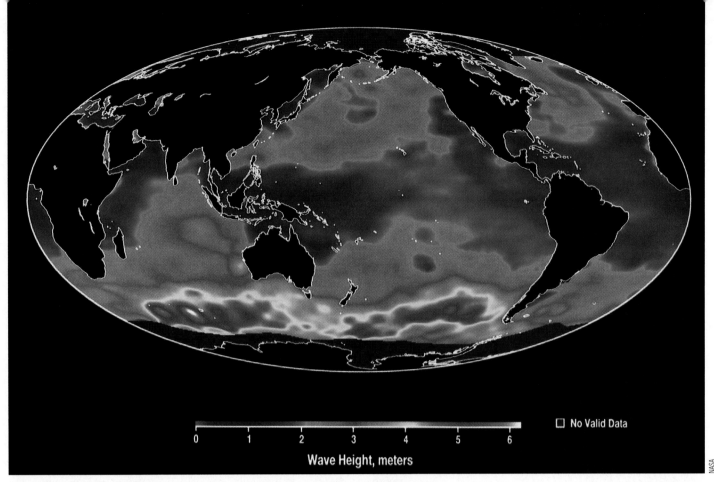

Figure 9.10 Global wave height acquired by a radar altimeter aboard the *TOPEX/Poseidon* satellite in October 1992. In this image, the highest waves occur in the southern ocean, where waves were more than 6 meters (19.8 feet) high (represented in white). The lowest waves (indicated by dark blue) are found in the tropical and subtropical ocean, where wind speed is lowest.

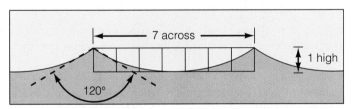

Figure 9.11 A wind wave of moderate size shown during its formation. The ratio of height to wavelength, called *wave steepness*, is 1:7; the crest angle does not exceed 120°.

waves made an indelible impression on all who witnessed them.

How big were these waves? When the executive officer had some time to spare, he did some calculations. **Figure 9.12** illustrates the attitude of the ship when the largest waves were measured. The height of the waves was determined by using a set of the ship's plans, a calculation of the height of the observer above the sea surface, the draft of the ship, and a sight to the horizon. The largest wave for which

Figure 9.12 How the great wave observed from the USS *Ramapo* was measured. An officer on the bridge was looking toward the stern and saw the crow's nest in his line of sight to the crest of the wave, which had just come in line with the horizon. Wave height was later calculated based on the geometry of the situation.

Figure 9.13 A NOAA research ship rides large waves in a storm off North Carolina.

a dependable observation had been made was 34 meters (112 feet) high, still a record!

Wavelength Records

9.10

But is that as big as wind waves can get? This is an interesting question and one not dispassionately answered. Calculating the wavelength from wave period (T in $S = L/T$), Bigelow and Edmondson (1947) reported that mariners have sighted waves with wavelengths of 451 meters (1,481 feet) off the west coast of Ireland, 583 meters (1,914 feet) off the Cape of Good Hope, and 829 meters (2,719 feet) in the equatorial Atlantic. (That last monster would have had a period of 25 seconds!) The nineteenth-century French admiral J. Mottez reported a wave with a wavelength of 790 meters (2,600 feet), a period of 23 seconds, and a speed of 123 kilometers (76 miles) per hour in the equatorial Atlantic west of Africa. The wavelength of the *Ramapo* wave was calculated at 360 meters (1,180 feet), so these sightings are perhaps exaggerated. Of course, much smaller waves can also add a great deal of excitement to a deck officer's day (**Figure 9.13**)!

Interference

9.11

The real situation of wind waves at sea is not as simple as has been suggested here. The ideal vision of one set of waves moving in one direction at one speed across an otherwise smooth surface is almost never observed in the ocean.

Independent wave trains exist simultaneously in the ocean most of the time. Since long waves outrun shorter ones, wind waves from different storm systems can overtake and interfere with one another. One wave doesn't crawl over the others when they meet; instead they add to (or subtract from) one another. Such interaction is known as **interference. In Figure 9.14a** a wave of one wavelength is represented as a blue line;

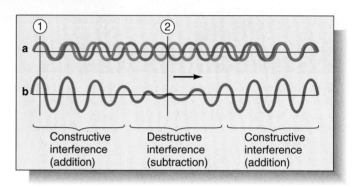

Figure 9.14 Interference. Two overlapping waves of different wavelength (**a**) generate both constructive and destructive interference (**b**).

a slightly longer wavelength wave as a green line. In the sea surface where these waves coincide (**Figure 9.14b**) you can see the alternation between addition (large crests and troughs) and subtraction (almost no waves at all). The cancellation effect of subtraction is termed **destructive interference,** not because of harm to lives or property, but because wave interference destroys or cancels the waves involved. **Constructive interference** is the additive formation of large crests or deep troughs. This pattern of constructive and destructive interferences explains why the ocean surface will be relatively calm for a time, rise to a few big waves, and then return to calm. Interference also explains why in some instances every ninth wave might be unusually large. As wavelengths change, though, every seventh wave might be large, or every fifth, or twelfth. Contrary to folklore, there is no set ratio.

Interference can have sudden unpleasant consequences on the open sea. In or near a large storm, wind waves at many wavelengths and heights may approach a single spot from different directions. If such a rare confluence of crests occurred at your position, a huge wave crest would suddenly erupt from

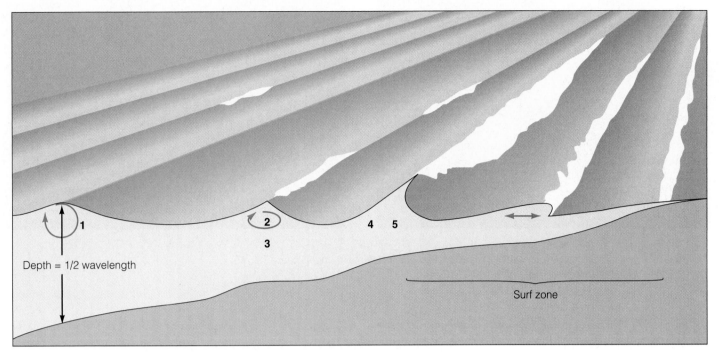

Figure 9.15 How a wave train breaks against the shore. (1) The swell "feels" bottom when the water is shallower than half the wavelength. (2) The wave crests become peaked because the wave's energy is packed into less water depth. (3) Constraint of circular wave motion by interaction with the ocean floor slows the wave, while waves behind it maintain their original rate. Therefore, wavelength shortens but period remains unchanged. (4) The wave approaches the critical 1:7 ratio of wave height to wavelength. (5) The wave breaks when the ratio of wave height to water depth is about 3:4. The movement of water particles is shown in red. Note the transition from a deep-water wave to a shallow-water wave.

a moderate sea to threaten your ship. The freak wave—called a **rogue wave**—would be much larger than any noticed before or after and would be higher than the theoretical maximum wave capable of being sustained in a fully developed sea. In such conditions one wave in about 1,175 is more than three times average height, and one in every 300,000 is more than four times average height! The huge wave described at the beginning of this chapter was a rogue wave.

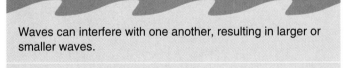

Waves can interfere with one another, resulting in larger or smaller waves.

Wind Waves Approaching Shore

Most wind waves eventually find their way to a shore and break, dissipating all their order and energy. The process begins with the transition of our now familiar deep-water wave to a shallow-water wave in water less than half a wavelength deep. **Figure 9.15** outlines the events that lead to the break:

1. The wave train moves toward shore. When the depth of the water is less than half the wavelength, the wave "feels" bottom.

2. The circular motion of water molecules in the wave is interrupted. Circles near the bottom flatten to ellipses. The wave's energy must now be packed into less water depth, so the wave crests become peaked rather than rounded.

3. Interaction with the bottom slows the wave. Waves behind it continue toward shore at the original rate. Wavelength therefore decreases but period remains unchanged.

4. The wave becomes too high for its wavelength, approaching the critical 1:7 ratio.

5. As the water becomes even shallower, the part of the wave below average sea level slows because of the restricting effect of the ocean floor on wave motion. When the wave was in deep water, molecules at the top of the crest were supported by the molecules ahead (thus transferring energy forward). This is now impossible because the *water* is moving faster than the *wave*. As the crest moves ahead of its supporting base, the wave breaks. The break occurs at about a 3:4 ratio of wave height to water depth (that is, a 3-meter wave will break in 4 meters of water). The turbulent mass of agitated water rushing shoreward during and after the break is known as **surf.** The **surf zone** is the region between the breaking waves and the shore.

a   b

Figure 9.16 Types of breaking waves. (**a**) Plunging waves form when the bottom slopes steeply toward shore. (**b**) Spilling waves form when the bottom slopes gradually.

Waves break against the shore in different ways, depending in part on the slope of the bottom. The break can be violent and toppling, leaving an air-filled channel (or "tube") between the falling crest and the foot of the wave. These **plunging waves** form when waves approach a steeply sloping bottom (**Figure 9.16a**). A more gradually sloping bottom generates a milder **spilling wave** as the crest slides down the face of the wave (**Figure 9.16b**).

Slope alone does not determine the position and nature of the breaking wave. The contour and composition of the bottom can also be important. Gradually shoaling bottoms can sap waves of their strength because of prolonged interaction against the bottom of the lowest elliptical water orbits.

Energy may be lost even more rapidly if the bottom is covered with loose gravel or irregular growths of coral. Masses of moving seaweeds or jostling chunks of sea ice can also extract energy from a wave. In a few rare cases the shore is configured in such a way that waves don't break at all—the waves have lost nearly all their energy by the time their remnants arrive at the beach.

Wave Refraction

What happens when a wave line approaches the shore at an angle, as it almost always does (**Figure 9.18**)? The line does not break simultaneously because different parts of it are in

Figure 9.17 *Neptune's Horse* by Walter Crane, a graphic representation of the power of breaking waves. The energy of a wave is proportional to the square of its height. Researchers report that each linear meter of a wave 2 meters (6.6 feet) above average sea level represents an energy flow of about 25 kilowatts (34 horsepower), enough to light 250 100-watt light bulbs; a wave twice as high would contain four times as much energy. A single wave 1.2 meters (4 feet) high striking the west coast of the United States may release as much as 50 million horsepower.

a

b

Figure 9.18 Wave refraction. (**a**) Diagram showing the elements that produce refraction. (**b**) Wave refraction around Maili Point, Oahu, Hawaii. Note how the wave crests bend almost 90° as they move around the point.

different depths of water. The part of the wave line in shallow water slows down, but the attached segment still in deeper water continues at its original speed, so the wave line bends (or refracts). The bend can be as much as 90° from the original direction of the wave train. This slowing and bending of waves in shallow water is called **wave refraction.** The refracted waves break in a line almost parallel to the shore.

Most waves change shape and speed as they approach shore. They may plunge or spill at the surf zone and bend to break nearly parallel to shore.

Long Waves 9.14

We now turn our attention to the longer waves generated by the low atmospheric pressure of large storms, by the sloshing of water in enclosed spaces, and by the sudden displacement of ocean water.

A Word About "Tidal Waves" 9.15

The popular media and general public tend to label *any* unusually large wave a "tidal wave" regardless of its origin, and the waves described next are prime candidates for this error. Press accounts of storms at sea usually list a rogue wave as a tidal wave, and very large sets of wind waves are called tidal waves by some yachtsmen or surfers. The sea waves associated with earthquakes are almost always called tidal waves in media damage reports. The waves caused by the approach of a tropical cyclone to land may also incorrectly be termed tidal waves. The term *tidal wave* is *not* synonymous with *large wave*, however. As we will see in the next chapter, the only true tidal waves are relatively harmless waves associated with the tides themselves.

Storm Surges 9.16

The abrupt bulge of water driven ashore by a tropical cyclone (hurricane) or frontal storm is called a **storm surge.** Its crest can temporarily add up to 7.5 meters (25 feet) to coastal sea level. Water can reach even greater heights when the surge is funneled into a confined bay or estuary.

Many factors contribute to the severity of a storm surge. The most important factor is the strength of the storm generating the surge. The low atmospheric pressure associated with a great storm will draw the ocean surface into a broad dome as much as 1 meter (about 3 feet) higher than average sea level. This dome of water accompanies the storm to shore, becoming much higher as the water gets shallower at the coast. There the water ramps ashore, driven forward by large storm-generated wind waves. A storm surge is a short-lived phenomenon. Technically it is not a progressive wave because it is only a crest; wavelength and period cannot be assigned to it.

Water in a storm surge does not come ashore as a single breaking wave, but rushes inland in what looks like a sudden, very high, wind-blown tide. Indeed, storm surges are sometimes called *storm tides* because the volume of water they force onshore is greatly increased if the surge arrives at the same time as a high tide. The wrong combination of low atmospheric pressure, strong onshore winds, high tide, and bottom contour can be especially dangerous if estuaries in the area have been swollen by heavy rainfall preceding the storm.

Storm surges have had catastrophic consequences. The frightful tropical storm of November 1970 in Bangladesh (described in Chapter 7) generated a storm surge up to 12 meters (40 feet) high, which caused the death of more than 300,000 people. Storms surges associated with extratropical cyclones (frontal storms) can also do tremendous damage. On 1 February 1953 a storm surge and high tide arrived simultaneously against the Dutch coast. Wind waves breached the dikes and flooded the low country, covering more than 3,200 square kilometers (800,000 acres) and drowning 1,783 people. The Dutch anticipate this coincidence of events will occur only about once in 400 years. The dikes have been rebuilt and the land reclaimed from the North Sea. On the opposite side of the North Sea, Londoners have spent $1.3 billion on a flood defense system at the mouth of the Thames River. The centerpiece of the project is an immense barrier against storm surges (**Figure 9.19**). Experts expect the barrier to prevent a devastating flood on an average of once every 50 years.

The coast of the United States is also at risk. In 1900 a storm surge topped the Galveston seawall, swept into the city, and killed more than 6,000 area residents. (In contrast, no lives were lost during a similar Texas storm in 1961 because coastal barriers had been constructed and advance warning permitted preparation and evacuation.) Hurricane Andrew's 1992 assault on the Florida coast was made even more lethal by its storm surge. Anyone who lives in a low-lying coastal area frequented by violent storms should be aware of the potential danger of storm surges.

a

Seiches

When disturbed, water confined to a small space (such as a bucket, a bathtub, or a bay) will slosh back and forth at a specific resonant frequency. The frequency changes with different amounts of water, or with different sizes or shapes of containers. If you carry a shallow container of water (like an ice cube tray) from one place to another, you're careful not to move the tray at its resonant frequency to avoid a spill. Most of the water's random motion quickly settles down after you place the tray on a table top, but the water in the tray may rock gently for some seconds at this one resonant frequency. That rocking is a **seiche** (pronounced "saysh").

The seiche phenomenon was first studied in Switzerland's Lake Geneva by eighteenth-century researchers curious about why the water level at the ends of the long, narrow lake rises and falls at regular intervals after wind storms. They found that constant breezes tend to push water into the downwind end of the lake. When the wind stops, the water is released to rock slowly back and forth at the lake's resonant frequency, completing a crest-trough-crest cycle in a little more than an hour. At the ends of the lake the water rises and falls a foot or two; at the center it moves back and forth without changing height. This kind of wave is called a **standing wave** because it oscillates vertically with no forward movement. The point (or line) of no vertical wave action in a standing wave—the place in the lake where the water moves only back and forth—is called a **node.** In Lake Geneva the wavelength of the seiche is twice the length of the lake itself; the node lies at the center (**Figure 9.20**). The lake acts like a large version of the ice cube tray in the example above.

Damage from seiches along most ocean coasts is rare. The wavelength may be tremendous, but seiche wave height in the open ocean rarely exceeds a few inches. Larger seiches can occur in harbors: Coastal seiches at Nagasaki, on the southern coast of Japan, occasionally reach 3 meters (10 feet). Seiches may disturb shipping schedules by interfering with the predicted arrival times of tides, or they may cause currents in harbors, which could snap mooring lines.

b

Figure 9.19 The Thames tidal barrier, near London. The funnel shape of the Thames estuary concentrates the surge as it nears London. The barrier sections can be raised to prevent flooding upstream (**a**). Each barrier section is controlled by hydraulic arms within towers. The great size of the $1.3-billion project can be seen in (**b**).

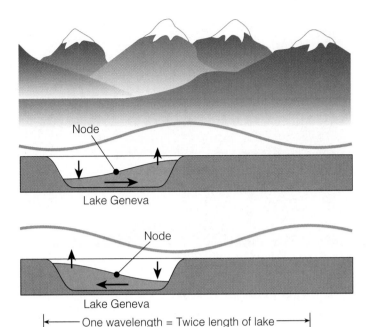

Figure 9.20 Seiche in Lake Geneva. After a wind storm, the water in the lake sloshes back and forth, forming a standing wave, or seiche, with a wavelength of 220 kilometers (132 miles), which is twice the length of the lake. The period of the seiche is about 72 minutes, and its height is about 0.5 meter (1.5 feet).

Tsunami and Seismic Sea Waves

Long-wavelength, shallow-water progressive waves caused by the rapid displacement of ocean water are called **tsunami,** a descriptive Japanese term combining *tsu* ("harbor") with *nami* ("wave"). The word is both singular and plural. Tsunami caused by the sudden vertical movement of Earth along faults (the same forces that cause earthquakes) are properly called **seismic sea waves.** Tsunami can also be caused by landslides, icebergs falling from glaciers, volcanic eruptions, and other direct displacements of the water surface. Note that all seismic sea waves are tsunami, but not all tsunami are seismic sea waves.

Origins

"Small" tsunami are caused by the displacement of surface water. Although less energy is released by landslides than by most seismic fractures, the resulting sea waves are still very destructive for people or structures near their point of origin. This is especially true if the wave is formed within a confined area.

Seismic sea waves originate on the seafloor when Earth movement along faults displaces seawater. **Figure 9.21** shows the birth of a seismic sea wave in the Aleutians. Rupture along a submerged fault lifts the sea surface above. Gravity pulls the crest downward, but the momentum of the water causes the crest to overshoot and become a trough. The oscillating ocean surface generates progressive waves that radiate from the epicenter in all directions. Waves would also form if the fault movement were downward. In that case a depression in the water surface would propagate outward as a trough. The trough would be followed by smaller crests and troughs caused by surface oscillation.

It seems strange to refer to tsunami—waves with wavelengths of up to 200 kilometers (125 miles)—as shallow-water waves. Yet half their wavelength would be 100 kilometers (62 miles), and even the deepest ocean trenches do not exceed 11 kilometers (7 miles) in depth. These immense

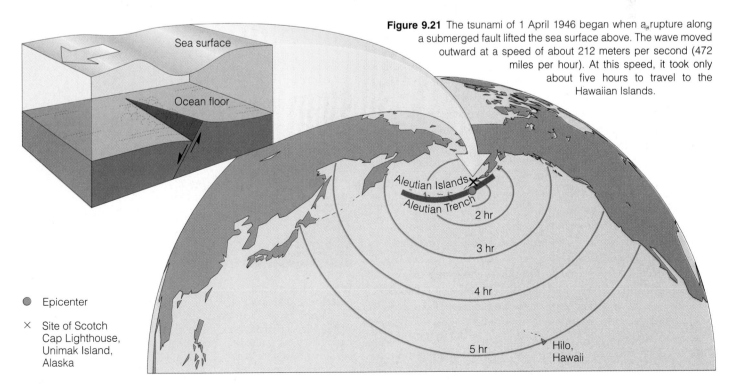

Figure 9.21 The tsunami of 1 April 1946 began when a rupture along a submerged fault lifted the sea surface above. The wave moved outward at a speed of about 212 meters per second (472 miles per hour). At this speed, it took only about five hours to travel to the Hawaiian Islands.

● Epicenter

✕ Site of Scotch Cap Lighthouse, Unimak Island, Alaska

http://info.brookscole.com/garrison

WAVES 173

waves are therefore never in water deeper than half their wavelength. Like any shallow-water wave, seismic sea waves are affected by the contour of the bottom and are commonly refracted, sometimes in unexpected ways.

Tsunami are always shallow-water waves. No ocean basin is deeper than half their wavelength.

Speed

The speed (*S*) of a tsunami is given by the formula for the speed of a shallow-water wave:

$$S = \sqrt{gd}$$

Because the acceleration due to gravity (*g*) is 9.8 meters (32.2 feet) per second per second and a typical Pacific abyssal depth (*d*) is 4,600 meters (15,000 feet), solving for *S* shows that the wave would move at 212 meters per second (470 miles per hour). At this speed a seismic sea wave will take only about five hours to travel from Alaska's seismically active Aleutians to the Hawaiian Islands.

Encountering Tsunami

We are familiar with the steepness of a wind wave and the short period of a few seconds between its crests. Tsunami are much different. Once a tsunami is generated, its steepness (ratio of height to wavelength) is extremely low. This lack of steepness, combined with the wave's very long period (5 to 20 minutes), enables it to pass unnoticed beneath ships at sea. A ship on the open ocean that encounters a tsunami with a 16-minute period would rise slowly and imperceptibly for about 8 minutes to a crest only 0.3 to 0.6 meter (1 or 2 feet) above average sea level. It would then ease into the following trough 8 minutes later. With all the wind waves around, such a movement would not be noticed.

As the tsunami crest approaches shore, however, the situation changes rapidly and often dramatically. The period of the wave remains constant, velocity drops, and wave height greatly increases. As the crest arrived at the coast, observers would see water surge ashore in the manner of a very high, very fast tide (**Figure 9.22**). In confined coastal waters relatively close to their point of origin, tsunami can reach a height of perhaps 30 meters (100 feet). The wave is a fast, onrushing flood of water, not the huge, plunging breaker of popular movies and folklore.

The wave energy spreads through an enlarging circumference as a tsunami expands from its point of origin. People on shore near the generating shock have reason to be concerned because the energy will not have dissipated very much. On 1 April 1946 a fracture along the Aleutian Trench generated a seismic sea wave that quickly engulfed the

a

b

c

Figure 9.22 Sequential photos showing the arrival of a tsunami at Laie Point, Oahu, Hawaii, on 9 March 1957. An 8.3 earthquake had occurred in the Aleutian Islands about five hours earlier. Note how the wave resembles a fast, onrushing flood of water, not the huge, plunging breaker of movies and folklore.

Figure 9.23 A tsunami in progress. The largest wave of the 1 April 1946 tsunami rushes ashore in Hilo. Terrified residents run for their lives; more than 150 died. V-shaped Hilo Bay is an especially dangerous place during a tsunami because its funnel-like outline concentrates the energy of the waves. Fourteen years after the 1946 disaster, a seismic disturbance in the subduction zone off western South America generated a tsunami that drowned 61 people in Hilo. Wave-cut scars on cliffs north of the town suggest that visits by large tsunami are not rare occurrences.

Scotch Cap Lighthouse on Unimak Island in the Aleutians. The lighthouse was completely destroyed and the five coastguardsmen operating the lighthouse died.

By the time the same seismic sea wave reached the Hawaiian Islands about five hours later, the wave circumference was enormous and its energy more dispersed. Even so, successive waves surged onto Hawaii's beaches at 15-minute intervals for more than two hours. One 9-meter (30-foot) wave struck the town of Hilo, and water rose to a height of 17 meters (56 feet) in an exposed valley near Polulu! At least 150 people were killed in Hawaii that morning, and property damage exceeded $25 million. **Figure 9.23** is a photograph taken in Hilo that morning.

Note that the destruction in Hawaii was not caused by one wave (as at Scotch Cap) but by a series of waves following one another at regular intervals. Some energy from the main tsunami wave was distributed into smaller waves ahead of or behind the main wave as it moved. If the epicenter of the displacement responsible for a tsunami is very far away, sea level at shore will rise and fall as these waves arrive. The interval between crests (the wave period) is usually about 15 minutes. Coastal residents far from a tsunami's origin can be lulled into thinking the waves are over; they return to the coastline only to be injured or killed by the next crest. This behavior contributed to loss of life in Hawaii.

Tsunami in History

Destructive tsunami strike somewhere in the world an average of once each year. An earthquake along the Peru–Chile Trench on 22 May 1960 killed more than 4,000 people, and the resulting tsunami that reached Japan, 14,500 kilometers (9,000 miles) away, killed 180 people and caused $50 million in struc-

tural damage. Los Angeles and San Diego Harbors were badly disrupted by seiches excited by the tsunami. In 1992 a tsunami struck the coast of Nicaragua and killed 170 people; 13,000 Nicaraguans were left homeless. In 1993 an earthquake in the Sea of Japan generated a tsunami that washed over areas 29 meters (97 feet) above sea level and killed 120 people (**Figure 9.24**). In 1998 a wave 7 meters (23 feet) high destroyed remote villages in New Guinea, killing more than 2,200 people. Sometimes even a relatively minor wave can cause considerable loss of life and property. On 17 February 1996, 53 people died at Biak, Indonesia, in a wave that advanced less than 9 meters (28 feet) past the normal high-tide line. Some recent lethal tsunami are described in **Figure 9.25.**

There have been much greater tsunami disasters. In 1703 at Awa, Japan, 100,000 people died. On 27 August 1883 the enormous volcanic explosion of Krakatoa in Indonesia generated 35-meter (115-foot) waves that destroyed 163 villages and killed more than 36,000 people. Evidence of even larger waves has been discovered. Researchers have found signs of a huge wave, perhaps as high as 91 meters (300 feet), that crashed against the Texas coast 66 million years ago. It may have been caused by a comet or asteroid striking the Gulf of Mexico near Yucatan (see Figure 12.14). The wave scoured the floor of the gulf; picked up sand, gravel, and shark's teeth; and deposited the material in what is now central Texas.

Either a crest or a trough of a tsunami can arrive first at a shore. Subsequent oscillations between crests and troughs will rise in amplitude and then subside.

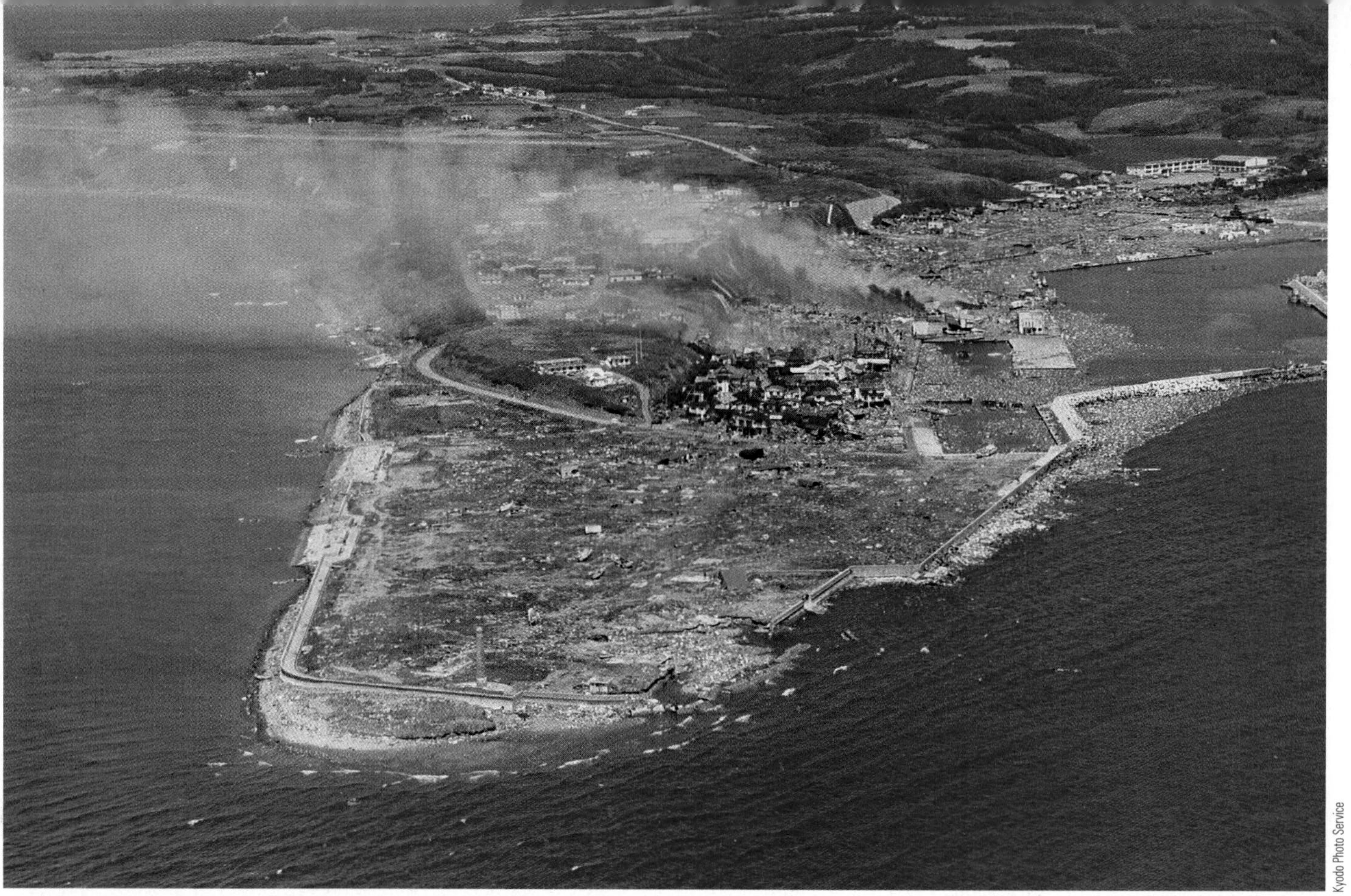
Kyodo Photo Service

Figure 9.24 Massive property damage in the fishing village of Aonae on the island of Okushiri, Japan, was caused by a tsunami generated by the Hokkaido Nansei-Oki earthquake in 1993. Maximum wave height was 31 meters (101 feet); 239 people died. This event has become the best-documented tsunami disaster in history.

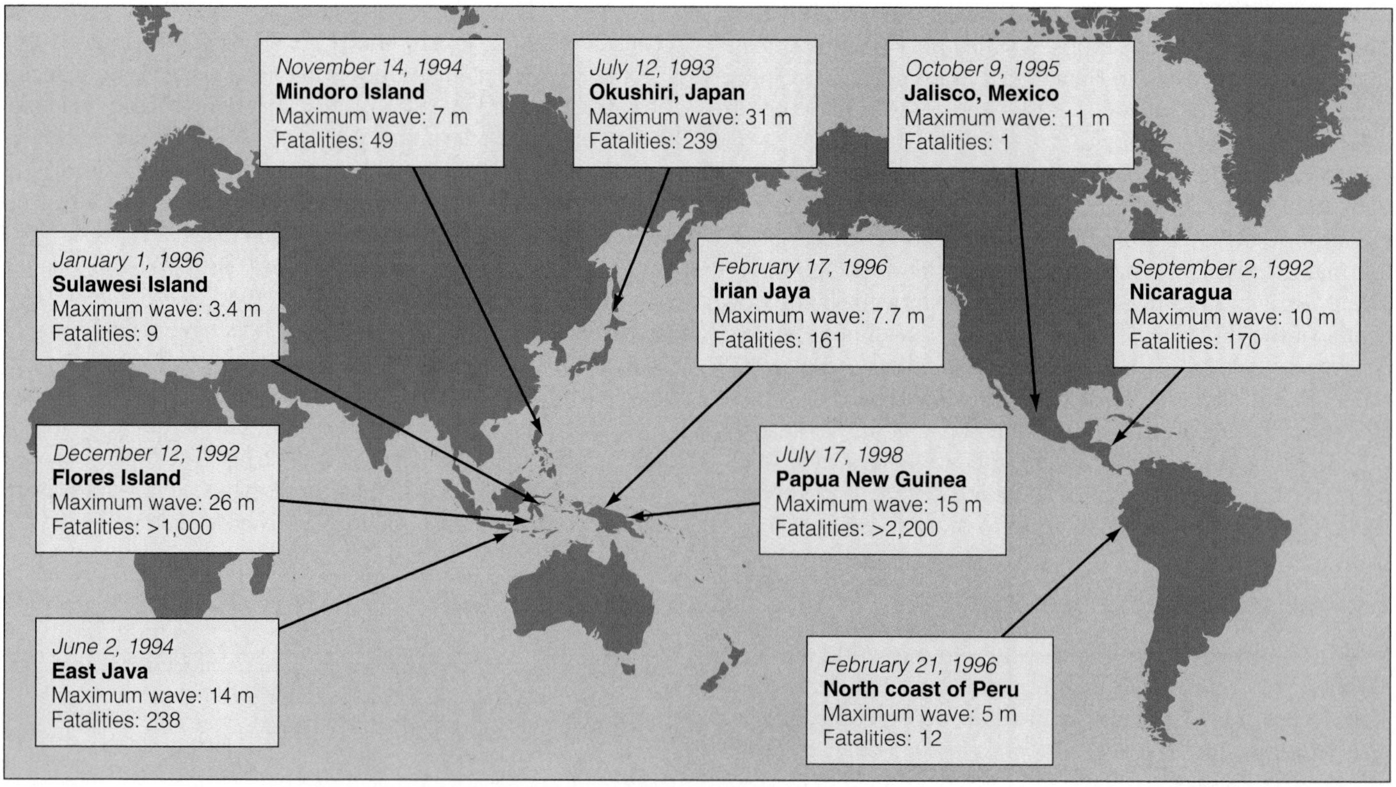
November 14, 1994
Mindoro Island
Maximum wave: 7 m
Fatalities: 49
July 12, 1993
Okushiri, Japan
Maximum wave: 31 m
Fatalities: 239
October 9, 1995
Jalisco, Mexico
Maximum wave: 11 m
Fatalities: 1
January 1, 1996
Sulawesi Island
Maximum wave: 3.4 m
Fatalities: 9
February 17, 1996
Irian Jaya
Maximum wave: 7.7 m
Fatalities: 161
September 2, 1992
Nicaragua
Maximum wave: 10 m
Fatalities: 170
December 12, 1992
Flores Island
Maximum wave: 26 m
Fatalities: >1,000
July 17, 1998
Papua New Guinea
Maximum wave: 15 m
Fatalities: >2,200
June 2, 1994
East Java
Maximum wave: 14 m
Fatalities: 238
February 21, 1996
North coast of Peru
Maximum wave: 5 m
Fatalities: 12

Figure 9.25 Ten destructive tsunami have claimed more than 4,000 lives since 1990.

a

b

Figure 9.26 (**a**) Tsunami hazard warning sign in a town on the Oregon coast. The configuration of the coast and the slope and instability of the bottom in this area make tsunami particularly dangerous. (**b**) An unfortunate effect of the tsunami warning network. Sightseers flocked to Oahu's Makapu'u Beach in Hawaii, awaiting a tsunami generated by a 6.5 earthquake centered in the Aleutian Trench on 7 May 1986.

The Tsunami Warning Network 9.23

Since 1948, an international tsunami warning network has been in operation around the seismically active Pacific to alert coastal residents of possible danger. Warnings must be issued rapidly because of the speed of these waves. Telephone books in coastal Hawaiian towns contain maps and evacuation instructions for use when the warning siren sounds.

The tsunami warning system was responsible for averting the loss of many lives after the great 27 March 1964 earthquake in Alaska (see Chapter 3). A 3.7-meter (12-foot) wave, probably the fourth crest to reach the coast, swept into Crescent City, California, about six hours after the quake. Though more than 300 buildings were destroyed or damaged, five gasoline storage tanks set ablaze, and 27 blocks of the city demolished, there were relatively few casualties.

It has been more than 30 years since a tsunami caused substantial damage in the United States. Some public safety experts suggest we have become complacent about the risks associated with these destructive waves. As can be seen in **Figure 9.26,** though, some coastal communities remind residents and visitors of the danger.

QUESTIONS FROM STUDENTS

1. If so much energy is expended as wind waves break, why doesn't water in the surf zone get hot?

The energy of wind waves is dissipated mostly as heat in the surf, but since water has such a high heat capacity (discussed in Chapter 6), the injection of heat into the surf zone doesn't sig-nificantly increase water temperature. The surf zone is also an area of vigorous mixing, so any heat released is quickly distributed through a large volume of water.

A number of methods have been proposed to take advantage of this energy before it is dissipated. An installation by Kvaerner Industrier A/S, a Norwegian energy company, is now producing power on the coast of Scotland. Its method of operation is shown in Figure 9.27.

2. What about wind waves and the Coriolis effect? Do wind waves turn right in the Northern Hemisphere as they move across the ocean surface?

The Coriolis effect has no influence on waves with periods shorter than about 5 minutes. Large shallow-water waves such as tsunami, seiches, and tides involve the mass movement of water and are influenced by Earth's rotation. Wind waves, however, with a period rarely exceeding 20 seconds, are not.

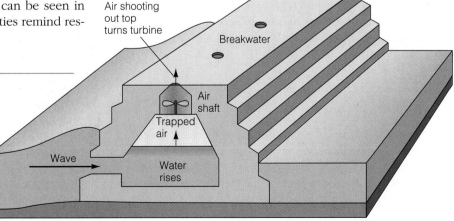

Figure 9.27 A device for using wave energy to generate electrical power.

3. I love to surf. Where should I plan to live?

The Pacific has the largest potential fetch distances, so the best chances for large wind waves are there. The biggest wind waves are made in the West Wind Drift, but temperatures there are much too cold for comfortable surfing. Besides, the sea state in areas of continuous high wind is chaotic, and surfing requires orderly waves. It is best to let the wave trains sort out. Hawaii is in the middle of the Pacific, so wind waves from polar storms in either hemisphere strike its shores only after much sorting by wavelength. Order is assured. The water is warm, too. I vote for Hawaii.

4. One of my friends has discovered I'm studying waves. He demands to know about the world's largest wave. What's the record, anyway?

If your friend classifies the largest wave by wavelength, the answer is easy: Tides take the prize, with a wavelength that is ideally half of Earth's circumference. Tides also win the speed race: Under ideal circumstances, their crests can move at speeds up to 1,600 kilometers (1,000 miles) per hour.

If your friend classifies the largest wave by height, things become more complicated. As you read in this chapter, the crew of *Ramapo* measured a wave 34 meters (112 feet) high in 1933. In general, the highest wind waves probably develop in the Antarctic Circumpolar Current, the nasty area of windy ocean ringing Antarctica. Multiple storms there, some with huge fetch distances, can cause very large waves and opportunity for constructive interference between them. In 1916, during an agonizing small-boat trip from an island on which their expedition was marooned, Frederic Worsley, captain of Ernest Shackleton's ship *Endurance*, encountered occasional tremendous waves greater than any he had seen in a lifetime at sea. He estimated a few were more than 30 meters (100 feet) high!

Some tsunami may be higher, however. Landslides in Lituya Bay, Alaska, have generated swash waves in the 50-meter (160-foot) range. And a wave of about 530 meters (1,740 feet) formed briefly as water surged up the opposite side of the bay on 9 July 1958. (It was more a titanic splash than a classic tsunami; one writer has likened the effect to dropping a sack of concrete into a half-filled bathtub.) The greatest height recently recorded for a tsunami in the ocean was 84 meters (278 feet) in 1971 off Ishigaki, Japan. But don't forget the 91-meter (300-foot) tsunami thought to have occurred 66 million years ago on the coast of what is now Texas after an asteroid struck the sea surface near the Yucatán Peninsula.

CHAPTER SUMMARY

Waves transmit energy, not water mass, across the ocean's surface. The speed of ocean waves usually depends on their wavelength, with long waves moving fastest. In order of wavelength from short to long (and therefore from slowest to fastest), ocean waves are generated by very small disturbances (capillary waves), wind (wind waves), rocking of water in enclosed spaces (seiches), seismic and volcanic activity or other sudden displacements (tsunami), and gravitational attraction (tides). The behavior of waves depends largely on the relationship between a wave's size and the depth of water through which it is moving. Waves can refract and reflect, break, and interfere with one another.

Unlike wind waves, the waves of very long wavelengths are always in "shallow water" (water less than half their wavelength deep). These long waves travel at high speeds. Some of the waves can be destructive, but their ability to cause damage is fortunately not proportional to their wavelengths. The waves with the longest wavelengths of all—the tides—are covered in the next chapter.

ON-LINE STUDY RESOURCES

The Web site for this book contains a wealth of helpful study aids. Log on to:

http://info.brookscole.com/garrison

and select *Essentials of Oceanography*, 3rd edition. Select Chapter 9 from the drop-down menu or click on one of these resource areas:

- For study and review, **Chapter at a Glance** gives you an outline of the chapter, **Chapter Summary** allows you to review the chapter's main ideas, and **Glossary** lists concepts and terms for the chapter along with their definitions.

- To test your mastery of the Terms and Concepts to Remember for this chapter, you can use the electronic **Flash Cards,** play the **Concentration** game, or work the **Crossword Puzzle.**

- For practice quizzing, try the multiple-choice **Tutorial Quiz,** the ten-question **True/False Quiz,** or the **Image Analysis Quiz,** which poses questions based on art and photos from the chapter.

TERMS AND CONCEPTS TO REMEMBER

capillary wave	$S = L/T$
constructive interference	$S = \sqrt{gd}$
deep-water wave	sea
destructive interference	seiche
disturbing force	seismic sea wave
fetch	shallow-water wave
fully developed sea	spilling wave
interference	standing wave
node	storm surge
orbit	surf
orbital wave	surf zone
plunging wave	swell
progressive wave	tsunami
restoring force	wave
rogue wave	wave crest

wave frequency   wave trough
wave height   wavelength
wave period   wind duration
wave refraction   wind strength
wave steepness   wind wave

STUDY QUESTIONS

1. How is an ocean wave different from a wave in a spring or a rope? How is it similar? How does it relate to a "stadium wave"—a wave form made by sports fans in a circular arena?

2. Draw a deep-water ocean wave and label its parts. Show the orbits of water particles. Include a definition of wave period. How would you measure wave frequency?

3. What factors influence the growth of a wind wave? What is a fully developed sea? Where would we regularly expect to find the largest waves? Are waves in fully developed seas always huge?

4. What happens when a wind wave breaks? What factors affect the break? How are plunging waves different from spilling waves?

5. Though they move across the deepest ocean basins, seiches and tsunami are referred to as "shallow-water waves." How can this be?

6. What causes tsunami? Are all seismic sea waves tsunami? Are all tsunami seismic sea waves? How fast do tsunami travel? Do they move in the same way or at the same speed in a confining bay as they would in the open ocean?

7. Are tsunami ever dangerous if encountered in the open sea? What happens when they reach shore?

Earth Systems Today CD Questions

1. Go to "Waves, Tides, and Currents," then to "Wave Properties" and review your understanding of terms used to describe progressive waves.

2. Go to "Waves, Tides, and Currents," then to "Wave Progression." Note in the animation the circular movement of water molecules in waves, and how the movement is affected by shallow water. This is a particularly effective demonstration of a difficult-to-understand concept.

3. Go to "Waves, Tides, and Currents," then to "Wind-Wave Relationships." Perform some experiments to determine the factors that influence a fully developed

sea. What is the best possible combination to generate truly large wind waves?

4. Go to "Earthquakes and Tsunami," and then to "Tsunami." Note the three choices: Prediction, Most Destructive, and Papau Simulation. Play each animation noting the time of propagation and the height of the waves near shore.

5. Go to "Earthquakes and Tsunami," and then to "Seismic Case Study, Alaska, 1964." Study each section of the case study, paying particular attention to the video and the speed of wave propagation through the whole Pacific Ocean. Did you see any references to 1964 in your research of question 4? Try your hand at finding the epicenter in the interactive part of this case study.

RESOURCES FOR FURTHER READING AND RESEARCH

The Web site for this book contains many ideas for further reading and research. Log on to:

http://info.brookscole.com/garrison

and select *Essentials of Oceanography*, 3rd edition. Select Chapter 9 from the drop-down menu or click on one of these resource areas:

- **Additional Readings** lists the major books and articles consulted in writing this chapter, along with comments from the author about their content and reading level.

- **Hypercontents** takes you to an extensive list of current links to Internet sites with news, research, and images related to individual subjects in the chapter. Just click on the icon that corresponds to a numbered section to see the links for that subject.

- **Internet Exercises** are critical thinking questions that involve research on the Internet with starter URLs provided.

- **InfoTrac College Edition Exercises** leads you to Critical Thinking Projects that use InfoTrac College Edition as a research tool.

- **Regional InfoTrac College Edition** articles are organized into East Coast, West Coast, and Gulf Coast regions, allowing you to study oceanography on a more local level.

 For more readings, go to InfoTrac College Edition, your on-line research library, at:

http://infotrac.thomsonlearning.com

Surfers ride a tidal bore as it moves up the Severn River near Gloucester in southwestern England. The tide wave has been funneled into the Severn estuary from the Celtic Sea and the Bristol Channel.

© Matt Hammersley/www.learn2climb.com

TIDAL BORES

Tidal waves are not as rare as you might think. When river mouths are exposed to very large tidal fluctuation, a tidal bore will sometimes form. Here is a true *tidal wave*—a steep wave moving upstream generated by the arrival of the tide crest in an enclosed river mouth. The funnel shape of the estuary confines the tide crest and forces it to move inland at a speed that exceeds the theoretical shallow-water wave speed for that depth. The forced wave then breaks to form a spilling wave front that moves upriver.

As many as 60 rivers experience tidal bores. Though most are less than 1 meter (3 feet) high, bores on China's Qiantang River may be as high as 8 meters (26 feet) and move at 40 kilometers (25 miles) per hour. The world's largest, this bore can be heard advancing more than 22 kilometers (14 miles) away!

A tidal bore's potential danger is lessened by its predictability. Tidal bores are common in the Amazon, the Ganges Delta, the Seine and Gironde in France, and England's Severn River. Some rivers in southern China may have three or four simultaneous bores at different places along the river's length.

10.1

Tides

Tides and the Forces That Generate Them

10.2

Tides are periodic, short-term changes in the height of the ocean surface at a particular place, caused by a combination of the gravitational force of the moon and sun and the motion of Earth. With a wavelength that can equal half of Earth's circumference, tides are the longest of all waves. Unlike the other waves we have met, these huge shallow-water waves are never free of the forces that cause them and thus are called *forced* waves. (After they are formed, wind waves, seiches, and tsunami are *free* waves; that is, they are no longer being acted upon by the force that created them and they do not require a maintaining force to keep them in motion.)

Tides are huge shallow-water waves—the largest waves in the ocean. Tides are caused by a combination of the gravitational force of the moon and sun and the motion of Earth.

The Greek navigator and explorer Pytheas first wrote of the connection between the position of the moon and the height of a tide around 300 B.C., but full understanding of tides had to await Isaac Newton's analysis of gravitation. Among many other things, Newton's brilliant *Mathematical Principles of Natural Philosophy* (1687) described the motions of planets, moons, and all other bodies in gravitational fields. A central finding: The pull of gravity between two bodies is proportional to the masses of the bodies but inversely proportional to the square of the distance between them. This means that heavy bodies attract each other more strongly than light bodies do, and that gravitational attraction quickly weakens as the distance grows larger.

While the main cause of tides is the gravity of the moon and sun acting on the ocean, the forces that actually generate the tides vary inversely with the *cube* of the distance from Earth's center to the center of the tide-generating object (the moon or sun). Distance is therefore even more important in this relationship. The sun is about 27 million times more massive than the moon, but the sun is about 387 times farther away than the moon, so the sun's influence on the tides is only about half that of the moon's.

The moon's influence on tides is about twice that of the sun's.

As we will see, Newton's gravitational model of tides—the *equilibrium theory*—deals primarily with the position and attraction of Earth, moon, and sun, and does not allow

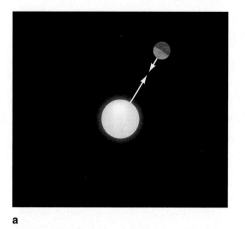
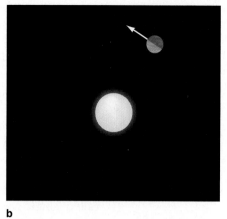
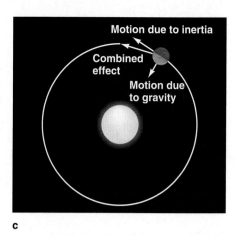

a

b

c

Figure 10.1 A planet orbits the sun in balance between gravity and inertia. (**a**) If the planet is not moving, gravity will pull it into the sun. (**b**) If the planet is moving, the inertia of the planet will keep it moving in a straight line. (**c**) In a stable orbit, gravity and inertia together cause the planet to travel in a fixed path around the sun.

for the influence on tides of ocean depth or the positions of continental landmasses. The equilibrium theory would accurately describe tides on a planet uniformly covered by water. A modification proposed by Pierre-Simon Laplace about a century later—the *dynamic theory*—takes into account the speed of the long-wavelength tide wave in relatively shallow water, the presence of interfering continents, and the circular movement or rhythmic back-and-forth rocking of water in ocean basins. We will explore the idealized situation of the equilibrium theory before moving to the real-world dynamic view.

The Equilibrium Theory of Tides

The **equilibrium theory** of the tides explains many characteristics of ocean tides by examining the balance and effects of the forces that allow a planet to stay in a stable orbit around the sun, or the moon to orbit Earth. The equilibrium theory assumes the ocean conforms instantly to the forces that affect the position of its surface; the ocean surface is pre-

sumed always to be in equilibrium (balance) with the forces acting on it.

The Moon and Tidal Bulges

We begin our examination of these forces by looking at the moon's effect on the ocean surface. Gravity tends to pull Earth and moon toward each other, but inertia—the tendency of moving objects to continue in a straight line—keeps them apart. (In this context, we sometimes call inertia *centrifugal force*, the "force" that keeps water against the bottom of a bucket when you swing it overhead in a circle.) Earth and the moon don't smash into each other (or fly apart) because they are in a stable orbit; their mutual gravitational attraction is exactly offset by their inertia (**Figure 10.1**).

Contrary to what you might think, the moon does not revolve around the center of Earth. Rather, the Earth–moon *system* revolves once a month (every 27.3 days) around the system's center of mass. Because Earth's mass is 81 times that of the moon, this common center of mass is located not in space, but 1,650 kilometers (1,023 miles) *inside* Earth. This center of mass is shown in **Figure 10.2**.

The moon's gravity attracts the ocean surface toward the moon. Earth's motion around the center of mass of the

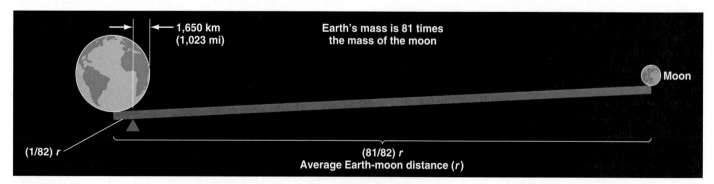

Figure 10.2 The moon does not rotate around the center of Earth. Earth and moon together—the Earth–moon *system*—rotate around a common center of mass about 1,650 kilometers (1,023 miles) beneath Earth's surface.

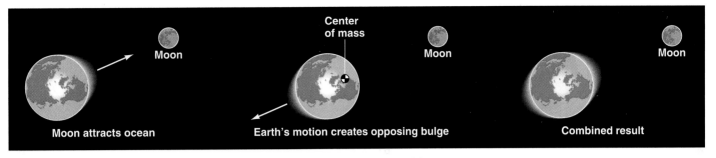

Figure 10.3 The moon's gravity attracts the ocean toward it. The motion of Earth around the center of mass of the Earth–moon system throws up a bulge on the side of Earth opposite the moon. The combination of the two effects creates two tidal bulges.

> Gravity and inertia cause the ocean surface to bulge. Tides occur as Earth rotates beneath the bulges.

Earth–moon system throws up a bulge on the opposite side of Earth. Two tidal bulges result (**Figure 10.3**).

How do these bulges cause the rhythmic rise and fall of the tides? In the idealized equilibrium model we are discussing, the bulges tend to stay aligned with the moon as Earth spins around its axis. **Figure 10.4** shows the situation from above the North Pole. As Earth turns eastward, an island on the equator is seen to move in and out of these bulges through one rotation (one day). The bulges are the crests of the planet-sized waves that cause **high tides. Low tides** correspond to the troughs, the area between bulges. Starting at 0000 (midnight) we see the island in shallow water at low tide. Around six hours later, at 0613 (6:13 A.M.), the island is submerged in the lunar bulge at high tide. At 1226 (about noon) the island is within the tide wave trough at low tide. At 1838 (6:38 P.M.) the island is again submerged,

this time in the opposite crest caused by inertia. About an hour after midnight (0050) on the next day, the island is back in shallow water, where it began.

The wave crests and troughs that cause high and low tides are actually very small: A 2-meter (7-foot) rise or fall in sea level is insignificant in comparison to the ocean's great size. Earth rotates beneath the bulges (tide wave crests) at about 1,600 kilometers (1,000 miles) per hour at the equator. The bulges appear to move across the ocean surface at this speed in an attempt to keep up with the moon. Theoretically, the wavelength of these tide waves is as long as 20,000 kilometers (12,500 miles)! The bulges tend to stay aligned with the moon as Earth spins around its axis. *The key to understanding the equilibrium theory of tides is to see Earth turning beneath these bulges.*

There are complications, of course. For example, **lunar tides**—tides caused by gravitational and inertial interaction of moon and Earth—complete their cycle in a tidal day (also called a lunar day). A complete tidal day is 24 hours 50 minutes long because the moon, which exerts the greatest tidal influence, rises 50 minutes later each day (**Figure 10.5**). Thus, the highest tide also arrives 50 minutes later each day.

Another complication arises from the fact that the moon does not stay right over the equator; each month, it moves from a position as high as 28½° above Earth's equator to 28½° below. When the moon is above the equator, the bulges are offset accordingly (**Figure 10.6**). When the moon is 28½° north of the equator, an island north of the equator will pass through the bulge on one side of Earth but miss the bulge on the other side. During one day the island passes through a very high tide, a low tide, a lower high tide, and another low tide (**Figure 10.7**).

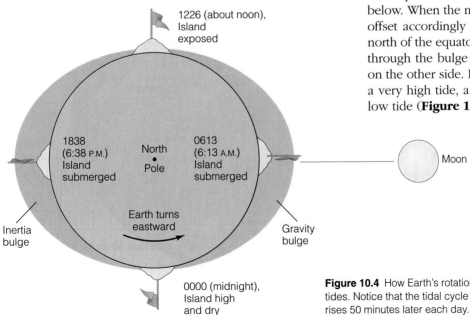

Figure 10.4 How Earth's rotation beneath the tidal bulges produces high and low tides. Notice that the tidal cycle is 24 hours, 50 minutes long because the moon rises 50 minutes later each day.

tide that low would expose intertidal organisms only rarely seen above water.

3. Are there tides in a glass of water?

Yes. They're too small to detect, but equilibrium tides do exist in very small bodies of water. Each molecule of water in the glass responds to the same planetary forces that affect molecules in the ocean.

4. You mentioned that the process of extracting power from the tides slows Earth's rotation. What would happen if all the world's electrical power needs could somehow be met by tidal generating plants?

If we assume it could actually be done, and if we assume a constant energy demand of today's requirement of 2×10^{20} Joules per year (somewhat unrealistic in light of the recent strong increase in electrical power generation), the length of a day would increase by an additional second in roughly a million and a half years. Clearly the "cashing in" of Earth's rotational kinetic energy to supply energy needs is not of great concern. Much more serious hazards would be disruptions of ocean currents, biological cycles, and esthetic values.

5. The news reporter said to expect "astronomical high tides" tonight. Should we pack our stuff and head for the hills?

Not necessarily. The reporter is calling attention to an alignment of the sun and moon that produces a high spring tide. "Astronomical" here refers to an alignment of heavenly bodies; it is not synonymous with "gigantic" or "spectacular."

CHAPTER SUMMARY

Tides have the longest wavelengths of the ocean's waves. They are caused by a combination of the gravitational force of the moon and the sun, the motion of Earth, and the tendency of water in enclosed ocean basins to rock at a specific frequency. Unlike the other waves, these huge shallow-water waves are never free of the forces that cause them and so act in unusual but generally predictable ways. Basin resonances and other factors combine to cause different tidal patterns on different coasts. The rise and fall of the tides can be used to generate electrical power, and they are important in many physical and biological coastal processes.

ON-LINE STUDY RESOURCES

The Web site for this book contains a wealth of helpful study aids. Log on to:

http://info.brookscole.com/garrison

and select *Essentials of Oceanography*, 3rd edition. Select Chapter 10 from the drop-down menu or click on one of these resource areas:

- For study and review, **Chapter at a Glance** gives you an outline of the chapter, **Chapter Summary** allows

you to review the chapter's main ideas, and **Glossary** lists concepts and terms for the chapter along with their definitions.

- To test your mastery of the Terms and Concepts to Remember for this chapter, you can use the electronic **Flash Cards,** play the **Concentration** game, or work the **Crossword Puzzle.**

- For practice quizzing, try the multiple-choice **Tutorial Quiz,** the ten-question **True/False Quiz,** or the **Image Analysis Quiz,** which poses questions based on art and photos from the chapter.

TERMS AND CONCEPTS TO REMEMBER

amphidromic point	mixed tide
astronomical tide	neap tide
diurnal tide	semidiurnal tide
dynamic theory of tides	slack water
ebb current	solar tide
equilibrium theory of tides	spring tide
flood current	tidal bore
high tide	tidal current
low tide	tidal datum
lunar tide	tidal range
mean sea level	tidal wave
meteorological tide	tide

STUDY QUESTIONS

1. What causes the rise and fall of the tides? What celestial bodies are most important in determining tides? Are there such things as "tidal waves"?

2. What are a high tide and a low tide? A spring tide and a neap tide? A tidal bore?

3. What are the most important factors influencing the heights and times of tides? What tidal patterns are observed? Are there tides in the open ocean? If so, how do they behave?

4. How does the latitude of a coastal city affect the tides there—or does it?

5. How is an astronomical tide different from a meteorological tide? Are the tides separate and independent of each other?

6. From what you learned about tides in this chapter, where would you locate a plant to generate electricity from tidal power? What are some advantages and disadvantages of using tides as an energy source?

 Earth Systems Today CD Questions

1. Go to "Waves, Tides, and Currents," to "High/Low Tides." Drag the moon to different elevations to see its effect on ocean tides.

2. Go to "Waves, Tides, and Currents," and to "Tidal Forces: Center of Mass." There are two interactive opportunities and one demonstration here. In "Center of Mass," you will see that the center of mass of the Earth–moon system is beneath the surface of Earth. In "Calculate Forces" and "Apply Forces," you can move the moon to find out how its influence changes with its proximity.

3. Go to "Waves, Tides, and Currents," then to "Tidal Shifts: Earth–Moon." Note the effect of the moon's position above or below Earth's equator on the types and heights of ocean tides. Now add the influence of sun by going to "Earth–Moon–Sun." Which has greater influence, moon or sun?

RESOURCES FOR FURTHER READING AND RESEARCH

The Web site for this book contains many ideas for further reading and research. Log on to:

http://info.brookscole.com/garrison

and select *Essentials of Oceanography*, 3rd edition. Select Chapter 10 from the drop-down menu or click on one of these resource areas:

- **Additional Readings** lists the major books and articles consulted in writing this chapter, along with comments from the author about their content and reading level.

- **Hypercontents** takes you to an extensive list of current links to Internet sites with news, research, and images related to individual subjects in the chapter. Just click on the icon that corresponds to a numberd section to see the links for that subject.

- **Internet Exercises** are critical thinking questions that involve research on the Internet with starter URLs provided.

- **InfoTrac College Edition Exercises** leads you to Critical Thinking Projects that use InfoTrac College Edition as a research fool.

- **Regional InfoTrac College Edition** articles are organized into East Coast, West Coast, and Gulf Coast regions, allowing you to study oceanography on a more local level.

 For more readings, go to InfoTrac College Edition, your on-line research library, at:

http://infotrac.thomsonlearning.com

CHAPTER 11

Sydney Harbour, Australia, a drowned river valley.

Coasts

"...THE FINEST HARBOUR IN THE WORLD"

On 20 January 1788 Captain Arthur Phillip led a fleet of ships into a shallow bay on the east coast of Australia. Discovered and explored by Captain James Cook eight years before, Botany Bay (as it was called) was insufficiently protected from the elements and offered little fresh water or fertile soil. Captain Phillip's ships were carrying more than a thousand people, most of them convicts and their jailers. His mission: Establish a prison colony as far from England as possible.

Phillip immediately knew Botany Bay would be inadequate to the task. A day after the landing he sent expeditions to look for a better location for the colony. Cook had mentioned a gap in the sandstone headlands to the north, and two small boats headed in that direction. Within two days the scouts returned with news of "... the finest harbour in the world." Less than a week later the whole colony moved to Sydney Cove (named after Lord Sydney, the British home secretary whose idea it was to transport criminals to this newly discovered continent).

Sydney Harbour seems an ideal coastal haven. But it was not always so. About 200 million years ago, layers of sediments were deposited in the huge delta of the Parramatta River system. The sands hardened to form the Hawkesbury Sandstones, which now form the local headlands and base rock. These were uplifted to near their present height by tectonic forces and eroded by the river. At the end of the last ice age, water released to the ocean from the melting ice caps caused a general rise in sea level. The river valleys were flooded to form the complex system of bays and promontories seen here. Phillip's sailing master wrote, "Here you are locked and it is impossible for the wind to do you the least damage."

11.1

Coastal areas join land and sea. The place where ocean meets land is usually called the **shore,** while the term **coast** refers to the larger zone affected by the processes that occur at this boundary. A sandy beach might form the shore in an area, but the coast (or coastal zone) includes the marshes, sand dunes, and cliffs just inland of the beach as well as the sandbars and troughs immediately offshore. The world ocean is bounded by about 440,000 kilometers (273,000 miles) of shore.

Because of its proximity to both ocean and land, a coast is subject to natural events and processes common to both realms. A coast is an active place. Here is the battleground on which wind waves break and expend their energy. Tides sweep water on and off the rim of land, rivers drop most of their sediments at the coasts, and ocean storms pound the continents. The *location* of a coast depends primarily on global tectonic activity and the volume of water in the ocean. The *shape* of a coast is a product of many processes: uplift and subsidence, the wearing-down of land by **erosion,** and the redistribution of material by sediment transport and deposition.

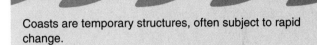

Coasts are temporary structures, often subject to rapid change.

The shape of a coast is a product of many processes: uplift and subsidence, the wearing down of land by erosion, and the redistribution of material by sediment transport and deposition.

Classifying Coasts

11.2

As we saw in Chapter 3, no area of geology has been left undisturbed by the revelations of plate tectonics. In the 1960s geologists began to classify coasts according to their tectonic position. *Active* coasts, near the leading edge of moving continental plates, were found to be fundamentally different from the more *passive* coasts near trailing edges. The shapes, compositions, and ages of coasts are better understood by taking plate movements into account. But as we'll see, the slow forces of plate movement are frequently obscured by the more rapid action of waves, by the erosion of land, and by the transport of sediments.

Another important consideration in the classification of coasts is long-term change in sea level. Five factors can cause sea level to change. Three of these factors are responsible for **eustatic change**—variations in sea level that can be measured all over the world ocean:

- The amount of water in the world ocean can vary. Sea level is lower during periods of global glaciation (ice ages) because there is less water in the ocean. It is higher during warm periods, when the glaciers are smaller. Periods of abundant volcanic outgassing can also add water to the ocean and raise sea level.

- The volume of the ocean's "container" may vary. High rates of seafloor spreading are associated with the expansion in volume of the mid-ocean ridges. This displaces the ocean's water, which climbs higher on the edges of the continents. Sediments shed by the continents during periods of rapid erosion can also decrease the volume of ocean basins and raise sea level.

- The water itself may occupy more or less volume as its temperature varies. During times of global warming, seawater expands and occupies more volume, raising sea level.

Of course, the continents rarely stay still as sea level rises and falls. Local changes are bound to occur, and two other factors produce variations in *local* sea level:

- Tectonic motions and isostatic adjustment can change the height and shape of the coast. Coasts can experience uplift as lithospheric plates converge or can be weighted down by masses of ice during a period of

widespread glaciation. The continents slowly rise when the ice melts. (See Figure 3.6 for a review of the principle of isostatic equilibrium.)

- Wind and currents, seiches, storm surges, an El Niño or La Niña event, and other effects of water in motion can force water against the shore or draw it away.

Sea level has been at its current elevation (give or take 0.5 meter, 1.5 feet) for only about 2,500 years. Over the past 2 million years, worldwide sea level has varied from about 6 meters (20 feet) above to about 125 meters (410 feet) below its present position. The recent low point occurred about 18,000 years ago at the height of the most recent glaciation. Indeed, sea level has been at the modern "high" only rarely in the past 2 million years—the dominant state for Earth is a much lower eustatic sea level position (**Figure 11.1**).

Changes in sea level produce major differences in the position and nature of coastlines, especially in areas where the edge of the continent slopes gradually or where the coast is rising or sinking. **Figure 11.2** shows an estimate of previous shore positions along the southeastern coast of the United States in the geologically recent past, and a prediction for the distant future should the present warming trend cause more of the polar ice to melt.

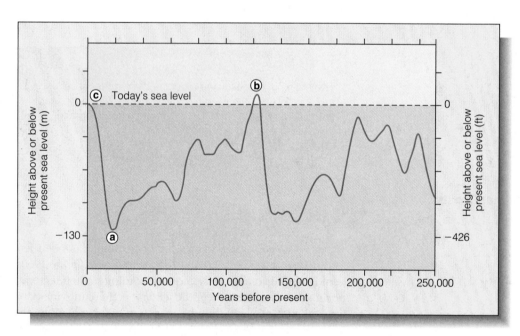

Figure 11.1 Changes in sea level over the last 250,000 years, as traced by data taken from ocean-floor cores. The rise and fall of sea level are largely due to the coming and going of ice ages—periods of increased and decreased glaciation, respectively. Water that formed the glaciers during these periods came from the ocean, and this caused sea level to drop. Point **a** indicates a low stand of −125 meters (−410 feet) at the climax of the last ice age some 18,000 years ago. Point **b** indicates a high stand of +6 meters (+19.7 feet) during the last interglacial period about 120,000 years ago. Point **c** shows the present sea level. The rate of sea-level rise has nearly leveled off, but sea level is rising slowly as we continue to emerge from the last period of glaciation.

Note that sea level has been considerably below its present position for nearly all of the last quarter-million years. Estuaries would have been less prevalent and continental shelves much narrower. Consider the implications for coastal civilizations. During the time between 18,000 and 8,000 years ago, sea level rose more than 100 meters (330 feet) at a rate of about 1 centimeter (½ inch) a year; more than 0.5 meter in a human lifetime. Could this have given rise to the flood legends common to many religions?

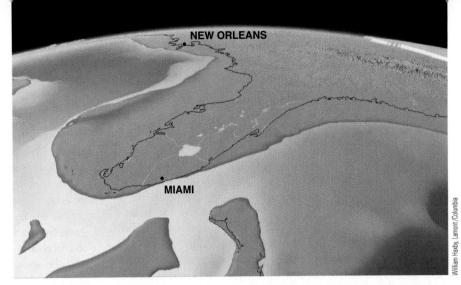

a

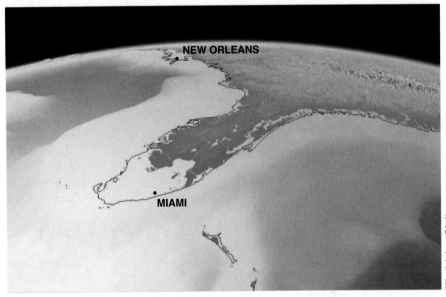

b

Figure 11.2 (**a**) The southeastern coast of the United States looked much different 18,000 years ago, during the last ice age. Because of lower sea level, the position of the gently sloping eastern coast has been as much as 200 kilometers (125 miles) seaward from the present shoreline, leaving much of the continental shelf exposed. (**b**) In the distant future, if the ocean were to expand and the polar ice caps were to melt because of global warming, sea level could rise perhaps 60 meters (200 feet), driving the coast inland as much as 250 kilometers (160 miles).

The location of a coast depends primarily on global tectonic activity and the volume of water in the ocean.

Change in sea level greatly influences coastal processes. For most of Earth's recent history, sea level has been lower than we find it today.

Because coasts are influenced by so many factors, perhaps the most useful scheme for classifying a coast is based on the predominant events that occur there: erosion and deposition. **Erosional coasts** are new coasts in which the dominant processes are those that *remove* coastal material. **Depositional coasts** are those coasts that are *steady or*

growing because of their rate of sediment accumulation or the action of living organisms (such as corals).

The rocky shores of Maine are erosional because erosion exceeds deposition; the sandy coastline from New Jersey to Florida is typically depositional because deposits of sediment tend to protect the shore from new erosion. The rocky central California coast is erosional, while the broad beaches of southern California are depositional. About 30% of the U.S. coastline is depositional and 70% is erosional. We will use the erosional–depositional classification scheme in the rest of this chapter.

Erosional coasts are new coasts in which the dominant processes are those that remove coastal material. Depositional coasts are those coasts that are steady or growing because of their rate of sediment accumulation or the action of living organisms.

Erosional Coasts

Land erosion and marine erosion both work to modify the nature of a rocky coast. Erosional coasts are shaped and attacked from the land by stream erosion, the abrasion of wind-driven grit, the alternate freezing and thawing of water in rock cracks, the probing of plant roots, glacial activity, rainfall, dissolution by acids from soil, and slumping.

Large storm surf routinely generates tremendous pressures. The crashing waves push air and water into tiny rock crevices. The repeated buildup and release of pressure within these crevices can weaken and fracture the rock. But it is not the hydraulic pressure of moving water alone that abrades the coasts. Tiny pieces of sand, bits of gravel, or stones hurled by waves toward the shore are even more effective at eroding the shore. Some indication of the violence of this activity may be inferred from **Figure 11.3.** The dissolving of minerals in the rocks by water contributes to the erosion of easily soluble coastal rocks such as limestone. Even the digging and scraping activities of marine organisms have an effect.

Figure 11.3 Attack from the sea is by waves and currents. The continuous onslaught of waves does most of the erosional work, with currents distributing the results of the waves' labor.

The rate at which a shore erodes depends on the hardness and resistance of the rock, the violence of the wave shock to which it is exposed, and the local range of tides. Hard rock resists wear. Coasts made of granite or basalt may retreat an insignificant amount over a human lifetime; the granite coast of Maine erodes only a few centimeters per decade. Coasts of soft sandstone or other weak (or soluble) materials, however, may disappear at a rate of a few meters per year.

Marine erosion is usually most rapid on **high-energy coasts,** areas frequently battered by large waves. High-energy coasts are most common adjacent to stormy ocean areas of great fetch and along the eastern edges of continents exposed to tropical storms. The coasts of Maine and British Columbia and the southern tips of South America and South Africa are typical high-energy coasts. **Low-energy coasts** are only infrequently attacked by large waves. Because of their generally protected location in the Gulf of Mexico, the U.S. Gulf states share a low-energy coast—at least between hurricanes!

Waves can affect the coast only where they strike, so erosion is concentrated near average sea level. A shore with little tidal variation can erode quickly because the wave action is concentrated near one level for longer times. Low-energy coasts protected by offshore islands usually erode slowly, as do areas below the low-tide line. Some erosion does occur below the surface because of the orbital motion of water in waves, but even the largest waves have little erosive effect at depths greater than about 15 meters (50 feet) below average sea level. Cliffs above shore are subject to pounding either directly from waves or by rocks hurled by waves.

Features of Erosional Coasts

Erosive forces can produce a wave-cut shore that shows some or all of the features illustrated in **Figure 11.4.** Note the complex, small-scale irregularities of this rocky coastline. **Sea cliffs** slope abruptly from land into the ocean, their steepness usually resulting from the collapse of undercut notches. The position of the sea cliffs marks the shoreward limit of marine erosion on a coast. The parade of waves cuts **sea caves** into the cliffs at local zones of weakness in the rocks. Most sea caves are accessible only at low tide. A blowhole can form if erosion follows a zone of weakness upward to the top of the cliff. When the tide is at just the right height, spray can blast from the fissure as waves crash into the cliff. Offshore features of rocky coasts can include natural arches, rock stacks, and a smooth, nearly level **wave-cut platform** just offshore, which marks the submerged limit of rapid marine erosion. Much of the debris removed from cliffs during the formation of these structures is deposited in the quieter water farther offshore, but some can rest at the bottom of the cliffs as exposed beaches. As we shall soon see, broad beaches are often features of depositional coasts.

Shore Straightening

The first effect of marine erosion on a newly exposed coast is to intensify the irregularity of the coastline. This happens because coastal rocks are usually not uniform in

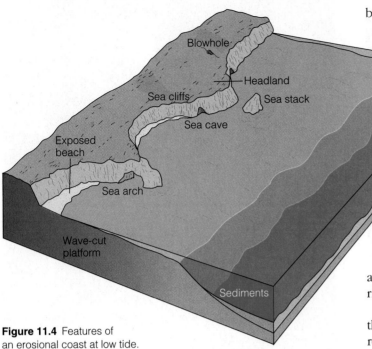

Figure 11.4 Features of an erosional coast at low tide.

beaches in the relatively calm bays. As erosion continues, the deposits may eventually protect the base of the shore cliffs from the waves. Coastal irregularities are thus smoothed with the passage of time. As you might expect, straightening occurs most rapidly on high-energy coasts.

Land Erosion Coasts

When sea level was lower during the last glaciations, rivers cut across the land and eroded sediment to form coastal river valleys. When higher sea level returned, the valleys were flooded, or *drowned*, with seawater. Sydney Harbour (see the chapter opener), Chesapeake Bay (**Figure 11.6**), and the Hudson River valley are examples of drowned river mouths.

Glaciers sometimes form in river valleys when rivers cut through the edges of continents at high latitudes. Deep, narrow bays known as **fjords** are often formed by tectonic forces and later modified by glaciers eroding valleys into deep, U-shaped troughs. Fjords are found in British Columbia, Greenland, Alaska, Norway, New Zealand, and other cold, mountainous places (**Figure 11.7**).

Volcanic Coasts

As we saw in Chapter 4, most islands that rise from the deep ocean are of volcanic origin. If the volcanism has been recent, the coasts of a volcanic island will consist of lobed lava flows extending seaward, common features in the Hawaiian Islands (**Figure 11.8**).

composition over long horizontal distances. Some hard rocks will resist erosion well, while softer rocks on the same coast may disappear almost overnight. (This explains the uneven character of the stacks, arches, and sea cliffs described above.)

Eventually, however, coastal erosion tends to produce a smooth shoreline. Because of wave refraction (see Figure 9.18), wave energy is focused onto headlands and away from bays by wave refraction (**Figure 11.5**). Sediment eroded from the headlands tends to collect as

Figure 11.5 Wave energy converges on headlands and diverges in the adjoining bays. The accumulation of sediment derived from the headland in the tranquil bays eventually smooths the contours of the shore.

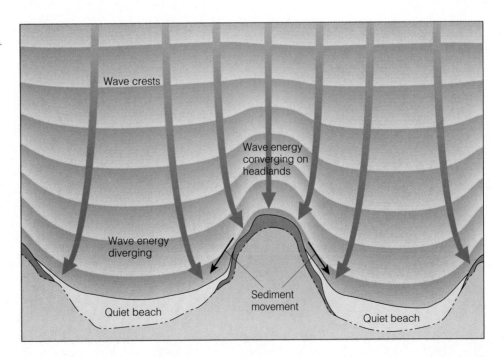

Figure 11.6 A false-color photograph of Chesapeake Bay taken from space. The complex bay is an example of a drowned river valley.

Figure 11.8 Lava flowing seaward from an eruption on the island of Hawaii forms a fresh coast exposed to erosion for the first time.

Figure 11.7 Milford Sound, a deep fjord on the western coast of New Zealand's South Island.

Depositional Coasts

The features found on depositional coasts are usually composed of sediments rather than rock. Accumulation and distribution of a layer of protective sediments along a coast can insulate that coast from rapid erosion; wave energy expended in churning overlying sediment particles cannot erode the underlying rock. So, with time, erosional shorelines can evolve into depositional ones. Unless the coast is rapidly rising or sinking, or unless other large-scale geological processes interfere, the inevitable process of erosion will tend to change the character of any coast from erosional to depositional.

Beaches

The most familiar feature of a depositional coast is the beach. A **beach** is a zone of loose particles that covers part or all of a shore. The landward limit of a beach may be vegetation, a sea cliff, relatively permanent sand dunes, or construction such as a seawall. The seaward limit occurs where sediment movement onshore and offshore ceases—a depth of about 10 meters (33 feet) at low tide. The continental United States has 17,672 kilometers (10,983 miles) of beaches, about 30% of the total shoreline (**Figure 11.9**).

Beaches result when sediment, usually sand, is transported to places suitable for deposition. Such places include the calm spots between headlands, shores sheltered by offshore islands, and regions with moderate surf or broad stretches of high-energy coasts. Sometimes the sediment is transported a very short distance—particles may simply fall from the cliff above and accumulate at the shoreline—but more often the sediment on a beach has been moved for long distances to its present location.

Wherever they are found, beaches are in a constant state of change. As we will see, they may be thought of as rivers of sand—zones of continuous sediment transport.

The Composition and Slope of Beaches

The material of beaches can range from boulders through cobbles, pebbles, and gravel to very fine silt. The rare black sand beaches of Hawaii are made of finely fragmented lava. Some beaches consist of shells and shell debris, or fragments of coral. Unfortunately, some also include large quantities of human junk: Glass or plastic beaches are not unknown. Cobble beaches can be very steep (occasionally with slopes in excess of 20°), but wide beaches of fine sand are sometimes nearly as flat as a parking lot.

In general, the flatter the beach, the finer the material from which it is made. The relationship between particle size and beach slope depends on wave energy, particle shape, and the porosity of the packed sediments. Water from waves washing onto a beach—the **swash**—carries particles onshore, increasing the beach's slope. If water returning to the ocean—the **backwash**—carries back the same amount of material as it delivered, the beach slope will be in equilibrium, so the beach will not become larger or steeper.

On fine-grain beaches, the ability of small, sharp-edged particles to interlock discourages water from percolating down into the beach itself, so water from waves runs quickly back down the beach, carrying surface particles toward the ocean. This process results in a very gradual slope. Broad, flat beaches also have a large area on which to dissipate wave energy, and they can provide a calm environment for the settling of fine sediment particles. In contrast, coarse particles

© Alan Hoelzle

Figure 11.9 A calm depositional shore—a beach scene on Florida's southwestern coast.

(gravel, pebbles) do not fit together well and readily allow water to drain between them. Onrushing water disappears *into* a beach made of coarse particles, so little water is left to rush down the slope, thereby minimizing the transport of sediments back to the ocean. Larger particles tend to build up at the back of the beach where they are thrown by large waves, increasing the steepness of the beach.

Beach Shape

Figure 11.10 shows a profile, or cross section, of a beach affected by small to moderate wave and tidal action. The scale is exaggerated vertically to show detail. The key feature of any beach is the **berm** (or berms), an accumulation of sediment that runs parallel to shore and marks the normal limit of sand deposition by wave action. The peaked top of the highest berm, called the **berm crest,** is usually the highest point on a beach. It corresponds to the shoreward limit of wave action during the most recent high tides. Inland of the berm crest, extending to the farthest point where beach sand has been deposited, is the **backshore.** The backshore is the relatively inactive portion of the beach, which may include windblown dunes and grasses. The **foreshore,** seaward of the berm crest, is the active zone of the beach, washed by waves during the daily rise and fall of the tides. It extends from the base of the berm—where a **beach scarp** (a vertical wall of variable height) is often carved by wave action at high tide—to the low-tide mark where the offshore zone begins. Below the low-tide mark, wave action, turbulent backwash, and longshore currents excavate a **longshore trough** parallel to shore. Irregular **longshore bars** (submerged or exposed accumulations of sand) complete the seaward profile.

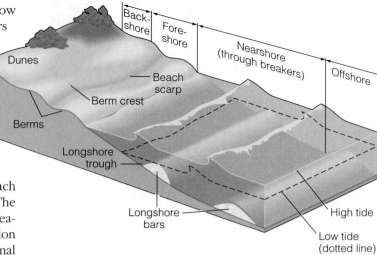

Figure 11.10 A typical beach profile.

This beach profile is only temporary, generated by the interplay of sediments, waves, and tides. Great storm waves can rearrange a beach in a day, transporting thousands of tons of sediment from the beach to hidden **sandbars** offshore. Most temperate-climate beaches undergo a seasonal transformation. Beaches are cut to a lower level in winter than in summer because higher waves accompany winter storms. Changes from summer to winter on a beach are shown in **Figure 11.11.**

Minor Beach Features

Small features are among the things that make beaches such interesting places. *Ripples* in the sand are caused by rushing currents (**Figure 11.12a**); similar structures can be seen in

Figure 11.11 The movement of sand on Carlsbad State Beach, California. (**a**) The beach in the summer of 1978. (**b**) The same beach in the spring of 1983 after a winter of severe storms. The sand has moved offshore, and coastal structures have been damaged.

a

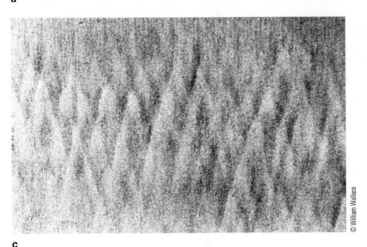

c

b

Figure 11.12 Minor features of beach sand. (**a**) Ripple marks. (**b**) The streamlike pattern of rill marks. (**c**) Backwash marks. Each V-shaped feature is caused by a small irregularity in the sand; each is about 15 centimeters (6 inches) long. (**d**) Layers of sediments are clearly seen in this freshly eroded beach face. The alternating layers were deposited by small waves during many ebb and flood tide cycles. Larger storm-driven waves have exposed the layers to view. The keys provide scale.

d

dry river bottoms or stream beds. *Rills* are small, branching surface depressions that channel water back to the ocean from a saturated beach during a falling tide (**Figure 11.12b**). Diamond-shaped *backwash marks* form when projecting shells, pebbles, or animals interrupt the backwash along the **low tide terrace** and cause uneven deposition or erosion of very fine sediment of a contrasting color (**Figure 11.12c**). Another result of uneven deposition can be seen in beach layering. A vertical slice into a calm beach will often reveal layer upon layer of sediments. The sediments are often separated by their density and may be of different colors and textures (**Figure 11.12d**).

Longshore Transport

If the submerged slope of the seafloor is steep, eroded sediments will soon drain to deeper waters. If the slope is not too steep, sediments will be transported along the coast by wave and current action. The movement of sediment (usually sand)

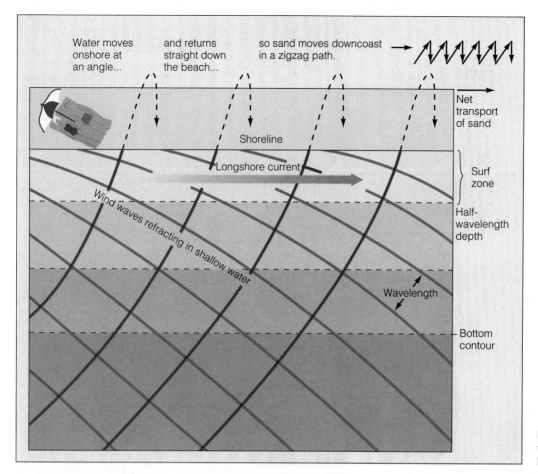

Figure 11.13 How wind-wave refraction and longshore current result in longshore transport of beach sand.

along the coast, driven by wave action, is referred to as **longshore drift.** Longshore drift occurs in two ways: the wave-driven movement of sand along the exposed beach, and the current-driven movement of sand in the surf zone just offshore (**Figure 11.13**).

Most wind waves approach at an angle and then refract in shallow water to break almost parallel to shore. Refraction is usually incomplete, however, and some angle remains when the waves break. If sediments have accumulated to form a beach, water from the breaking wave will rush up the beach at a slight angle but return to the ocean by running straight downhill under the influence of gravity. The sand grains disturbed by the wave will follow the water's path, moving up the beach at an angle but retreating down the beach straight down the slope. Net transport of the grains is longshore, parallel to the coast, away from the direction of the approaching waves. Net sand flow along the Pacific and Atlantic coasts of the United States is usually to the south because the waves that drive the transport system usually approach from the north, where storms most commonly occur.

Sediments are also transported in the surf zone in a **longshore current.** The waves breaking at a slight angle distribute a portion of their energy away from their direction of approach. This energy propels a narrow current in which sediment already suspended by wave action can be transported downcoast. The speed of the longshore current can approach 4 kilometers (about 2½ miles) per hour.

Sand moving in the wash of waves along the beach and sediments propelled in the longshore current just offshore are often joined by much greater loads of sediment brought to the coast by rivers. Net southward transport of all this material along the central California coast exceeds 230,000 cubic meters (300,000 cubic yards) per year. Typical figures for the Atlantic coast of the United States are about two-thirds of this value. Transport is to the south because southern-moving waves from northern storms provide most of the beach energy on the Atlantic and Pacific coasts.

Depositional coasts often support beaches, accumulations of loose particles. The shape and volume of beaches change as a function of wave energy and the balance of sediment input and removal.

Coastal Cells

Most new sand on a coast is brought in by rivers. The sand is moved parallel to the beach by longshore drift, and it is moved onshore and offshore at right angles to the beach as the seasons change. If a beach is stable in size, neither growing nor shrinking, the amount of new sand entering must be

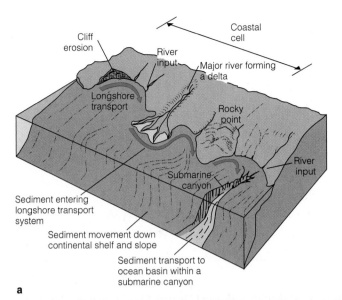

a

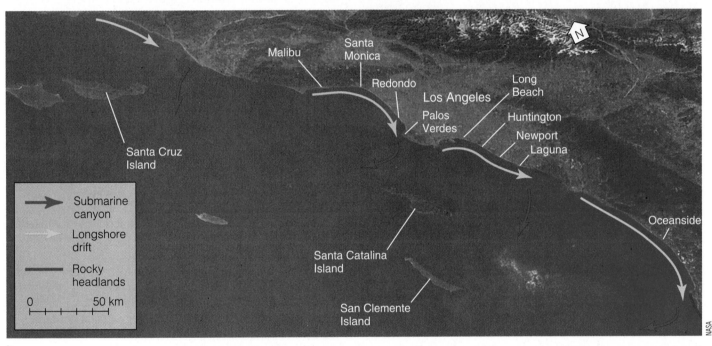

b

Figure 11.14 Coastal sediment transport cells. **(a)** The general features of coastal cells, in which sand is introduced by rivers, transported southward by the longshore drift, and trapped within the nearshore heads of submarine canyons. **(b)** A series of coastal cells in southern California. The yellow arrows show sand flowing toward the submarine canyons (shown in red).

balanced by the amount of old sand being removed. Sand that drifts below the reach of wave action is lost from the coast and may migrate farther out on the continental shelf. Some sand is driven by longshore currents into the nearshore heads of submarine canyons. Sand moving away from shore in these canyons sometimes forms impressive sandfalls (see Figure 4.16 for an example) and is lost from the beaches above. The bulk of this material is transported by gravity down the axis of the canyon and ultimately deposited on a submarine fan at the base of the slope.

The natural sector of a coastline in which sand *input* and sand *outflow* are balanced may be thought of as a **coastal cell.** The main features of such a cell are illustrated in **Figure 11.14a.** Coastal cells are usually bounded by submarine canyons that conduct sediments to the deep sea. Their size varies greatly. They are often very large along the relatively smooth, tectonically passive trailing edges of continents;

coastal cells along the southeastern coast of the United States, for example, are hundreds of kilometers long. On the active leading edge of the continent, they are smaller. Four cells exist in the 360 kilometers (225 miles) between southern California's Point Conception and the Mexican border. Each terminates in a submarine canyon at the downcoast end (**Figure 11.14b**).

Large-Scale Features of Depositional Coasts

Aside from beaches, depositional coasts exhibit some other large-scale features resulting from the deposit of sediments. Some of these features are illustrated in **Figure 11.15.**

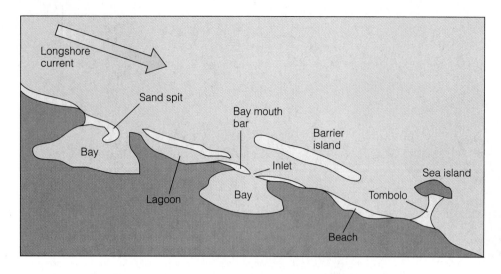

Figure 11.15 A composite diagram of the large-scale features of an imaginary depositional coast. Not all these features would be found in such close proximity on a real coast.

Sand Spits and Bay Mouth Bars

Sand spits are among the most common of these features. A sand spit forms where the longshore current slows as it clears a headland and approaches a quiet bay. The slower current in the mouth of the bay is unable to carry as much sediment, so sand and gravel are deposited in a line downcurrent of the headland. As can be seen in Figure 11.19, sand spits often have a curl at the tip. This is caused by the current-generating waves being refracted around the tip of the spit.

A **bay mouth bar** forms when a sand spit closes off a bay by attaching to a headland adjacent to the bay. The bay mouth bar protects the bay from waves and turbulence and encourages the accumulation of sediments there. An inlet—a passage to the ocean—may be cut through a bay mouth bar by tidal action, by water flowing from a river emptying into the bay, or by heavy storm rains. A bay mouth bar is shown in **Figure 11.16.**

Barrier Islands and Sea Islands

Depositional coasts can also develop narrow, exposed sandbars that are parallel to but separated from land. These are known as **barrier islands** (**Figure 11.17**). About 13% of the world's coasts are fringed with barrier islands.

Barrier islands can form when sediments accumulate on submerged rises parallel to the shoreline. Some islands off the Mississippi–Alabama coast developed in this way. Larger barrier islands are thought to form in a different way, however. Near the end of the last major rise in sea level, about 6,000 years ago, coastal plains near the edge of the continental shelf were fronted by lines of sand dunes. Rising sea level caused the ocean to break through the dunes and form a **lagoon**—a long, shallow body of seawater isolated from the ocean. The high lines of coastal dunes became islands. As sea level continued to rise, wave action caused the islands and lagoons to migrate landward. Most of the barrier islands off the southeastern coast of the United States originated in this way. They are still migrating slowly landward as sea level continues to rise.

Every year severe storms generate waves intense enough to erode barrier island beaches. The largest of these storms can generate waves that overwash the low islands. Runoff from rivers swollen by rains, coupled with water driven by wind waves and storm surge, can rapidly flood a lagoon and cut new inlets through barrier islands.

Figure 11.16 A bay mouth bar. The inlet is now closed, but increased river flow (from inland rainfall) or large waves combined with very high tides could break the bar.

Figure 11.17 Barrier islands off the Texas Gulf coast. (This photo, taken from space, is on a much larger scale than Figure 11.16.)

Despite these dangers, about 70 of the barrier islands have been commercially developed, and millions of people live on them. The most famous barrier islands include Atlantic City, New Jersey; Ocean City, Maryland; Miami Beach and Palm Beach, Florida; and Galveston, Texas. Roughly once every hundred years a winter storm has catastrophic effects on populated areas of Atlantic barrier islands, and the southeastern Atlantic and Gulf coasts must contend with occasional large hurricanes. The continuing subsidence of these passive coasts (combined with changes caused by commercial development and the ongoing rise in sea level) will undoubtedly cost lives and destroy property. **Figure 11.18** suggests the extent of the threat.

Unlike barrier islands, **sea islands** contain a firm central core that was part of the mainland when sea level was lower. The rising ocean separated these high points from land, and sedimentary processes surrounded them with beaches. Hilton Head, South Carolina, and Cumberland Island, Georgia, are sea islands. If the island is close to shore, a bridge of sediments called a **tombolo** may accumulate to connect the island to the mainland. Tombolos can also connect offshore rocky outcrops or volcanoes to the mainland. A sea island and a tombolo are shown in Figure 11.15.

Depositional coasts, especially along subsiding continental margins, often exhibit characteristic large-scale features.

Deltas

In a few places, sediments washing off the land have built out the coasts extensively. The shoreline in such places is much different from its configuration at the end of the last ice age. The most important of these coastal features are **deltas.**[1]

Deltas do not form at the mouth of every sediment-laden river. A broad continental shelf must be present to provide a platform on which sediment can accumulate. Tidal range is usually low, and waves and currents generally mild. There are no large deltas along the Atlantic coast of the United States because sediments that arrive at the coast are deposited in the sunken river mouths or dispersed by tides and currents. Also, there are no large deltas along the western margins of North and South America because these coasts are converging margins, where an oceanic plate is being subducted and the continental shelf is very narrow; sediment that would form a delta is swept down the continental slope or dispersed along the coast by waves. Deltas are most common on the low-energy shores of enclosed seas (where the tidal range is not extreme) and along the tectonically stable trailing edges of some continents. The largest deltas are those of the Gulf of Mexico (the Mississippi, **Figure 11.19a**), the Mediterranean Sea (the Nile), the Ganges–Brahmaputra river system in the Bay of Bengal (**Figure 11.19b**), and the huge

---

[1] The term is derived from the triangular shape of the capital Greek letter delta (Δ).

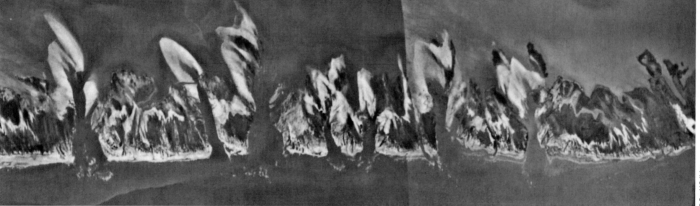

Figure 11.18 Barrier island modification: actual and potential. **(a)** The beach extending along the Matagorda Peninsula (Texas) barrier in September 1960. **(b)** The same area six days after the passage of Hurricane Carla in September 1961. The beach and island have been breached, and washover deltas are clearly seen. **(c)** Ocean City, Maryland, a developed barrier island. Host to 8 million visitors a year, this city (and others similarly situated) has no effective protection against flooding and damage from severe storms.

deltas formed by the rivers of China that empty into the South China Sea.

The shape of a delta represents a balance between the accumulation of sediments and their removal by the ocean. For a delta to maintain its size or grow, the river must carry enough sediment to keep marine processes in check. The combined effects of waves, tides, and river flow determine the shape of a delta. *River-dominated deltas* are fed by a strong flow of fresh water and continental sediments, and they form in protected marginal seas. They terminate in a

well-developed set of *distributaries*—the split ends of the river—in a characteristic bird's-foot shape (as shown in the Mississippi). In *tide-dominated deltas*, freshwater discharge is overpowered by tidal currents that mold sediments into long islands parallel to the river flow and perpendicular to the trend of the coast. The largest tide-dominated delta has formed at the mouths of the Ganges–Brahmaputra river system. *Wave-dominated deltas* are generally smaller than either tide- or river-dominated deltas and have a smooth shoreline punctuated by beaches and sand dunes. Instead of a bird's-

Deltas form on broad continental shelves where rivers deposit sediment, tidal range is low, and wave and current action is generally mild.

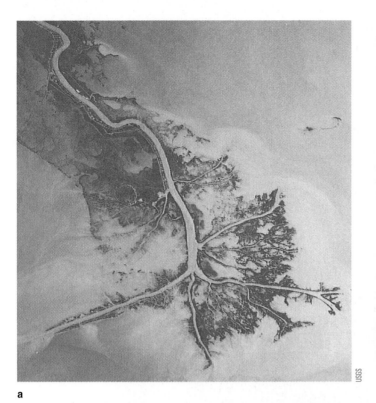

a

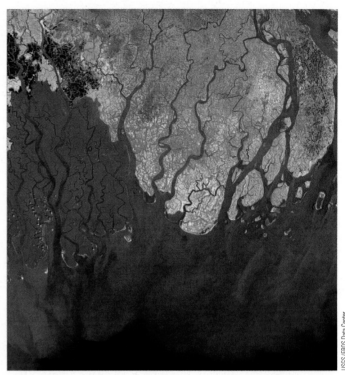

b

USGS

USGS/EROS Data Center

Figure 11.19 (**a**) River deltas form at places where sediment-laden rivers enter enclosed or semi-enclosed seas, where wave energy is limited. The bird's-foot shape of the Mississippi Delta is seen clearly in this photograph. Lobed and bird's-foot deltas form where deposition overwhelms the processes of coastal erosion and sediment transportation. The sediment-laden water looks brown or tan in this photograph taken from low orbit. (**b**) The mouths of the Ganges–Bramaputra river system on 28 February 2000. This tide-dominated delta, home to about 120 million people, is routinely flooded during cyclones and monsoon rains. Note the sediment (milky purple color) flowing from the delta into the Bay of Bengal.

foot pattern of distributaries, a wave-dominated delta has one main exit channel.

Coasts Formed by Biological Activity

Coasts can be extensively modified by the activities of animals and plants. The most dramatic modifications occur in the tropics, where coral polyps form reefs around volcanic islands or along the margin of a continent. The greatest of all reefs is the Australian Great Barrier Reef (**Figure 11.20**), which begins in the Torres Strait separating New Guinea and Australia and runs down the northeastern coast of Australia for 2,500 kilometers (1,500 miles). The reef is not a single object, but a composite of more than 3,000 individual coral reefs covering 350,000 square kilometers (135,000 square miles)—collectively the largest structure made by living organisms on Earth.

The Florida Keys—a series of low islands extending south of the tip of Florida—are an excellent example of a coral reef coast in the continental United States. How can a coral coast extend above sea level? The Keys are relatively high because they were formed during a time between glaciations when sea level stood about 6 meters (20 feet) higher than it does today. Some coral reef islands of the Pacific and Indian Oceans are much lower—a great storm can submerge and fracture them. Those that extend above sea level do so because chunks of reef margin are thrown toward the center of these islands by storm waves and winds. The accumulated blocks are cemented together by the limestone (calcium carbonate) they contain.

Other coasts have been formed by mangroves, trees that can grow in salt water. The coast of southwestern Florida has been extended and shaped by the activity of mangroves, whose root systems trap and hold sediments around the plant (**Figure 11.21**). The root complex forms an impenetrable barrier and safe haven for organisms around the base of the trees.

Coasts can be extensively modified by the actions of living organisms.

Figure 11.20 A small section of the Great Barrier Reef, Queensland, Australia. This coast has been extensively modified by biological activity.

Figure 11.21 A mangrove coast in Florida. Mangrove trees trap sediments, building and stabilizing the coast.

Estuaries

An **estuary** is a body of water partially surrounded by land, where fresh water from a river mixes with ocean water. Estuaries are areas of remarkable biological productivity and diversity. The coasts of the United States contain about 15,150 square kilometers (5,850 square miles) of estuarine waters. Chesapeake Bay, San Francisco Bay, and Puget Sound are all estuaries.

Classification of Estuaries

Estuaries are classified into four types depending on their origins:

- Drowned river mouths
- Fjords
- Bar-built
- Tectonic

All four types are shown in **Figure 11.22.**

Estuaries formed at drowned river mouths are common throughout the world, particularly along the Atlantic coast of the United States. Remember, sea level has risen about 125 meters (410 feet) since the end of the last major period of

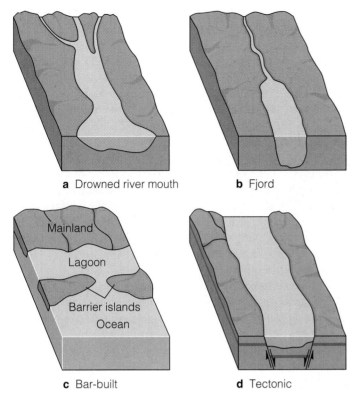

a Drowned river mouth   **b** Fjord

Mainland

Lagoon

Barrier islands

Ocean

c Bar-built   **d** Tectonic

Figure 11.22 Estuaries classified by their origins. (**a**) Drowned river mouths: Chesapeake Bay and the mouths of the James, York, and Susquehanna Rivers. (**b**) Fjords: Milford Sound (see Figure 11.7) and the Strait of Juan de Fuca in Washington state. (**c**) Bar-built: Albemarle and Pamlico Sounds in North Carolina. (**d**) Tectonic: San Francisco Bay and Tomales Bay.

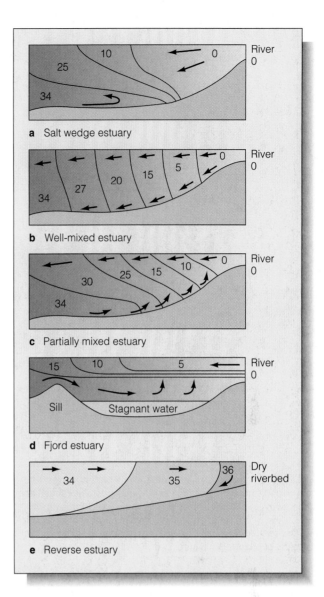

a Salt wedge estuary

b Well-mixed estuary

c Partially mixed estuary

d Fjord estuary

e Reverse estuary

Figure 11.23 Types of estuaries in vertical cross sections. The salinity values show the amount of mixing between fresh water (0‰) and seawater (34‰) in the various types. (**a**) Salt wedge estuary. (**b**) Well-mixed estuary. (**c**) Partially mixed estuary. (**d**) Fjord estuary. (**e**) Reverse estuary, in which evaporation plays a major role.

glaciation some 18,000 years ago, and this has resulted in the incursion of seawater into river mouths. The mouths of the York, James, and Susquehanna Rivers, and Chesapeake Bay, are examples of this type of estuary.

As Figure 11.7 suggests, fjords are steep, glacially eroded, U-shaped troughs. They are often about 300 to 400 meters (1,000 to 1,300 feet) deep, but typically terminate in a shallow lip, or sill, of terminal glacial deposits. In fjords with shallow sills, little vertical mixing occurs below the sill depth, and the bottom waters can become stagnant (look ahead to Figure 11.23d). In fjords with deeper sills, the bottom waters mix slowly with adjacent oceanic waters. While fjords are common in Norway, Greenland, New Zealand, Alaska, and western Canada, they are not common in the lower 48 states. The Strait of Juan de Fuca in Washington is a good example.

Bar-built estuaries form when a barrier island or a barrier spit is built parallel to the coast above sea level. Since these estuaries are shallow and usually have only a small inlet connecting them to the ocean, tidal action is limited. Waters in bar-built estuaries are mainly mixed by the wind. Albemarle and Pamlico Sounds in North Carolina, and Chincoteague Bay in Maryland, are bar-built estuaries.

Estuaries produced by tectonic processes are coastal indentations formed by faulting and local subsidence. Fresh water and seawater both flow into the depression, and an estuary results. San Francisco Bay is, in part, a tectonic estuary.

Characteristics of Estuaries

Three factors determine the characteristics of estuaries: the shape of the estuary, the volume of river flow at the head of the estuary, and the range of tides at the estuary's mouth. The mingling of waters of different densities, the rise and fall of the tide, and the variations in river flow—along with the actions of wind, ice, and the Coriolis effect—guarantee that patterns of water circulation in an estuary will be complex.

Estuaries are categorized by their circulation patterns. The simplest circulation patterns are found in **salt wedge estuaries,** which form where a rapidly flowing large river enters the ocean in an area where tidal range is low or moderate. The exiting fresh water holds back a wedge of intruding seawater (**Figure 11.23a**). Note that density differences

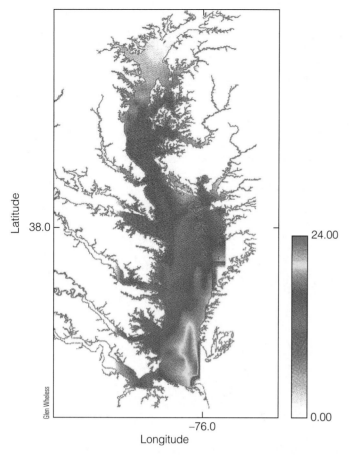

38.0 — Latitude

Glen Wheless

−76.0 — Longitude

24.00

0.00

Figure 11.24 Cheasapeake Bay, an example of a partially mixed estuary. The typical distribution of surface salinity in the estuary ranges from 28‰ at the mouth to 1‰ near the upper reaches. The Coriolis effect forces the inflowing salt water against the right (eastern) bank. (Notice how the 20‰ contour lines trend toward the right bank.) Compare this diagram to the photograph in Figure 11.6.

cause fresh water to flow over salt water. The seawater wedge retreats seaward at times of low tide or strong river flow, and it returns landward as the tide rises or when river flow diminishes. Some seawater from the wedge joins the seaward-flowing fresh water at the steeply sloped upper boundary of the wedge, and new seawater from the ocean replaces it. Nutrients and sediments from the ocean can enter the estuary in this way. Examples of salt wedge estuaries are the mouths of the Hudson and Mississippi Rivers.

A different pattern occurs where the river flows more slowly and the tidal range is moderate to high. As their name implies, **well-mixed estuaries** contain differing mixtures of fresh and salt water through most of their length. Tidal turbulence stirs the waters together as river runoff pushes the mixtures to sea. A well-mixed estuary is illustrated in **Figure 11.23b.** The mouth of the Columbia River is an example.

Deeper estuaries exposed to similar tidal conditions but greater river flow become **partially mixed estuaries.** Partially mixed estuaries share some of the properties of salt wedge and well-mixed estuaries. Note in **Figure 11.23c** the influx of seawater beneath a surface layer of fresh water flowing seaward; mixing occurs along the junction. Energy for mixing comes from both tidal turbulence and river flow.

England's Thames River, San Francisco Bay, and Chesapeake Bay are examples.

In well-mixed and partially mixed estuaries in the Northern Hemisphere, the incoming seawater will press against the right side of the estuary because of the Coriolis effect. Outflowing river water will also trend to the right of its direction of travel. This rightward drift can be seen in the contour of lines representing surface salinity in Chesapeake Bay (**Figure 11.24**).

Fjord estuaries form where glaciers have gouged steep U-shaped valleys below sea level. Typically, fjord estuaries have small surface areas, high river input, and little tidal mixing. River water tends to flow seaward at the surface with little contact with the seawater below (**Figure 11.23d**). In fjord estuaries with steep sills, a layer of stagnant water—cold water containing little oxygen and few nutrients—can form above the floor.

Reverse estuaries can form along arid coasts when rivers cease to flow. The evaporation of seawater in the uppermost reaches of these estuaries will cause water to flow from the ocean into the estuary, producing a gradient of *increasing* salinity from the ocean to the estuary's upper reaches (**Figure 11.23e**). Reverse estuaries are common on the Pacific coast of Mexico's Baja Peninsula and along the U.S. Gulf coast.

Estuaries can form at river mouths where fresh water mixes with seawater. Estuaries are among the most complex and biologically productive coasts.

The Value of Estuaries

Some of the oldest continuous civilizations have flourished in estuarine environments. The lower regions of the Tigris and Euphrates Rivers, the Po River Delta region of Italy, the Nile Delta, the mouths of the Ganges, and the lower Hwang Ho Valley have supported dense human habitation for thousands of years. Estuaries continue to be irresistibly attractive to developers. In areas of high population density, estuaries are routinely dredged to provide harbors, marinas, and recreational resources, and filled to make space for homes and agricultural land.

Estuaries often support a tremendous number of living organisms. The easy availability of nutrients and sunlight, protection from wave shock, and the presence of many habitats permit the growth of many species and individuals. Estuaries are frequently nurseries for marine animals; several species of perch, anchovy, and Pacific herring take advantage of the abundant food in estuaries during their first weeks of life. Unfortunately for their inhabitants, the high demand for development is incompatible with a healthy estuarine ecosystem. More than half the nation's estuaries and other wetlands have been lost. Of the original 870,000 square kilometers (215 million acres) of wetlands that once existed in the lower 48 states, only about 360,000 square kilometers (90 million acres) remain.

Characteristics of U.S. Coasts

Plate tectonic forces have had immense influence on the margins of continents, and the edges of the United States are no exception. The results of plate movement on the Pacific coast differ greatly from those on the Atlantic and Gulf coasts, primarily because the Pacific coast is near an active plate margin and the Atlantic and Gulf coasts are not.

The Pacific Coast

The Pacific coast is an actively rising margin on which volcanoes, earthquakes, and other indications of recent tectonic activity are easily observed. Pacific coast beaches are typically interrupted by jagged rocky headlands, volcanic intrusions, and the effects of submarine canyons. Wave-cut terraces are found as much as 400 meters (1,300 feet) above sea level in a number of places, evidence that tectonic uplift has exceeded the general rise in sea level over the past million years (**Figure 11.25**).

Most of the sediments on the Pacific coast originated from erosion of relatively young granitic or volcanic rocks of nearby mountains. The particles of quartz and feldspar that constitute most of the sand were transported to the shore by flowing rivers. The volume of sedimentary material transported to Pacific coast beaches from inland areas greatly exceeds the amount originating at the coastal cliffs. Deltas tend not to form at Pacific coast river mouths because the

continental shelf is narrow, river flow is generally low (except for the Columbia River), and beaches are usually high in wave energy. The predominant direction of long-shore drift is to the south because northern storms provide most of the wave energy.

The Atlantic Coast

The Atlantic coast is a passive margin, tectonically calm and subsiding because of its trailing position on the North American Plate. Subsidence along the coast has been considerable—3,000 meters (10,000 feet) over the last 150 million years. A deep layer of sediment has built up offshore, material that helped produce today's barrier islands. Relatively recent subsidence has been more important in shaping the present coast, however. Except for the coast of Maine (which is still in isostatic rebound after the recent departure of the glaciers), coastal sinking and rising sea level have combined to submerge some parts of the Atlantic coast at a rate of about 0.3 meter (1 foot) per century. This process has formed the huge flooded valleys of Chesapeake and Delaware Bays, the landward-migrating barrier islands, and the shrinking lowlands of Florida and Georgia.

Rocks to the north (in Maine, for example) are among the hardest and most resistant to erosion of any on the continent, so beaches are uncommon in Maine. But from New Jersey southward, the rocks are more easily fragmented and weathered, and beaches are much more common. As on the Pacific coast, sediments are transported coastward by rivers from eroding inland mountains, but the transported material is trapped in estuaries and therefore plays a less important role on beaches. Eastern beaches are typically formed of sediments from shores eroding nearby, or from the shoreward movement of offshore deposits laid down when the sea level was lower. The amount of sand in an area thus depends

Figure 11.25 Wave-cut terraces on San Clemente Island off the California coast.

in part on the resistance or susceptibility of nearby shores to erosion. Sand moves generally south on these beaches just as it does on the Pacific coast, but the volume of moving sand is less in the East.

As we have seen, glaciers have also contributed to the shaping of the northern part of the Atlantic coast: Large portions of Long Island and all of Cape Cod are remnants of debris deposited by glaciers.

The Gulf Coast

The Gulf coast experiences a smaller tidal range and—hurricanes excepted—a smaller average wave size than either the Pacific or Atlantic coast. Reduced longshore drift and an absence of interrupting submarine canyons allow the great volume of accumulated sediments from the Mississippi and other rivers to form large deltas, barrier islands, and a long raised "super berm" that prevents the ocean from inundating much of this sinking coast.

These are fortunate conditions because the rate of subsidence in the Gulf coast is greater than that for most of the Atlantic coast. Subsidence here is not the result of tectonic activity but rather due to sediment compaction, de-watering, and the removal of oil and natural gas. Sediment starvation and dredging have made the situation worse around some large cities. At Galveston, Texas, for example, sea level appears nearly 64 centimeters (25 inches) higher than it was a century ago, and parts of New Orleans are now 2 meters (6.6 feet) below sea level. As we have seen, the results of hurricanes at such places can be tragic. The protective natural berm can easily be breached, and floodwaters can surge far inland.

Human Interference in Coastal Processes

Beaches exist in a tenuous balance between accumulation and destruction. Human activity can tip the balance one way or the other. For example, consider the rocky **breakwater** shown in **Figure 11.26.** The breakwater interrupts the progress of waves to the beach, weakening the longshore current and allowing sand to accumulate there. Without dredging, the beach will eventually reach the breakwater and fill the small boat anchorage the breakwater was built to provide. This is a minor example of human alteration of a beach, yet it serves to introduce the growing problem of human influences on coastal processes.

a

b

Figure 11.26 Growth of a beach protected by a breakwater: Santa Monica, California. (**a**) The shoreline as it appeared in 1931. (**b**) The same shoreline in 1949 after the breakwater was built. The boat anchorage formed by the breakwater is filling with sand deposited by disruption of the longshore current.

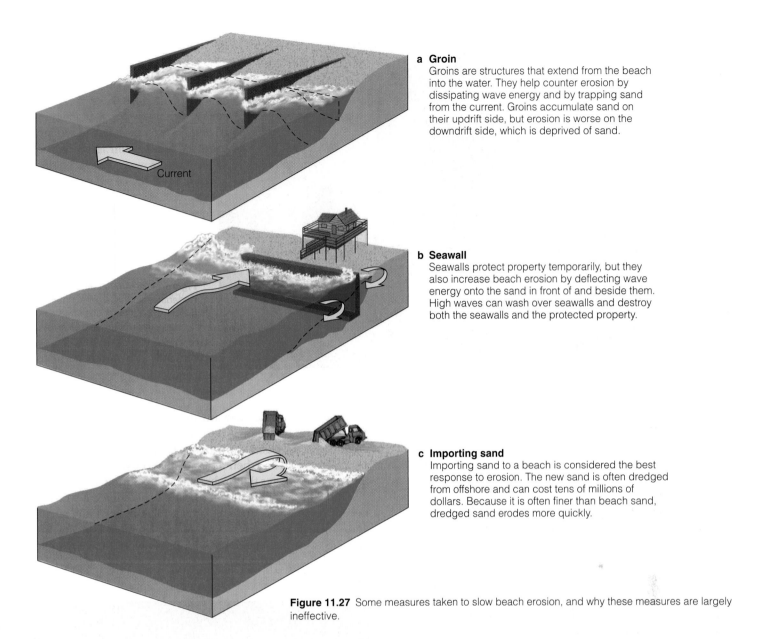

a Groin
Groins are structures that extend from the beach into the water. They help counter erosion by dissipating wave energy and by trapping sand from the current. Groins accumulate sand on their updrift side, but erosion is worse on the downdrift side, which is deprived of sand.

Current

b Seawall
Seawalls protect property temporarily, but they also increase beach erosion by deflecting wave energy onto the sand in front of and beside them. High waves can wash over seawalls and destroy both the seawalls and the protected property.

c Importing sand
Importing sand to a beach is considered the best response to erosion. The new sand is often dredged from offshore and can cost tens of millions of dollars. Because it is often finer than beach sand, dredged sand erodes more quickly.

Figure 11.27 Some measures taken to slow beach erosion, and why these measures are largely ineffective.

We often divert or dam rivers, build harbors, and develop property with surprisingly little understanding of the impact our actions will have on the adjacent coast. Our role then becomes that of powerless observers. Residents of eroding coasts can only accept the inevitable loss of their property to the attack of natural forces, but residents of coasts in which deposition exceeds erosion are sometimes presented with alternatives. The choices are almost never simple. For example, should rivers be dammed to control devastating floods? If the dams are built, they will trap sediments on their way from mountains to coast. Beaches within the coastal cell fed by the dammed river will shrink because the sand on which they depend (to replenish losses at the shore) is blocked. Alarmed coastal residents will then take steps to hang onto whatever sand remains. They may try to trap "their" beaches by erecting **groins**—short extensions of rock or other material placed at right angles to longshore drift, to stop the longshore transport of sediments. This temporary expedient usually accelerates erosion downcoast (**Figure 11.27a**). Diminished beaches then expose shore cliffs to accelerated erosion. Wind-wave energy that would have harmlessly churned sand grains now speeds the destruction of natural and artificial structures. Seawalls don't help either (**Figure 11.27b**). They increase beach erosion by deflecting wave energy onto the sand. Churning by this increased energy eventually undermines the seawall, causing it to collapse. The importation of sand trapped behind dams (or from other sources) is also only a temporary—and very expensive—expedient (**Figure 11.27c**). Scenes like that shown in **Figure 11.28** will be more common.

What are the implications of these unlooked-for sand movements? Douglas Inman, Director of the Center for Coastal Studies at Scripps Institution of Oceanography, believes that *at least 20% of the beach-bounded coastline of the United States is in danger of serious or catastrophic alteration*. On the West Coast a 30-year period of relatively mild weather may be ending. During this time, people felt it was safe to build close to the shore. Increased dam building and breakwater,

shares the same basic underlying life processes; life almost certainly began in the ocean. Life and Earth have changed together, generation by generation, over some 4 billion years.

There are perhaps 100 million different species (kinds) of living things on Earth. Despite their astonishing diversity in form and lifestyle, all species share the same underlying mechanisms for capturing and storing energy, manufacturing proteins, and transmitting information between generations. Biologists know there is nothing special about the atoms or energy of life, no way to distinguish the physical components of life from their nonliving counterparts. What *does* distinguish life from nonlife is the ability of living things to capture, store, and transmit energy—and the ability to reproduce.

Life on Earth is notable for both its unity and diversity: *diversity* because there may be as many as 100 million different species (kinds) of living things on Earth, *unity* because each species shares the same underlying mechanisms for basic life processes.

Energy and Marine Life

The energy most marine organisms need to function comes directly or indirectly from the sun. The sun produces enormous quantities of energy—some in the form of visible light, a tiny portion of which strikes Earth. Only about 1 part in 2,000 of the light that reaches Earth's surface is captured by organisms, but that "small" input of energy powers nearly all the growth and activity of living things. Light energy from the sun is trapped by chlorophyll in organisms called *producers* (certain bacteria, algae, and green plants) and changed into chemical energy. The chemical energy is used to build simple carbohydrates and other organic molecules—**food**—which are then used by the producer or eaten by animals (or other organisms) called *consumers*. Because light energy is used to synthesize molecules rich in stored energy, the process is called **photosynthesis.** Here is a general formula for photosynthesis:

$$6CO_2 + 6H_2O \xrightarrow{\text{sunlight}} C_6H_{12}O_6 + 6O_2$$

$6CO_2$		$6H_2O$			$C_6H_{12}O_6$		$6O_2$
6 molecules of carbon dioxide	+	6 molecules of water	(yields by photo-synthesis)		**a molecule of glucose (food, a carbohydrate)**	+	6 molecules of oxygen

The energy is released when food such as glucose is used for growth, repair, movement, reproduction, and the other functions of organisms. The breakdown of food eventually produces waste heat, which flows away from Earth into the coldness of space. This one-way flow of energy is shown in **Figure 12.1.**

Photosynthesis is the dominant method of binding energy into carbohydrates, but there is another. **Chemosynthesis,** employed by a few relatively simple forms of life, is the pro-

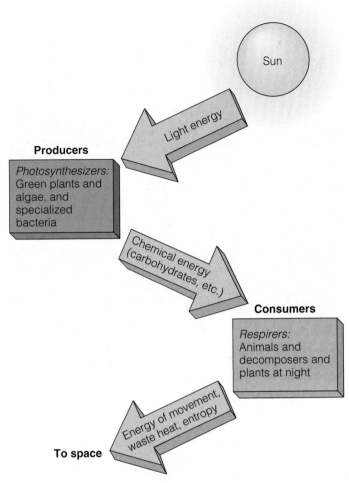

Figure 12.1 The flow of energy through living systems. At each step, energy is degraded (transformed into a less useful form).

duction of usable energy directly from energy-rich inorganic molecules available in the environment rather than from the sun. As we will see in Chapter 14, some unusual forms of marine life depend on chemosynthesis. Overall, chemosynthetic production of food in the ocean is very small in comparison with photosynthetic production.

Producers assemble food molecules using energy from the sun (photosynthesis) or from energy-rich inorganic molecules (chemosynthesis).

Primary Productivity **12.3**

The synthesis of organic materials from inorganic substances by photosynthesis or chemosynthesis is called **primary productivity** (**Figure 12.2**). Primary productivity is expressed in *grams of carbon bound into organic material per square meter of ocean surface area per year* ($gC/m^2/yr$). The organic

material produced is usually glucose, a carbohydrate. The source of carbon for glucose is dissolved CO_2. *Phytoplankton*—minute, drifting photosynthetic organisms you will meet in Chapter 13—produce 90% to 96% of oceanic carbohydrates. Seaweeds—larger marine plants discussed in Chapter 14—contribute 2% to 5% of the ocean's primary productivity. Chemosynthetic organisms probably account for 2% to 5% of the total. Though estimates vary widely, recent studies suggest that total ocean productivity ranges from 75 to 150 grams of carbon bound into carbohydrates per square meter of ocean surface per year (75 to 150 $gC/m^2/yr$). (For comparison, a well-tended alfalfa field produces about 1,600 $gC/m^2/yr$.)

Productivity is expressed in grams of carbon bound into carbohydrates (food) per square meter of ocean surface per year ($gC/m^2/yr$).

How does marine net productivity compare with terrestrial net productivity? Recent research suggests the global net productivity in *marine* ecosystems is 35 to 50 billion metric tons of carbon bound into carbohydrates per year; global net *terrestrial* productivity is roughly similar at 50 to 70 billion metric tons per year.[1] However, the total producer **biomass** (the mass of living tissue) in the ocean is only 1 to 2 billion metric tons, compared with 600 to 1,000 billion metric tons of living biomass on land! Clearly, marine producers are *much* more efficient in their production of food than are their land-based counterparts.

The total mass of a **primary producer** is assumed to be about ten times the mass of the carbon it has bound into carbohydrates. So, a primary productivity of 100 $gC/m^2/yr$ represents the yearly growth of about 1,000 grams of primary producers for each square meter of ocean surface (**Figure 12.3**). Between 35 and 50 billion metric tons of carbon is believed to be bound into carbohydrates in the ocean each year, so between 350 and 500 billion metric tons of marine plants and plantlike organisms are produced annually. Each year this vast bulk is consumed by the metabolic activity of the producers themselves and by the consumers that graze on them. The component atoms are then reassembled by photosynthesis into carbohydrates in a continuous solar-powered cycle.

Primary producers—autotrophs—are organisms that synthesize food from inorganic substances by photosynthesis and chemosynthesis.

---

[1] 1 billion metric tons = 1.1 billion tons

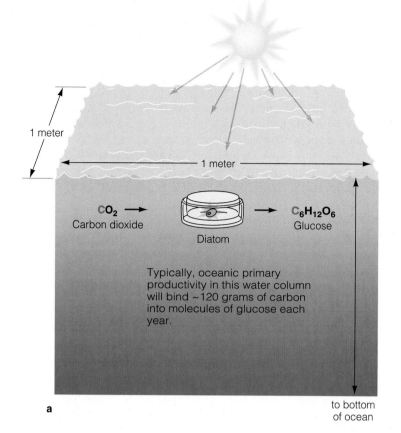

CO_2
Carbon dioxide

Diatom

$C_6H_{12}O_6$
Glucose

1 meter

1 meter

Typically, oceanic primary productivity in this water column will bind ~120 grams of carbon into molecules of glucose each year.

to bottom of ocean

a

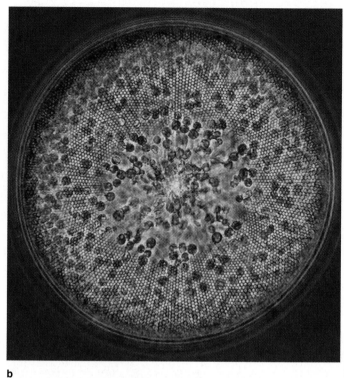

b

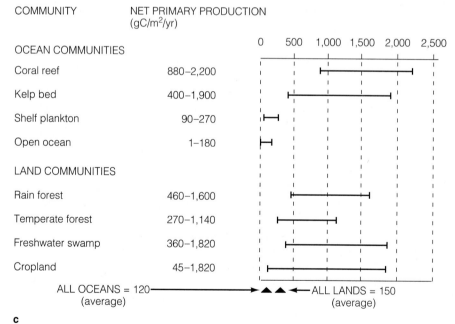

COMMUNITY	NET PRIMARY PRODUCTION (gC/m²/yr)
OCEAN COMMUNITIES	
Coral reef	880–2,200
Kelp bed	400–1,900
Shelf plankton	90–270
Open ocean	1–180
LAND COMMUNITIES	
Rain forest	460–1,600
Temperate forest	270–1,140
Freshwater swamp	360–1,820
Cropland	45–1,820

0   500   1,000   1,500   2,000   2,500

ALL OCEANS = 120 (average)   ALL LANDS = 150 (average)

c

Figure 12.2 (**a**) Oceanic productivity—the incorporation of carbon atoms into carbohydrates—is measured in grams of carbon bound into carbohydrates per square meter of ocean surface area per year (gC/m²/yr). (**b**) The diatom *Coscinodiscus*, an important marine primary producer. In a very bright light this single-celled plantlike organism would barely be visible to the unaided eye. (**c**) Net annual primary productivity in some marine and terrestrial communities.

Wim van Egmond

Feeding (Trophic) Relationships

Photosynthetic and chemosynthetic organisms can be called either primary producers or **autotrophs** because they make their own food. The bodies of autotrophs are rich sources of chemical energy for any organisms capable of consuming them. **Heterotrophs** are organisms that must consume other organisms because they are unable to synthesize their own

food molecules. Some heterotrophs consume autotrophs, and some consume other heterotrophs.

We can label organisms by their positions in a "who eats whom" feeding hierarchy called a **trophic pyramid.** The primary producers shown at the bottom of the pyramid in **Figure 12.4** are mostly chlorophyll-containing photosynthesizers. The animal heterotrophs that eat them are called **primary consumers** (or herbivores), the animals that eat them

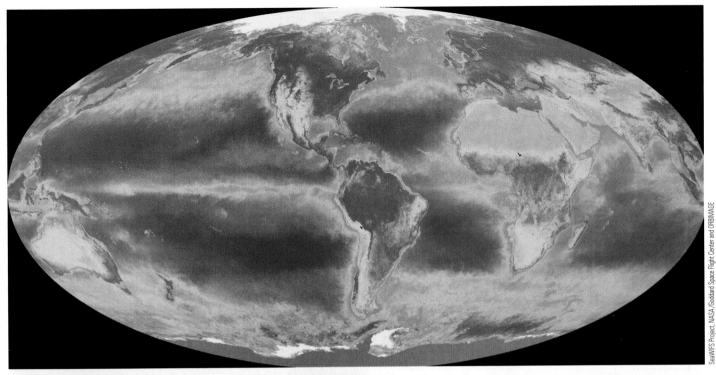

SeaWiFS Project, NASA /Goddard Space Flight Center and ORBIMAGE

Chlorophyll a Concentration mg/m3

Figure 12.3 Oceanic productivity can be observed from space. NASA's *SeaWiFS* satellite, launched in 1997, can detect the amount of chlorophyll in ocean surface water. Chlorophyll content provides an estimate of productivity. Red, yellow, and green areas indicate high primary productivity; blue areas indicate low. This image was derived from measurements made from September 1997 through August 1998.

Trophic Level

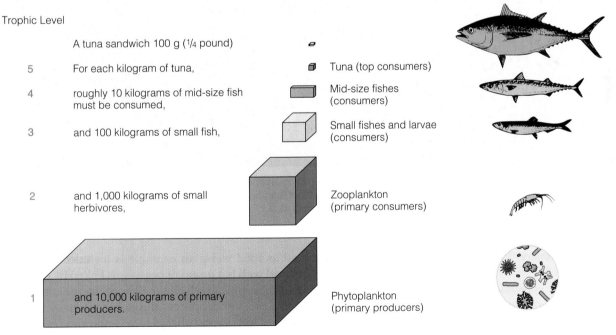

	A tuna sandwich 100 g (¼ pound)	
5	For each kilogram of tuna,	Tuna (top consumers)
4	roughly 10 kilograms of mid-size fish must be consumed,	Mid-size fishes (consumers)
3	and 100 kilograms of small fish,	Small fishes and larvae (consumers)
2	and 1,000 kilograms of small herbivores,	Zooplankton (primary consumers)
1	and 10,000 kilograms of primary producers.	Phytoplankton (primary producers)

Figure 12.4 A generalized trophic pyramid. How many kilograms of primary producers are necessary to maintain 1 kilogram of tuna, a top carnivore? What is required for an average tuna sandwich? Using the trophic pyramid model shown here, you can see that 1 kilogram of tuna at the fifth trophic level (the fifth feeding step of the pyramid) is supported by 10 kilograms of mid-size fish at the fourth, which in turn is supported by 100 kilograms of small fish at the third, which have fed on 1,000 kilograms of zooplankton (primary consumers) at the second, which have eaten 10,000 kilograms of phytoplankton (small autotrophs, primary producers) at the first. The quarter-pound tuna sandwich has a long and energetic history. (These figures have been rounded off to illustrate the general principle. The actual measurements are difficult to make and quite variable.)

interact at a particular location. A **population** is a group of organisms of the same species occupying a specific area. The location of a community, and the populations that comprise it, depend on the physical and biological characteristics of that living space. In the next two chapters we will survey the organisms in the ocean's two great realms: the pelagic and benthic environments. Pelagic organisms live suspended in the water; benthic organisms live on or in the ocean bottom.

The largest marine community—and the most sparsely populated—is the pelagic community lying within the uniform mass of permanently dark water between the sunlit surface and the deep bottom. Few animals live there because so little food is available, but those organisms that survive are among the strangest in the ocean. Opportunities for feeding in the deep open-ocean community are few and far between, so some animals are able to consume prey larger than themselves should the occasion arise. Because so few animals are present, mating is also a rare event; in a few species males and females become permanently bonded during their first encounter, the male burrowing into the female's body for a life-long free ride.

In contrast, the smallest obvious marine communities may be those benthic communities established against solitary rocks on an otherwise flat, featureless seabed. Drifting larvae will colonize the place; the established community can seem an oasis of life and activity in an otherwise static sedimentary desert. Seaweeds will grow, worms will burrow, snails will scrape algae from the hard surfaces, and small fishes will nestle among crevices. Hundreds of small plants and animals can live their lives within a meter of one another, interacting in a compact solitary community with no similar environment available for thousands of meters. The larvae of the next generation drift away with little chance of finding a suitable place to carry on their lives. Microscopic communities also exist—interacting populations can exist on a single grain of sand or on one decomposing fish scale.

Organisms are distributed throughout the marine environment in specific communities—groups of interacting producers, consumers, and recyclers that share a common living space.

Organisms Within Communities

There are many different places to live and many different "jobs" for organisms within even a simple community. A **habitat** is an organism's "address" within its community, its physical *location*. Each habitat has a degree of environmental uniformity. An organism's **niche** is its "occupation" within that habitat, its relationship to food and enemies, an expression of what the organism is *doing*. For example, the small fishes living among the coral heads in a coral reef commu-

nity share the same habitat, but each species has a slightly different niche. Each population in the community has a different "job" for which its shape, size, color, behavior, feeding habits, and other characteristics particularly suit it.

Competition

The availability of resources such as food, light, and space within a community determines the number and composition of the populations of organisms within that community. Competition for the necessities of life may occur within the community between members of the *same* population or between members of *different* populations. Subtle swings in physical or biological factors may give one population the advantage for a time, then shift to favor another.

When *members of the same population* (all members of the same species) compete with each other, some individuals will be larger, stronger, or more adept at gathering food, avoiding enemies, or mating. These animals tend to prosper, forcing their less successful relatives to emigrate, fight, or die in the course of competition. This kind of competition continually adjusts the characteristics of individuals in a population to their environment.

When *members of different populations* compete, one population may be so successful in its "job" that it eliminates competing populations. In a stable community, two populations cannot occupy the same niche for long. Eventually the more effective competitor overwhelms the less effective one. For example, the little barnacle *Chthamalus* lives on the uppermost rocks in many intertidal communities, while the larger barnacle *Balanus* lives on the lower rocks (**Figure 12.13**). Planktonic larvae of both species can attach themselves to rocks anywhere in the intertidal zone and begin to grow. In the lower zone the faster-growing *Balanus* push the weaker *Chthamalus* off the rocks, while at higher positions *Balanus* cannot survive because it is less resistant to drying

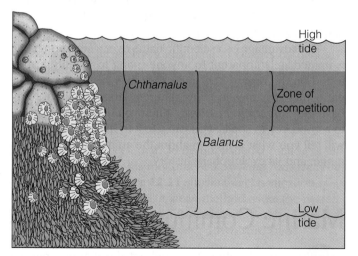

Figure 12.13 Competition between two species of barnacles prevents either from occupying as much of the intertidal zone as possible. The central zone of overlap indicates the area where *Chthamalus* and *Balanus* compete for food and space.

and exposure than the tough little *Chthamalus*. At the top and bottom of their distribution, the two species do not compete for food or space. The competition at the intersection of their ranges prevents each species from occupying as much of the habitat as might otherwise be possible.

As we have seen, physical and biological factors affect the numbers and positions of organisms in a community. The number of individuals per unit area (or volume) is known as the **population density**—rare individuals have a much lower population density than dominant ones. In general, *more* different species exist in benign habitats where physical factors stay near optimal values (like a coral reef or rain forest), and *fewer* species exist in rigorous habitats where physical factors range to extremes (like a beach or a desert). That is, "easy" habitats typically have high **species diversity** (they contain more species in more niches within a given area) and harsh habitats usually have lower species diversity. Relatively few species can cope with the stressful environment of the polar ocean, for example, but many species have adapted to the relatively benevolent environment of a tropical reef.

Change in Marine Communities

Like the organisms that comprise them, communities change over time. The slow changes associated with seafloor spreading, climate cycles, atmospheric composition, or newly evolved species have shaped this generally slow evolution. As on land, the species, community composition, and location of a marine community are changed by the environmental factors to which members of the community are exposed. Communities themselves can gradually modify the physical aspects of their environment: A coral reef is an extreme example. The massive accumulation of coral and sediments within the reef can alter current patterns, influence ocean temperature, and change the proportions of dissolved gases.

But rapid changes can also occur in marine communities. A natural catastrophe—a volcano eruption, a landslide that blocks a river, or the collision of an asteroid with Earth, for example—can disrupt a community. Similarly, human activities—such as altering an estuary by damming a river, dumping excess nutrients into a nearshore area, or stressing organisms with toxic wastes—can cause rapid, disruptive changes. The composition of offshore communities changes abruptly near new sewage outfalls, for example.

A stable, long-established community is known as a **climax community.** This self-perpetuating aggregation of species tends not to change unless disrupted by severe external forces such as violent storms, significant changes in current patterns, epidemic diseases, or an influx of large amounts of fresh water or pollutants. A disrupted climax community can be re-established through the process of **succession,** the orderly changes of a community's species composition from temporary inhabitants to long-term inhabitants. Disruption makes the environment more hostile to the original species, but destruction of species in the original community leaves open habitats and niches. A few highly tolerant species will move into the area, eventually drawing in other species that depend on them. If the environment is permanently changed by the disruption, a climax community different from the previous one will be established.

Mass Extinctions

The environment for life changes as time passes, and the changes are not always gradual. The biological history of Earth has been interrupted—catastrophically—at least six times in the last 450 million years. In these events, known as **mass extinctions,** a great many species died off simultaneously (in geological terms). Scientists are not certain of the causes of the mass extinctions, but leading candidates for a couple of these events include the collision of Earth with an asteroid or comet.

Flocks of asteroids orbit the sun along with Earth and the other planets (**Figure 12.14a**). Some of these asteroids have orbits that cross our own, and meetings are inevitable. Earth and its neighbors are pocked with impact craters as evidence of these meetings (**Figure 12.14b**). The consequences to Earth of a collision with even a small asteroid are all but unimaginable. An asteroid only 10 kilometers (6 miles) in diameter would strike Earth with an energy equivalent to the explosion of half a trillion tons of TNT (**Figure 12.14c**). More than 100 million metric tons of Earth's crust would be thrown into the atmosphere, obscuring the sun for decades and causing acid rain that would pollute the planet's surface. Concussive shock waves would shatter structures, crush large organisms, and trigger earthquakes for a radius of hundreds of kilometers. If the impact occurred in the Atlantic Ocean—say, 1,600 kilometers (1,000 miles) east of Bermuda—the resulting tsunami would wash away the resort islands and swamp most of Florida. Boston would be struck by a 100-meter (300-foot) wall of water. A hole more than 25 kilometers (16 miles) wide and perhaps 10 kilometers (6 miles) deep would mark the point of impact. Clouds of fine particles, accelerated to escape velocity, would travel around the sun in orbits that would intersect Earth's; a steady rain of fine debris might fall for tens of thousands of years.

These kinds of cataclysmic events have been disquietingly common in Earth's past. Where are the craters? As you may recall from the discussion of plate tectonics in Chapter 3, much of the ocean floor has been recycled by the movement of lithospheric plates, so any undersea impact craters more than about 100 million years old have disappeared. The distortion and erosion of continents have obscured the outlines of ancient craters on land, although they are more

a

Figure 12.14 (**a**) What one asteroid looks like, courtesy of the *Galileo* spacecraft en route to study Jupiter. This asteroid would extend from Washington, D.C., halfway to Baltimore. Asteroids are rocky, metallic bodies with diameters ranging from a few meters to 1,000 kilometers (600 miles). Most were swept into the planets during their formation, but about 6,000 large asteroids are still orbiting the sun in a belt between Mars and Jupiter. Unfortunately, the orbits of many dozens of others take them across Earth's orbit. (**b**) What an impact crater looks like. Aerial view of Manicougan Crater, Québec, where an asteroid struck Earth some 210 million years ago. The crater is about the size of Rhode Island. (**c**) An artist's conception of a large cometary nucleus, 10 kilometers (6 miles) across, striking Earth 65 million years ago. The cataclysmic explosion is thought to have propelled shock waves and huge clouds of seabed and crust all over Earth, producing a time of cold and dark that contributed to the extinction of many species, including the dinosaurs.

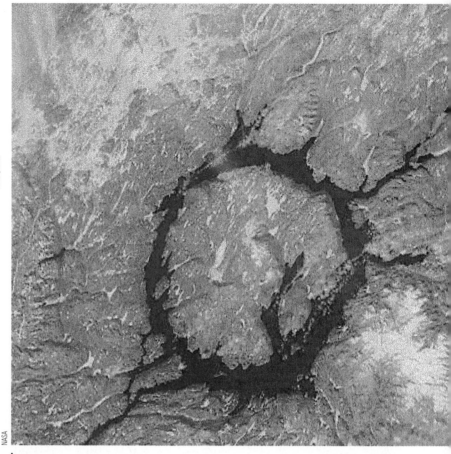

b

c

Table 12.2 The Six Great Mass Extinctions			
Geological Period in Which Extinction Occurred	Millions of Years Ago	Percentage of Marine Extinctions for:[a]	
		Families	Genera
Late Ordovician	435	27	57
Late Devonian	365	19	50
Late Permian	245	57	83
Late Triassic[b]	220	23	48
Late Cretaceous[b]	65	17	50
Late Eocene	35	2	16

[a]Rough estimate of percentage of all families and genera of marine animals with hard parts (so we have fossil evidence of their existence) rendered extinct; numbers are given to the nearest 5 million years (see column 2).
[b]Mass extinctions for which asteroid or comet strikes may be responsible.
Source: C. Sagan and A. Druyan, *Comet* (New York: Random House, 1985).

readily visible from space than from the surface—especially if we know what to look for (as in Figure 12.14b). Evidence for one massive impact has been bolstered by the discovery of a thin, worldwide layer of iridium-rich continental rock dated at the boundary between the Cretaceous and Tertiary periods (see Appendix II). Iridium is rare on Earth but common in asteroids. The thin iridium-rich layer may have formed from the dust settling after a collision some 65 million years ago.

As can be seen in **Table 12.2,** vast numbers of marine organisms have perished in mass extinctions. (Extinctions of land families and genera are thought to have been roughly comparable.) At the end of the Cretaceous period, about 65 million years ago, almost one of five families, half the genera, and three-quarters of the species disappeared. Many scientists believe the asteroid impact triggered the mass extinction. Clearly these tumultuous interruptions do not represent biological business as usual. In a few instances, rocky visitors from space appear to have massively disrupted the environment for life on this planet. The animals and plants, bacteria and single-celled organisms we see on Earth are descendants of the survivors.

Marine communities change as time passes, sometimes slowly growing or shrinking, and sometimes disappearing catastrophically.

QUESTIONS FROM STUDENTS

1. Total terrestrial primary productivity appears to be roughly the same as total marine primary productivity. But the total biomass of producers in the ocean is at least 300 times smaller (and maybe 1,000 times smaller) than the total biomass of producers on land! How can that be? How can terrestrial and marine productivity be so similar?

Because of the astonishing efficiency of small marine autotrophs (phytoplankton), the turnover time of molecules is exceedingly fast—nutrients and carbohydrates are cycled with speed and efficiency. There may not be nearly as great a biomass of producers in the ocean, but they appear to be *very* busy indeed!

2. What would terrestrial productivity and food chains be like if 99% of the land environment were too dark for successful photosynthesis? In other words, what if land plants had to contend with an environment as dark as the ocean?

Productivity would plummet, of course. If the land were lighted as the ocean is, Milne (1995) estimated that all land animals would be dependent on the plant growth in a lighted area the size of the United States east of the Mississippi River. Life on land would be much less abundant than at present, and animals would be concentrated in or around the lighted region. Because land plants are less efficient than aquatic ones (in part because of the infrastructure needed for support, to pump juices around their bodies, and to hold leaves to the light and air), total land productivity would be reduced to less than 1% of present values.

3. If humans have a fluid much like seawater bathing their cells, why can't we drink seawater and survive?

Human cells function in an environment hypotonic to seawater; that is, blood plasma is less saline than seawater. Drinking seawater therefore causes water to leave the intestinal walls, flood the intestine, and leave the body. There is a net loss via intestine or kidneys even if the seawater is diluted with fresh water before drinking. Moral: Never drink seawater in a survival situation at sea, and never dilute fresh water with seawater (in any proportion) to stretch your supply.

4. What proportion of total world productivity is achieved by chemosynthesis?

An interesting and controversial question! Chemosynthesis occurs in the marine environment primarily at hydrothermal vents associated with the 65,000-kilometer (41,000-mile) network of mid-ocean ridges. Specialized communities of organisms have evolved in the zones of warm or hot mineral-rich water percolating through the seafloor in these places. Once thought to be rare and isolated, vent communities have now been discovered in broadly separate locations and may be quite common to mid-ocean ridges. The trophic pyramid in these deep communities is based on chemosynthetic bacteria. Some biological oceanographers have suggested that if vent communities are abundant on ridges, or if bacteria can grow at warmer temperatures than previously thought, chemosynthesis there may account for a much larger proportion of total oceanic productivity than was previously thought. And recent discoveries have shown chemosynthetic communities deep *within* the seabed itself!

5. What's the difference between the photic zone and the euphotic zone?

The photic zone is the sunlit uppermost layer of the ocean. The euphotic zone is part of the photic zone. Within the euphotic zone, autotrophic organisms receive enough sunlight to make enough food (by photosynthesis) for their lives to continue. During daylight hours light is available below the euphotic zone, but it is not bright enough to allow the photosynthetic machinery of autotrophs to produce enough food to sustain them indefinitely. Unless they rise into the euphotic zone, they will eventually die.

CHAPTER SUMMARY

Life on Earth is notable for both its unity and diversity: *diversity* because there may be as many as 100 million different species (kinds) of living things on Earth, *unity* because each species shares the same underlying mechanisms for basic life processes. The atoms in living things are no different from the atoms in nonliving things, and the energy that powers living things is the same energy found in inanimate objects.

Because of its watery nature and origin, in a sense all life on this planet is marine. The oceanic environment is a relatively easy place for cells to live. In part at least, life in the ocean is so successful—*total* marine productivity is as high as it is—because of the ocean's physical characteristics, but these characteristics may also be limiting factors for an organism.

Primary producers—autotrophs—are organisms that synthesize food from inorganic substances by photosynthesis and chemosynthesis. Autotrophic marine organisms transform energy from the sun (or from certain inorganic molecules) into chemical energy to power their own growth, and in turn they are consumed by heterotrophic organisms. The feeding relationships in a community resemble complex webs.

A variety of physical factors affect the density, variety, and success of the life-forms in each marine habitat. These factors include water's transparency, temperature, dissolved nutrients, salinity, dissolved gases, hydrostatic pressure, acid–base balance, and others.

Marine organisms are naturally classified by their physical characteristics and by the degree to which they resemble other organisms. The various marine environments populated by marine life may be classified by their physical characteristics. Organisms are distributed throughout the marine environment in specific communities—groups of interacting producers, consumers, and recyclers that share a common living space. The location of a community, and the variety of organisms that comprise it, depend on the physical and biological characteristics of that living space.

ON-LINE STUDY RESOURCES

The Web site for this book contains a wealth of helpful study aids. Log on to:

http://info.brookscole.com/garrison

and select *Essentials of Oceanography*, 3rd edition. Select Chapter 12 from the drop-down menu or click on one of these resource areas:

- For study and review, **Chapter at a Glance** gives you an outline of the chapter, **Chapter Summary** allows you to review the chapter's main ideas, and **Glossary** lists concepts and terms for the chapter along with their definitions.

- To test your mastery of the Terms and Concepts to Remember for this chapter, you can use the electronic **Flash Cards,** play the **Concentration** game, or work the **Crossword Puzzle.**

- For practice quizzing, try the multiple-choice **Tutorial Quiz,** the ten-question **True/False Quiz,** or the **Image Analysis Quiz,** which poses questions based on art and photos from the chapter.

TERMS AND CONCEPTS TO REMEMBER

abyssal zone	mass extinctions
artificial system of	metabolic rate
classification	natural system of
autotroph	classification
bathyl zone	neritic zone
benthic zone	niche
biomass	nutrient
chemosynthesis	oceanic zone
climax community	pelagic zone
community	photosynthesis
disphotic zone	physical factor
ectothermic	population
endothermic	population density
euphotic zone	primary consumer
food	primary producer
food web	primary productivity
habitat	scientific name
hadal zone	species diversity
heterotroph	sublittoral zone
hierarchy	succession
hydrostatic pressure	taxonomy
kingdom	top consumer
limiting factor	trophic pyramid
Linnaeus (Carl von Linné)	zone
littoral zone	

STUDY QUESTIONS

1. What is the ultimate source of the energy used by most living things?

2. What do primary producers produce? How is productivity expressed?

3. What is an autotroph? A heterotroph? How are they similar? How are they different?

4. What is a trophic pyramid? What is the relationship of organisms in a trophic pyramid? Does this have anything to do with food webs?

5. Name and briefly discuss five physical factors of the marine environment that affect living organisms. How is each different in the ocean from on the land?

6. What is a limiting factor? Can you think of some examples not given in the text?

7. How is the marine environment classified? Which scheme is most useful? Justify your answer.

8. How does a natural system of classification differ from an artificial system? Can you give an example of each? Was the hierarchy-based system invented by Linnaeus natural or artificial? What *is* a hierarchy-based system?

9. What are communities? How are marine organisms distributed in communities, and what factors influence who lives where? What is a climax community?

10. Would you support the expenditure of government funds to search for asteroids or other bodies on a collision course with Earth? What would the public's response be to the discovery of a serious threat?

RESOURCES FOR FURTHER READING AND RESEARCH

The Web site for this book contains many ideas for further reading and research. Log on to:

http://info.brookscole.com/garrison

and select *Essentials of Oceanography*, 3rd edition. Select Chapter 12 from the drop-down menu or click on one of these resource areas:

- **Additional Readings** lists the major books and articles consulted in writing this chapter, along with comments from the author about their content and reading level.

- **Hypercontents** takes you to an extensive list of current links to Internet sites with news, research, and images related to individual subjects in the chapter. Just click on the icon that corresponds to a numbered section to see the links for that subject.

- **Internet Exercises** are critical thinking questions that involve research on the Internet with starter URLs provided.

- **InfoTrac College Edition Exercises** leads you to Critical Thinking Projects that use InfoTrac College Edition as a research tool.

- **Regional InfoTrac College Edition** articles are organized into East Coast, West Coast, and Gulf Coast regions, allowing you to study oceanography on a more local level.

 For more readings, go to InfoTrac College Edition, your on-line research library, at:

http://infotrac.thomsonlearning.com

CHAPTER 13

© Paul Ward

A wandering albatross banks over the sea surface in search of food. The bird's wingspan may be nearly 3.6 meters (12 feet).

MASTERS OF THE STORM

Foraging for months across the ocean in conditions of strong winds and high waves, seeking no shelter during storms, enduring tropical heat and polar sleet, soaring continuously through air with a gliding efficiency exceeding that of the most perfectly built human sailplane—the albatrosses are true masters of the sky.

The magnificent wandering albatrosses (genus *Diomedea*) are the largest of these birds. Their wingspan reaches 3.6 meters (12 feet), their weight 10 kilograms (22 pounds). The key to the success of the albatross is its aerodynamically efficient wing—a very long, thin, narrow, cupped and pointed structure ideal for high-speed soaring and gliding. This specialized wing allows albatrosses to cover huge distances in search of food with very little expense of energy. Using the uplift from wind deflected by ocean waves to stay aloft, they soar in long looping arcs. Albatrosses rarely flap their wings, which lock at elbow and shoulder for nearly effortless trans-

Pelagic Communities

port. Flying is their natural state; the heart of an albatross beats more slowly in flight than while it is sitting calmly on the ocean surface.

Satellite tracking data indicate that wandering albatrosses can cover 15,000 kilometers (9,300 miles) between visits to feed their chicks, reaching speeds of 80 kilometers (50 miles) per hour. High speeds over long distances are their specialty. One bird was observed to travel 808 kilometers (502 miles) at an *average* speed of 56 kilometers (35 miles) per hour.

Albatrosses can find schools of fish by the odor of fish oil wafted tens of kilometers downwind. They catch fish (and squid) by dipping their bill into the water during flight or while resting on the surface when the water is calm. Albatrosses take shore leave only during breeding season. They mate for life. Chicks are hatched and raised on remote islands to which albatrosses regularly migrate from nearly any point over the world ocean. They are astonishing animals, and no one who has seen one sweeping over the sea surface ever forgets the experience.

Pelagic organisms live suspended in seawater. They are immensely varied, but all have common problems of maintaining their vertical position, producing or obtaining food, and surviving long enough to reproduce. They can be divided into two broad groups based on their lifestyle: The **plankton** drift or swim weakly, going where the ocean goes, unable to move consistently against waves or current flow. The **nekton** are pelagic organisms that actively swim. **Figure 13.1** shows some representative pelagic organisms.

Pelagic organisms live suspended in the water. Planktonic organisms drift or swim weakly; nektonic organisms actively swim.

Plankton

The pelagic organisms that constitute plankton are as important as they are inconspicuous. The word is derived from the Greek word *planktos*, which means "wandering."

The diversity of planktonic organisms is astonishing: There are giant drifting jellyfish with tentacles 8 meters (25 feet) long, small but voracious arrow worms, many single-celled creatures that glow brightly when disturbed, mollusks with slowly beating flaps that resemble butterfly wings, crustaceans that look like microscopic shrimp, miniature jet-propelled animals that live in jellylike houses and filter food from water, and shimmering crystal-shelled algae. The only feature common to all plankton is their inability to move consistently laterally through the ocean. However, many can and do move vertically in the water column.

Plankton include many different plantlike species and virtually every major group of animals. The term *plankton* is not a collective natural category like mollusks or algae, which implies an ancestral relationship between the organisms; instead it describes a basic ecological connection. Members of the plankton community, informally referred to as *plankters*, can and do interact with one another: Grazing, predation, parasitism, and competition occur among members of this dynamic group. The organisms within the ovals in Figure 13.1 are plankton.

The term *plankton* is not a collective natural category; it is a description of a lifestyle. The plankton include many plantlike species and nearly every major group of animals.

a

Figure 13.1 (**a**) Pelagic communities: representative plankton and nekton of the subtropical Atlantic Ocean. (**b**) Key. Deep-sea fishes and the sargassum fish (8) are shown near their actual size. Surface-dwelling fishes are at least 20 times larger in life than shown in the figure. This stylized representation shows organisms to be much more crowded than they are in real life.

1 dolphins, *Delphinus*	26 mullet larva, *Mullus*
2 tropic birds, *Phaëthon*	27 sea butterfuly, *Clione*
3 paper nautilus, *Argonauta*	28 copepods, *Calanus*
4 anchovies, *Engraulis*	29 assorted fish eggs
5 mackerel, *Pneumatophorus*, and sardines, *Sardinops*	30 stomatopod larva
6 squid, *Onykia*	31 hydromedusa, *Hybocodon*
7 *Sargassum*	32 hydromedusa, *Bougainvilli*
8 sargassum fish (*Histro*)	33 salp (pelagic tunicate), *Doliolum*
9 pilot fish (*Naucrates*)	34 brittle star larva
10 white-tipped shark *Carcharhinus*	35 copepod, *Calocalaus*
11 pompano, *Palometa*	36 cladoceran, *Podon*
12 ocean sunfish, *Mola mola*	37 foraminifer, *Hastigerina*
13 squids, *Loligo*	38 luminescent dinoflagellates, *Noctiluca*
14 rabbitfish, *Chimaera*	39 dinoflagellates, *Ceratium*
15 eel larva, *Leptocephalus*	40 diatom, *Coscinodiscus*
16 deep sea fish	41 diatoms, *Chaetoceras*
17 deep sea angler, *Melanocetus*	42 diatoms, *Cerautulus*
18 lantern fish, *Diaphus*	43 diatom, *Fragilaria*
19 hatchetfish, *Polyipnus*	44 diatom, *Melosira*
20 "widemouth," *Malacosteus*	45 dinoflagellate, *Dinophysis*
21 euphausid shrimp, *Nematoscelis*	46 diatoms, *Biddulphia regia*
22 arrowworm, *Sagitta*	47 diatoms, *B. arctica*
23 amphipod, *Hyperoche*	48 dinoflagellate, *Gonyaulax*
24 sole larva, *Solea*	49 diatom, *Thalassiosira*
25 sunfish larva, *Mola mola*	50 diatom, *Eucampia*
	51 diatom, *B. vesiculosa*

b

Collecting and Studying Plankton

13.3

The first large-scale, systematic study of plankton was carried out by biologists aboard the research vessel *Meteor* during the German Atlantic Oceanographic Expedition of 1925–26. Many of the tools and techniques they pioneered are still in use today. **Plankton nets** of the type perfected aboard *Meteor* are essential to plankton studies (**Figure 13.2**). These conical nets are customarily made of nylon or Dacron cloth woven in a fine interlocking pattern to assure consistent

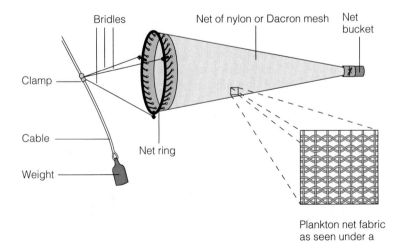

Plankton net fabric as seen under a microscope

a

Figure 13.2 Plankton nets. (**a**) The standard conical net is made of fine mesh and has a mouth up to 1 meter (3.3 feet) in diameter. The net is towed behind a ship for a set distance. The number of organisms present in the water can be estimated if the trapped organisms are counted and the volume of sampled water is known. (**b**) The net shown here has a somewhat coarser mesh because its target organisms, shrimplike crustaceans known as *krill*, are relatively large.

b

spacing between threads. The net is hauled slowly for a known distance behind a ship, or cast to a set depth, and then reeled in. Trapped organisms are flushed to the net's pointed end and gently removed for analysis. Quantitative analysis of plankton requires identification of the organisms and an estimate of the sampled volume of water.

Very small plankton can slip through a plankton net. Their capture and study require concentration by centrifuge or entrapment by a plankton filter through which water is drawn. The filter is later disassembled and the plankton studied in place.

Measurements of physical ocean conditions such as dissolved carbon dioxide and oxygen content, pH, temperature, and light intensity at the time and place of sampling are important in interpreting the samples.

Phytoplankton: The Autotrophs

Autotrophic plankton are generally called **phytoplankton,** a term derived from the Greek word *phyton,* meaning "plant." A huge, nearly invisible mass of phytoplankton drifts within the sunlit surface layer of the world ocean. Phytoplankton are critical to all life on Earth because of their great contribution to food webs and their generation of large amounts of atmospheric oxygen through photosynthesis. Planktonic autotrophs are thought to bind *at least* 35 billion metric tons of carbon into carbohydrates each year, at least 40% of the food made by photosynthesis on Earth! These easily overlooked, mostly single-celled, drifting photosynthesizers are much more important to marine productivity than the larger and more conspicuous seaweeds.

There are at least eight major types of phytoplankton, of which the most prominent are the diatoms and dinoflagellates. In the past decade, however, the small coccolithophores and silicoflagellates, and the extremely small cyanobacteria and autotrophic picoplankton, have been found to be more important in world productivity than most researchers could have imagined.

DIATOMS   The dominant photosynthetic organisms in the plankton—and in the world—are the **diatoms.** More than 5,600 species of diatoms are known to exist. The larger species are barely visible to the unaided eye. Most are round, but some are elongated, branched, or triangular.

Typical diatoms are shown in **Figure 13.3.** The name means "cut through," a reference to the patterns of perforations through the diatom's rigid cell wall, or **frustule.** As much as 95% of the mass of the frustule consists of silica (SiO_2), giving this covering the optical, physical, and chemical characteristics of glass—an ideal protective window for a photosynthesizer. Magnification reveals that the frustule consists of two closely fitting halves, or **valves,** which fit together like a well-made gift box, the top valve adhering tightly over the lip of the bottom one. The pattern of perforations, slits, striations, dots, and lines on the surface of the valves is different for each diatom species (**Figure 13.3b**).

When diatoms die, their valves can drift to the seafloor to accumulate as layers of siliceous ooze (see Chapter 5).

Inside the diatom's tailored valves lies an extraordinary photosynthetic machine. Fully 55% of the energy of sunlight absorbed by a diatom can be converted into the energy of carbohydrate chemical bonds, one of the most efficient energy conversion rates known. Excess oxygen not needed in the cell's respiration is released through the perforations in the frustule into the water. Some oxygen is absorbed by marine animals, some is incorporated into bottom sediments, and some diffuses into the atmosphere. Most of the oxygen we breathe has moved recently through the many glistening pores of diatoms.

Diatoms store energy as fatty acids and oils, compounds that are lighter than their equivalent volume of water and assist in flotation. As you might guess, flotation is a potential problem for diatoms because the weight of their heavy silica frustule seems at odds with their requirement for staying near the sunlit ocean surface. Oil floats, glass sinks, and a balanced amount of both produces neutral buoyancy. Not all diatoms need to float, however. Many nonplanktonic species lie on shallow bottoms where light and nutrients are able to support photosynthesis. These benthic species are nearly always elongated (or *pennate*) in shape.

DINOFLAGELLATES   Most **dinoflagellates** are single-celled autotrophs (**Figure 13.4a**). A few species live within the tissues of other organisms, but the great majority of dinoflagellates live free in the water. Most have two whiplike projections called **flagella** in channels grooved in their protective outer covering of cellulose. One flagellum drives the organism forward, while the other causes it to rotate in the water. Their flagella allow dinoflagellates to adjust their orientation and vertical position to make the best photosynthetic use of available light.

Some species of dinoflagellates can become so numerous that the water turns a rusty red as light reflects from the accessory pigments within each cell. These species are responsible for the phenomenon of *red tide* (**Figure 13.4b**). During times of such rapid growth (usually in the spring), the concentration of these microscopic organisms may briefly reach 6 million per liter (23 million per gallon)! At night, the huge numbers of dinoflagellates in a red tide can cause breaking waves to glow a bright blue, a phenomenon known as bioluminscence.

Red tide can be dangerous because some dinoflagellate species synthesize potent toxins as by-products of metabolism. Among the most effective poisons known, these toxins may affect nearby marine life or (if the dinoflagellates are dried on the beach and blown inland) even humans. Some of the toxins are similar in chemical structure to the muscle relaxant curare but are tens of times more powerful. Humans should avoid eating certain species of clams, mussels, and other filter feeders during the summer months when toxin-producing dinoflagellates are abundant in the plankton. If shellfish from a particular area are unsafe, a state agency will issue an advisory that may remain in effect for six weeks or longer until the danger is past.

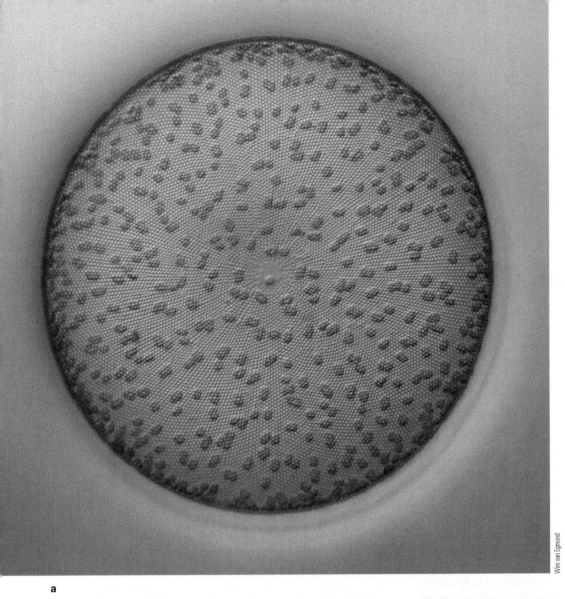

a

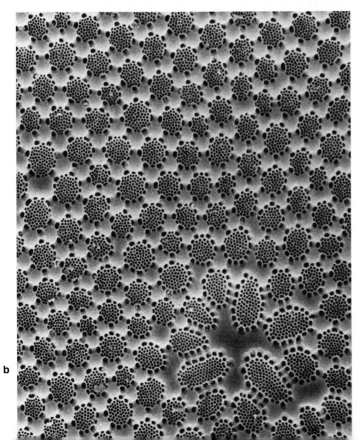

b

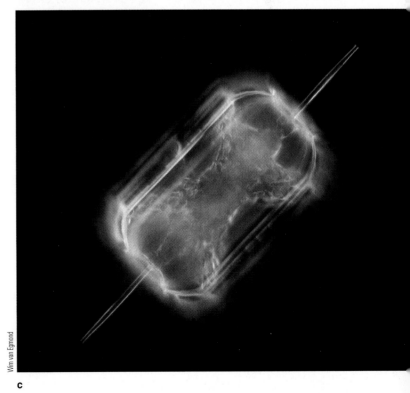

c

Figure 13.3 Diatoms. (**a**) The transparent glass valves, or "shell," of the diatom *Coscinodiscus* as shown with a light micrograph. The many small perforations that give diatoms their name are clearly visible. Note also the many green chloroplasts—cell organelles responsible for photosynthesis. (**b**) A closer view using an electron microscope shows the perforations to be groups of still smaller holes, each small enough to exclude bacteria and some marine viruses. The holes allow the diatom to pass gases, nutrients, and waste products through the otherwise impermeable silica covering. (From C. Shih and R. G. Kessel, "Living Images," 1982 Sudbury, MA: Jones & Bartlett Publishers. www.jbpub.com. Reprinted with permission.) (**c**) *Ditylum*, a diatom, photographed in visible light. Note the junction between the frustules. The cells in this figure are about the size of the period at the end of this sentence.

Wim van Egmond

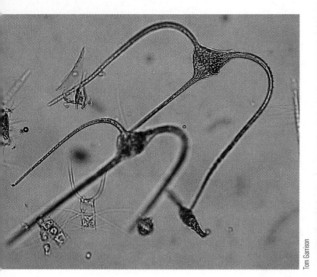

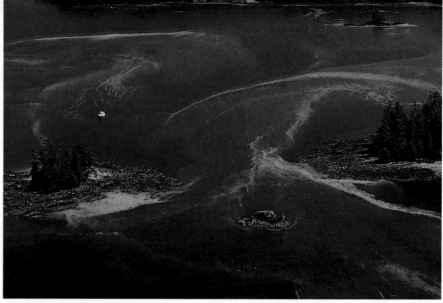

a   b

Figure 13.4 (a) Two representatives of the genus *Ceratium*, a common dinoflagellate. Two flagella beat in opposing grooves in the armor-plated body. One groove is visible at the equator of the upper cell, between the two "horns." These specimens are about 0.5 millimeter (0.02 inch) across. **(b)** During a red tide, the presence of millions of dinoflagellates turns seawater brownish-red.

COCCOLITHOPHORES AND OTHER PHYTOPLANKTON Most other types of phytoplankton are extraordinarily small and so are called **nanoplankton.** The **coccolithophores,** for example, are tiny single cells covered with discs of calcium carbonate (coccoliths) fixed to the outside of their cell walls (**Figure 13.5a**). Coccolithophores live near the ocean surface in brightly lit areas. The translucent covering of coccoliths may act as a sunshade to prevent absorption of too much light. In areas of high coccolithophore productivity, most notably in temperate coastal areas, their numbers occasionally become so great that the water appears milky or chalky (**Figure 13.5b**). Coccoliths can also build seabed deposits of ooze. The famous White Cliffs of Dover in England and the extensive chalk deposits of northern Texas consist largely of fossil coccolith ooze deposits uplifted by geological forces.

The plantlike organisms that make up the phytoplankton are responsible for most of the ocean's primary productivity. Of these, diatoms are the most productive and efficient.

Phytoplankton Productivity

Where is phytoplankton productivity the greatest? This question is among the most important in biological oceanography. Since phytoplankton form the base of nearly all oceanic food webs, the biological characteristics of any ocean area will depend heavily on the presence and success of phytoplankton.

With some exceptions, the distribution of phytoplankton corresponds to the distribution of nutrients. Because of coastal upwelling and land runoff, nutrient levels are highest near the continents. Plankton are most abundant there, and productivity is highest. The water above some continental shelves sustains productivity in excess of 1 $gC/m^2/day$! (Please see Figure 12.2 for a review of primary productivity notation and estimates.) But what of the open ocean? Where is productivity greatest away from land?

In the tropics? The open tropical oceans have abundant sunlight and CO_2 but are generally deficient in surface nutrients because the strong thermocline discourages the vertical mixing necessary to bring nutrients from the lower depths. The tropical oceans away from land are therefore oceanic deserts nearly devoid of visible plankton. The typical clarity of tropical water underscores this point. In most of the tropics, productivity rarely exceeds 30 $gC/m^2/yr$, and seasonal fluctuation in productivity is low.

Tropical coral reefs—benthic habitats—are exceptions to this general rule. Reef areas, which account for less than 2% of the tropical ocean surface, are productive places because autotrophic dinoflagellates live *within* the tissues of coral animals and don't drift as plankton. Nutrients are made available by coastal upwelling and by the coral's own metabolism. These nutrients are cycled tightly through the reef and not lost to sinking.

In the polar regions? At very high latitudes the low sun angle, reduced light penetration due to ice cover, and weeks or months of darkness in winter severely limit productivity. At the height of summer, however, 24-hour daylight, a lack of surface ice, and the presence of upwelled nutrients can lead to spectacular plankton blooms. The surface of some sheltered bays can look like tomato soup because dinoflagellates and other plankton are so abundant. This bloom cannot last because nutrients are not quickly recycled and because the sun is above the critical angle for a few weeks at best. The

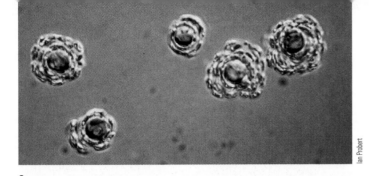

a

Figure 13.5 (**a**) A rare light microscope photograph of *Emiliania huxleyi*, a coccolithophore. The tiny calcified plates (coccoliths) covering the cells are 6 micrometers (120 millionths of an inch) across. Photosynthetic pigments give the cells a golden or yellow-green color. The coccoliths (not the organic cells themselves) act like mirrors suspended in the water, and can reflect a significant amount of the incoming sunlight. The reflectance from the blooms can be picked up by satellites in space, allowing the extent of the blooms of this species to be distinguished in fine detail. Figure 5.7c is a scanning electron microscope image of the same species that shows the individual coccoliths in more detail. (**b**) A coccolithophore bloom is clearly visible south of Ireland in this natural color satellite image.

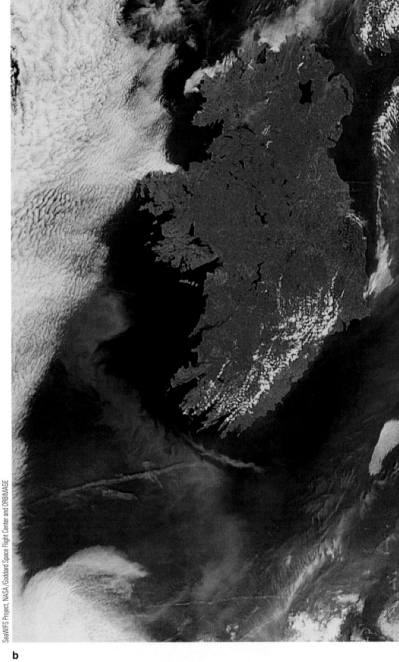

b

short-lived summer peak does not compensate for the long, unproductive winter months.

Average productivity at very high northern latitudes tends to be lower than at high southern latitudes, averaging less than 25 gC/m²/yr. The Arctic Ocean is almost surrounded by landmasses, which limit water circulation and therefore nutrient upwelling. The southern ocean, on the other hand, is enriched by water upwelling to replace sinking Antarctic Bottom Water. This rich mixture is stirred by the West Wind Drift. The Antarctic accounts for a much greater share of high-latitude production than the Arctic because more nutrients are available.

In the temperate and subpolar zones? With the tropics generally out of the running for reasons of nutrient deficiency and the north polar ocean suffering from slow nutrient turnover and low illumination, the overall productivity prize is left to the temperate and southern subpolar zones. Thanks to the dependable light and the moderate supply of nutrients, annual production over temperate continental shelves and in southern subpolar ocean areas is the greatest of any open ocean area. Typical productivity in the temperate zone is about 120 gC/m²/yr. In ideal conditions southern subpolar productivity can approach 250 gC/m²/yr!

Figure 13.6 shows the levels of productivity in tropical, temperate, and northern polar ocean areas. Note that *nearshore* productivity is almost always higher than *open ocean* productivity, even in the relatively productive temperate and south subpolar zones.

Curiously, the open ocean area with the greatest annual productivity is an exception to the general picture developed in this section. The slender, cold finger of high productivity pointing west from South America along the equator is a result of wind-propelled upwelling due to Ekman transport on either side of the geographical equator. Look for this area in Figure 12.3.

Does productivity change with the seasons? **Figure 13.7** shows the relationship of phytoplankton biomass to season

and latitude. The low, flat line representing annual tropical productivity contrasts with the high, thin peak representing the Arctic summer. The higher of the two peaks for the temperate zone indicates the plankton bloom of northern spring caused by increasing illumination, while the smaller peak representing the northern fall is caused by nutrients returning to the surface.

Most primary productivity occurs above the continental shelves of the temperate zones and in the south subpolar zones.

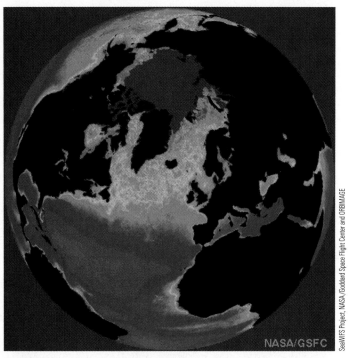

Figure 13.6 Measuring the concentration of chlorophyll in ocean water by satellite. This false-color image from the Coastal Zone Color Scanner aboard the *Nimbus 7* satellite represents average conditions in late spring in the Northern Hemisphere. It shows the concentration of chlorophyll in the upper layer of the ocean, with higher amounts indicated by green, orange, and red. Note the high phytoplankton concentrations induced by increased nutrient availability along the coasts. The centers of the oceanic gyres contain relatively few phytoplankton, as shown by their purple hue.

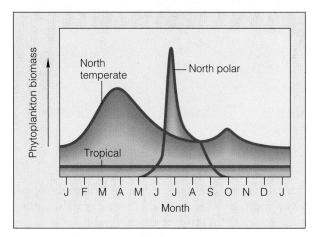

Figure 13.7 Variation in the biomass of phytoplankton by season and latitude. (Note that the area under each curve represents total productivity.)

Zooplankton: The Heterotrophs

Heterotrophic plankton—the planktonic organisms that eat the primary producers—are collectively called **zooplankton.** Zooplankters are the most numerous primary consumers of the ocean. They graze on the diatoms, dinoflagellates, and other phytoplankton at the bottom of the trophic pyramid in

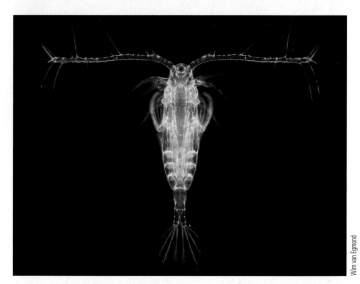

Wim van Egmond

Figure 13.8 A zooplanktonic copepod. Copepods probably are the most abundant and widely distributed animal in the world. The species shown here reaches a maximum size of about 0.5 millimeter (about 0.02 inch).

a way similar to cows grazing on grass. The variety of zooplankton is surprising; nearly every major animal group is represented. Each is expert at the painstaking concentration of food from the water. The most abundant zooplankters, accounting for about 70% of individuals, are tiny shrimplike animals called *copepods* (**Figure 13.8**). Copepods are crustaceans, a group that also includes crabs, lobsters, and shrimp.

Not all zooplankters are small, however. Many are in the 1–2 centimeter (½–1 inch) size range. The largest drifters are giant jellyfish of the genus *Cyanea*; their bells may be more than 3.5 meters (12 feet) in diameter! We have a special term for plankton larger than about 1 centimeter (½ inch) across: *macroplankton.*

Most zooplankters spend their whole lives in the plankton community, so we call them **holoplankton.** But some planktonic animals are the juvenile stages of crabs, barnacles, clams, sea stars, and other organisms that will later adopt a benthic or nektonic lifestyle. These temporary visitors are called **meroplankton.** Most animal groups are represented in the meroplankton; even the powerful tuna serves a brief planktonic apprenticeship. These useful categories can be applied to phytoplankton as well as to zooplankton. Holoplanktonic organisms are by far the more numerous forms of both phytoplankton and zooplankton.

One of the ocean's most important zooplankters is the pelagic arthropod known as **krill** (genus *Euphausia*, **Figure 13.9**), the keystone of the Antarctic ecosystem. This thumbsized shrimplike crustacean mostly grazes on the abundant diatoms of the southern polar ocean. In turn, krill are eaten in tremendous numbers by seabirds, squids, fishes, and whales. Some 500 to 750 million metric tons (550 to 825 million tons) of krill inhabit the Antarctic Ocean, with the greatest concentrations in the productive upwelling currents of the Weddell Sea. Krill travel in huge schools that can extend over several square miles, and collectively they exceed the biomass of Earth's entire human population!

© Uwe Kils

Figure 13.9 Krill (*Euphausia superba*). These shrimplike crustaceans, shown here about twice its actual size, occur throughout the world ocean. They are particularly numerous in Antarctic seas.

The great diversity of members of the plankton community is perhaps best illustrated by the informally named "jellies." Common to all oceans, these diaphanous animals range in size from microscopic to immense, and employ a wonderful range of adaptations to reduce weight, retard sinking, and snare prey. A few of these animals are shown in **Figure 13.10.**

Small but important, planktonic **foraminifera** are related to amoebas (**Figure 13.11**). Like amoebas, they extend long protoplasmic filaments to snare food. Most foraminifera have calcium carbonate shells. As we saw in Chapter 5, extensive white deposits of calcareous ooze have builtup on the seabed from their skeletons. As was the case with some phytoplankton sediments, some of these layers have been uplifted and can be found on land.

Plankton and Food Webs **13.7**

Zooplankton and other animals eat phytoplankton, and still other animals eat these primary consumers. The mass of zooplankton is typically about 10% that of phytoplankton.

It is interesting to note that the largest marine animals, such as whale sharks (fish) and baleen whales (mammals), do not expend their energy tracking down and attacking big animals. Instead these largest of all feeders concentrate zooplankton from the water and consume it in vast quantities. The zooplankton they eat are usually not the primary consumers but are the somewhat larger secondary consumers, usually crustaceans such as krill that have themselves fed on the microscopic primary consumers. In this way whales and other large filter feeders get around the inevitable energy losses of a long food web by harvesting energy closer to the source and gaining the advantages of efficiency and quantity.

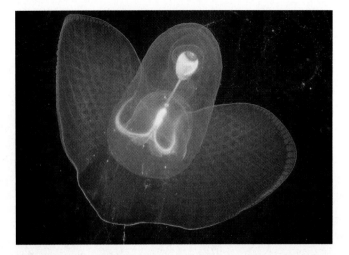

a

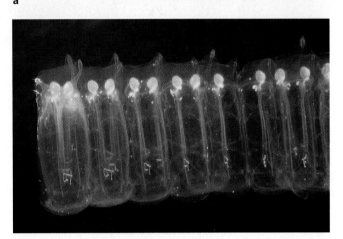

b

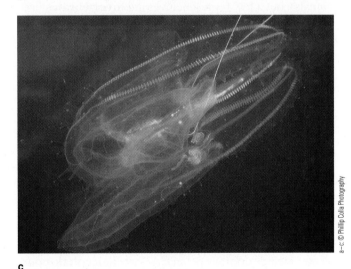

a–c: © Phillip Colla Photography

c

Figure 13.10 Jellies. (**a**) Unlikely as it may seem, the zooplankton includes gliding snails (pteropods). This example, genus *Corolla,* is about 8 centimeters (3 inches) across, and feeds by capturing smaller drifting organisms in a mucus net. (**b**) Chains of pelagic tunicates (salps of genus *Pegea*) drift with the current. Each animal looks like a small orange pea at the top of an individual gelatinous "house" used to filter the food from the water. Individual "houses" in the colony are about 6 centimeters (2.375 inches) high. (**c**) Sometimes called "comb jellies" because of the rows of iridescent cilia along their surfaces, ctenophores reel in small organisms and food particles on sticky tentacles. Genus *Leucothea*, pictured here, is typically 10 centimeters (4 inches) long.

Figure 13.11 A foraminiferan. The word means "bearers of windows." Light streams through thin parts of the shell.

Wim van Egmond

bite the flesh with horny beaks. Squids swim in groups and usually prey on small fishes. Squids can also confuse predators with clouds of ink. Some kinds of squids eject a dummy of coagulated ink that's a rough duplicate of their size and shape. The squid is long gone by the time the attacker discovers the deception! At least one species of squid living below the euphotic zone produces a sparkling luminous ink instead of black ink (which, of course, would be ineffective in the darkness). Squids can grow to surprising sizes (**Figure 13.12**)—the record length, including tentacles, is 18 meters (59 feet)! Most are much smaller.

Shrimps and Their Relatives

Arthropoda, the animal category that includes the lobsters, shrimps, crabs, krill (Figure 13.9), and barnacles, is the most successful animal group on Earth. They occupy the greatest variety of habitats, consume the greatest quanti-

Nekton

Pelagic animals that swim actively are known as **nekton.** Most nektonic animals are **vertebrates** (animals with backbones, such as fishes, reptiles, marine birds, and marine mammals), but a few representatives are **invertebrates** (animals without backbones, such as squids and nautiluses and some species of shrimplike arthropods).

> Important nektonic animals include cephalopods, shrimps, fishes, seabirds, and marine mammals. Each has a continuously evolving suite of adaptations that allows it to meet the challenges of the marine environment.

Squids and Nautiluses

The most highly evolved of the **mollusks**—a category of animals that includes clams and snails—are the **cephalopods,** a group of marine predators containing squids, nautiluses, and octopuses. These animals have a head surrounded by a foot divided into tentacles. The nautiluses retain a large, coiled external shell, but squids have only a thin vestige of the shell within their bodies, and octopuses (nearly all of which are benthic organisms) have none at all. Pelagic cephalopods move by swimming with special fins or by squirting jets of water from an interior cavity.

Most cephalopods catch prey with stiff adhesive discs on their tentacles that function as suction cups, and they tear or

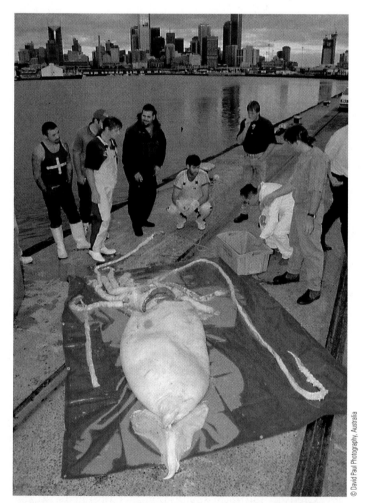

Figure 13.12 This large squid, genus *Architeuthis*, was captured by commercial fishermen in New Zealand. It weighs 114 kilograms (250 pounds) and is 7.6 meters (25 feet) long. The world's largest preserved specimen, it was placed on display in New York City's American Museum of Natural History in October 1999.

© David Paul Photography, Australia

Figure 13.13 The great white shark, genus *Carcharodon*, a predator of seals, sea lions, and large fish. The great white is one of the most dangerous of the sharks encountered by human swimmers. In this extraordinary sequence taken off the South African coast, a large great white shark rockets from the ocean with a freshly caught sea lion in its jaws.

ties of food, and exist in almost unimaginable numbers. Their most important innovation is an **exoskeleton.** Unlike the often cumbersome shells of some other marine animals, the exoskeleton of an arthropod fits and articulates like a finely crafted suit of armor. Its three layers serve to waterproof the covering, tint it a protective color, and make it resilient and strong. Muscles within the animal are attached to the exoskeleton to move the appendages.

The 2,000-plus species of shrimps, most of which are pelagic, range from about 1 centimeter to more than 20 centimeters (½ inch to more than 8 inches) in length. The larger species are often called *prawns*. Their semitransparent bodies are flattened from side to side, and swimming species normally move by the rhythmic waving of their legs. Rapid movement in an emergency is accomplished by a quick flex of the abdomen and tail.

Fishes

Fishes are vertebrates that usually live in water and possess gills for breathing and fins for swimming. There are more species of fishes, and more individuals, than species and individuals of all other vertebrates *combined*. This is not a surprising fact given the vast oceanic habitat the planet provides. Fishes range in adult length from less than 10 millimeters to more than 20 meters (0.4 inch to 60 feet) and weigh from about a 0.1 gram to about 41,000 kilograms (0.004 ounce to 45 tons). Some fishes are capable of short bursts of speed in excess of 113 kilometers (70 miles) per hour; some species hardly ever move.

Fishes live near the surface and at great depth, in warm water and cold, even frozen within ice or dried in balls of mud. Like other ectothermic (cold-blooded) organisms, the great majority of fishes are incapable of generating and maintaining a steady internal temperature from metabolic heat, so the internal body temperature of a fish is usually the same as that of the surrounding environment. About 40% of the 30,000-plus fish species live at least part of their lives in fresh water; 60% live exclusively in seawater. Fishes have evolved to fit almost every conceivable watery habitat but are most numerous on the bottom or in productive seawater over the continental shelves. Some species have a "sixth sense," an ability to detect small changes in the electrical field sur-

rounding their bodies that assists in the detection of prey or avoidance of predators. Some electric eels, catfish, and rays can use internally generated electricity for defense and offense.

Fishes are divided into two major groups based on the material that forms their skeletons: cartilaginous fishes and bony fishes.

The Cartilagenous Fishes

All members of the class **Chondrichthyes**—the group that includes sharks, skates, rays, and chimaeras—have a skeleton made of a tough, elastic tissue called **cartilage.** Though there is some calcification in the cartilaginous skeleton, true bone is entirely absent from this group. These fish have jaws with teeth, paired fins, and often active lifestyles. Sharks and rays tend to be larger than bony fishes, and except for some whales, sharks are the largest living vertebrates.

About 350 species of sharks and 320 species of rays are known to exist. Nearly all are marine, though a few species inhabit estuaries and a very few are permanent inhabitants of fresh water. Although there are many exceptions, sharks tend to favor swimming through open water, while rays tend to be found on or near the bottom.

Sharks have an undeservedly bad reputation. More than 80% of shark species are less than 2 meters (6.6 feet) long as adults, and only a few of the remaining 20% are aggressive toward humans. Like other cartilaginous fishes, sharks are not very intelligent and certainly don't hold grudges or behave in the malignant ways so vividly portrayed in popular novels and movies. Still, some sharks are indeed dangerous to humans, and the great white sharks in the genus *Carcharodon* are perhaps the most dangerous of all (**Figure 13.13**).[1] These swimmer's nightmares attain lengths of 7 meters (23 feet) and weigh up to 1,400 kilograms (3,000 pounds). Great whites are not white, but grayish brown or blue above and creamy on the lower half. A dangerous relative, the mako shark, reaches lengths of 4 meters (13 feet) and is known

---

[1] Some perspective: Worldwide, sharks are responsible for about six known human fatalities per year. Each year, more people are killed in the United States by dogs than have been killed by sharks in the last century. For every human killed by a shark, humans kill more than 16 million sharks, mostly for food and medicines.

Figure 13.14 A diver hitches a ride on a whale shark. Unless he is struck by the fish's tail as he dismounts, the diver is in no danger. Large animals (like seals and divers) are not part of the diet of this type of shark.

also to attack small boats. These and other predatory species, such as tiger sharks and hammerhead sharks, are attracted to prey by vibrations in the water, which they detect with sensitive organs arrayed in lines beneath the surface of their skin. Smell also plays an important role in hunting their prey, which is usually fish and marine mammals.

Though most famous, the man-eaters are not the largest shark species. This honor goes to the immense warm-water whale sharks in the genus *Rhineodon*, which reach sizes in excess of 18 meters (60 feet) and 41,000 kilograms (90,000 pounds) (**Figure 13.14**). Whale sharks and their somewhat smaller relatives, the basking sharks, are docile and present little threat to people. These greatest of fishes swim slowly near the surface with their huge mouths open, feeding on plankton. They may filter as much as 2,200 cubic meters (2,000 tons) of water per hour through a fine mesh of gill rakers. Accumulated plankton is periodically backflushed into the mouth where it is concentrated for swallowing.

The Bony Fishes

The 27,000-plus species of bony fishes, members of class **Osteichthyes,** owe much of their success to the hard, strong, lightweight skeleton that supports them. These most numerous of fish—and most abundant, most diverse, and successful of all vertebrates—are found in almost every marine habitat from tidepools to the abyssal depths. Their numbers include the air-breathing lungfishes and the lobe-finned coelacanths, whose ancient relatives broke from the path of fish evolution to establish the dynasties of land vertebrates.

About 90% of all living fishes are contained within the osteichthyan order **Teleostei,** which contains the cod, tuna, halibut, goldfish, and other familiar species. Within this large category are varieties of fishes with independently

movable fins for well-controlled swimming, great speed for pursuit or avoidance of predators, highly effective camouflage, social organization, orderly patterns of migration, and other advanced features (**Figure 13.15**). Their economic importance is great: Some 92 million metric tons (101 million tons) of bony fishes are taken annually from the ocean to help satisfy the human demand for protein.

The Problems of Fishes

Seawater may seem to be an ideal habitat, but living in it does present difficulties. Water is about a thousand times more dense than air and a hundred times more viscous, and it impedes motion effectively at low speeds. How can a fish best move through it? How can a fish maintain its vertical position in the water column? Must a fish swim constantly to offset the weight of muscle and bone? And how about breathing? Can oxygen and carbon dioxide be exchanged efficiently under water? How can predators be thwarted? There seem to be many problems, but these most successful vertebrates have evolved structures and behaviors to cope.

MOVEMENT, SHAPE, AND PROPULSION Active fish usually have streamlined shapes that make their propulsive efforts more effective. A fish's resistance to movement, or **drag,** is determined by its frontal area, body contour, and surface texture. Drag increases geometrically with increasing speed. Faster-swimming fish are therefore more highly modified to minimize the slowing effects of the dense, relatively sticky medium in which they live. The most effective antidrag shape is the tapering, torpedolike body plan.

A fish's forward thrust comes from the combined effort of body and fins. Muscles within slender flexible fish (such as eels) cause the body to undulate in S-shaped waves that

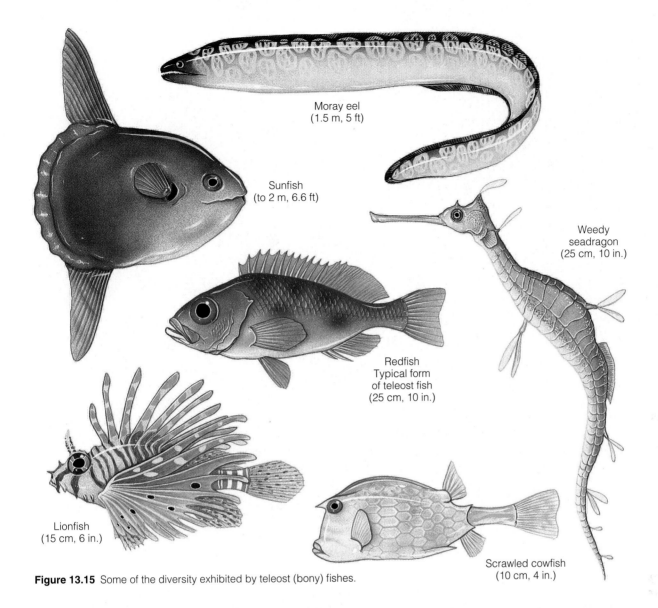

Moray eel
(1.5 m, 5 ft)

Sunfish
(to 2 m, 6.6 ft)

Weedy
seadragon
(25 cm, 10 in.)

Redfish
Typical form
of teleost fish
(25 cm, 10 in.)

Lionfish
(15 cm, 6 in.)

Scrawled cowfish
(10 cm, 4 in.)

Figure 13.15 Some of the diversity exhibited by teleost (bony) fishes.

pass down the body from head to tail in a snakelike motion. The eel pushes forward against the water much as a snake pushes against the ground. This type of movement is not very efficient, however. More advanced fishes have a relatively inflexible body, which undulates rapidly through a shorter distance, and a hinged tail to couple muscular energy to the water. The fish's body can be shorter and can face more squarely in the direction of travel; the drag losses are lower.

MAINTENANCE OF LEVEL   Fish tissue is typically more dense than the surrounding water, so fishes will sink unless their weight is offset by propulsive forces or by buoyant gas- or fat-filled **swim bladders.** Cartilaginous fishes have no swim bladders and must work continuously to maintain their position in the water column. Sharks generate lift with a high tail and fins that act like airplane wings. Bony fishes that appear to hover motionless in the water usually have well-developed swim bladders just below their spinal columns. The volume of gas in these structures provides enough buoyancy to offset the animal's weight. The quantity of gas is controlled by secre-

tion and absorption of gas from the blood, and by muscular contraction of the swim bladder to compensate for temporary changes in depth.

GAS EXCHANGE   How can fish breathe underwater? **Gas exchange,** the process of bringing oxygen into the body and eliminating carbon dioxide (CO_2), is essential to all animals. At first glance the task may seem more difficult for water breathers than for air breathers, but air breathing animals add an extra step. We air breathers must first dissolve gases in a thin film of water in our lungs before they can diffuse across a membrane.

Fish take in water containing dissolved oxygen at the mouth, pump it past fine **gill membranes,** and exhaust it through rear-facing slots. The higher concentration of free oxygen dissolved in the water causes oxygen to diffuse through the gill membranes into the animal; the higher concentration of CO_2 dissolved in the blood causes CO_2 to diffuse through the gill membranes to the outside. The gill membranes themselves are arranged in thin filaments and plates

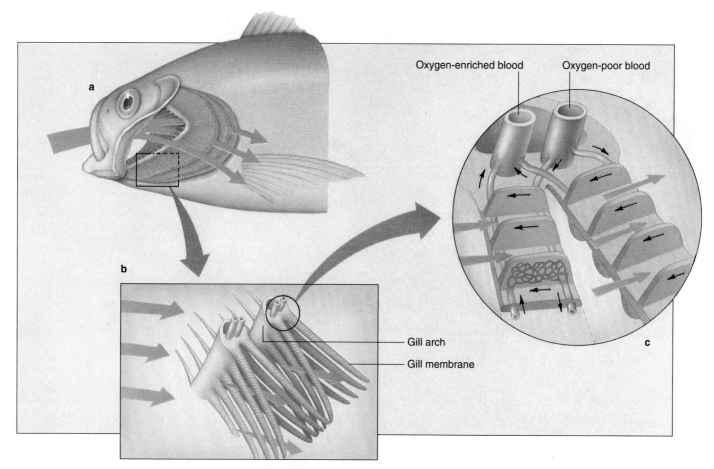

Oxygen-enriched blood   Oxygen-poor blood

Gill arch

Gill membrane

a

b

c

Figure 13.16 Cutaway of a mackerel, showing the position of the gills (**a**). Broad arrows in (**b**) and (**c**) indicate the flow of water over the gill membranes of a single gill arch. Small arrows in (**c**) indicate the direction of blood flow through the capillaries of the gill filament in a direction opposite to that of the incoming water. This mechanism is called *countercurrent flow.*

efficiently packaged into a very small space (**Figure 13.16**). Water and blood circulate in opposite directions—in a countercurrent flow—which increases transfer efficiency.

An active fish like a mackerel requires so much oxygen and generates so much waste CO_2 that its gill surface area must be ten times its body surface area! (Sedentary fish have proportionally less gill area.) With their large gill area and countercurrent flow, active fish extract about 85% of the dissolved oxygen in water flowing past their gills. Air-breathing vertebrates, by contrast, extract only about 25% of the oxygen from the air entering their lungs.

FEEDING AND DEFENSE Competitive pressure among the large number of fish species has given rise to a wonderful variety of feeding and defense tactics. Sight is very important to most fishes, enabling them to see their prey or avoid being eaten. Even some deep-water fishes that live below the photic zone have excellent eyesight for seeing luminous cues from potential mates or meals. Hearing is also well developed, as is the ability to detect low-frequency vibrations. The bizarre flattened crossbar that gives the hammerhead shark its name may provide a kind of stereo smell to sense differing amounts

of interesting substances in the water. Salmon smell their way to their home streams after years at sea by detecting faint chemical traces characteristic of the water from the stream in which they hatched.

About a quarter of all bony fish species exhibit **schooling** behavior at some time during their life cycle. A fish school is a massed group of individuals of a single species and size class packed closely together and moving as a unit. There is no leadership in fish schools, and the movement of fish within them seems to be controlled automatically by direct interaction between lateral-line sensors and the locomotor muscles themselves. I can personally attest to the effectiveness of schooling as a means of defense. On a few diving trips I've noticed a large moving mass just beyond the limit of clear visibility. Is it a fish school, or is it a single large animal? Many predators might not stay around long enough to find out! Schools have the added benefits of reducing chance detection by a predator, providing ready mates at the appropriate time, and increasing feeding efficiency.

Fishes that live in the twilight world at the bottom of the photic zone use bioluminescence in feeding, avoiding being eaten, and attracting mates. Some members of this sparsely

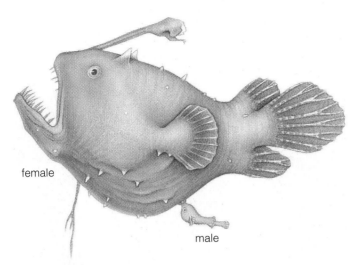

female

male

Figure 13.17 Some species of deep-sea anglerfish have bioluminescent lures. Victims curious about the lure are quickly eaten by the predator. This female of genus *Melanocoetus* is about 10 centimeters (4 inches) long and carries a permanently attached male.

Figure 13.18 A green sea turtle, *Chelonia mydas*.

populated community have built-in luminescent organs that cast dim blue light downward; this light masks their own shadows, so they have less chance of being detected and eaten. Other deep-swimming organisms attract their infrequent meals with a luminous lure (**Figure 13.17**). These animals also use patterns of glowing spots or lines to identify themselves to members of the same species, a necessary first step in mating. Some use flashes of light to dazzle or frighten potential predators.

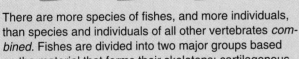

There are more species of fishes, and more individuals, than species and individuals of all other vertebrates *combined*. Fishes are divided into two major groups based on the material that forms their skeletons: cartilagenous and bony.

Marine Reptiles

13.15

Each of the three main groups of reptiles has marine representatives: turtles, sea snakes and marine lizards (iguanas), and marine crocodiles. Like all reptiles, marine reptiles are ectothermic, breathe air with lungs, are covered with scales and a relatively impermeable skin, and are equipped with special **salt glands** to concentrate and excrete excess salts from body fluids. Except for one widely ranging species of turtle, all marine reptiles require the warmth of tropical or subtropical waters.

The best-known and most successful living marine reptiles are the eight species of sea turtles. Unlike land turtles,

sea turtles have relatively small streamlined shells without enough interior space to retract head or limbs. The shell provides an effective passive defense, and adult sea turtles have no predators except humans. Their forelimbs are modified as flippers and provide propulsive power, their hindlimbs act as rudders. The two species of green sea turtles (genus *Chelonia*, **Figure 13.18**) are the most abundant and widespread species. Green turtles range over great distances looking for the marine algae, turtle grass, and other plants on which they feed.

Sea turtles are justly famous for their remarkable feats of navigation. Sea turtles return at two-, three-, or four-year intervals to lay eggs on the beaches at which they were hatched. Homing behavior can be a great advantage to any animal: If the parent survived its earliest childhood at this location, it will probably be a suitable place for hatching the next generation. The navigation of green turtles to tiny Ascension Island, an emergent point of the Mid-Atlantic Ridge between Brazil and Africa, has been extensively studied. Researchers have found that the turtles use solar angle (to derive latitude), wave direction, smell, and visual cues first to find the island and then to discover the spot on the beach where they hatched perhaps 20 years before!

Marine Birds

13.16

Birds probably evolved from small, fast-running dinosaurs about 160 million years ago. Their reptilian heritage is clearly visible in their scaly legs and claws, and in the configuration of their internal organs and skeletons. The success of the 8,600 living species of birds is due in large part to the evolution of feathers (derivatives of reptilian scales) used to insulate the body and to provide aerodynamic surfaces for flight. Birds (and mammals) are *endothermic*; they generate and regulate metabolic heat to maintain a constant internal temperature that is generally higher than the temperature of their surroundings.

Flying birds have light, thin, hollow bones without fatty insulation; they have forsaken the heavy teeth and jaws of reptiles for a lightweight beak. Their highly efficient respiratory system can accept great quantities of oxygen, and their large four-chambered heart circulates blood under high pressure. All birds lay eggs on land, and most incubate them and provide care for the young. Some seabirds may stay at sea for years, but all must eventually return to land to breed.

Only about 270 kinds of birds, about 3% of known bird species, qualify as seabirds. Most seabirds live in the Southern Hemisphere. Like the marine reptiles, seabirds have special salt-excreting glands in their heads to eliminate the excess salt taken in with their food. Salty brine from these glands may sometimes be seen dripping from the tips of their beaks. Marine birds are voracious feeders, and the ocean will usually be teeming with life wherever they are found. True seabirds generally avoid land unless they are breeding, obtain nearly all their food from the sea, and seek isolated areas for reproduction.

Of the four groups of seabirds, the gulls and the pelicans may be most familiar to us because there are many species and because they spend much of their time near shore. But the groups best adapted to the pelagic world are the tubenoses (albatrosses, petrels) and the penguins.

The 100 species of tubenoses are the world's most oceanic birds. Their prosaic common name does not convey any sense of their beauty and grace; it refers to the plumbing in their beak responsible for sensing air speed, detecting smells, and ducting saline water from the salt glands. As you read in this chapter's opener, the supremely competent albatrosses—and their relatives, the petrels and shearwaters—are masters of the ocean's sky.

Penguins have completely lost the ability to fly, but they use their reduced wings to swim for long distances and with great maneuverability. Their flightlessness makes it practical to have fatty insulation, greasy peglike feathers, stubby appendages, and large size and weight; indeed, such heat-conserving adaptations are critical to the survival of aquatic organisms in very cold climates. Their neutral buoyancy is an advantage as they forage for food underwater. Emperor penguins, the largest of the living penguin species, may dive to depths of 265 meters (875 feet) and stay submerged for ten minutes or more. Penguins feed on fish, large zooplankters, bottom-dwelling mollusks or crustaceans, and squids. A few of the 18 species spend two uninterrupted years at sea between breedings.

Penguins are native only to the Southern Hemisphere and range from the size of a large duck to a height of more than 1 meter (3.3 feet) and weight exceeding 36 kilograms (80 pounds). They are thought to consume about 86% of all food taken by birds in the southern ocean—about 34 million metric tons (37 million tons) per year, mostly larger zooplanktonic crustaceans. The small Galápagos penguin lives a comparatively easy life fishing the cold, nutrient-rich Humboldt Current at the equator, but its Antarctic relatives lead what must surely be the most rigorous existence of any sea-

Figure 13.19 Emperor penguins, one of the larger species. Emperors subsist on a diet of mainly fish and squid.

bird. For example, Emperor penguins breed and incubate during the bitterly cold Antarctic winter (**Figure 13.19**).

Seabirds are indicators of oceanic productivity and are second only to marine mammals in consumption of large planktonic crustacea. The four groups of seabirds are the gulls, the pelicans, the even more pelagic tubenoses (albatrosses, petrels), and the penguins.

Marine Mammals

The class **Mammalia,** to which humans belong, is the most advanced vertebrate group. About 4,300 species of mammals are known. The three living groups of marine mammals are the porpoises, dolphins, and whales of order **Cetacea;** the seals, sea lions, walruses, and sea otters of order **Carnivora;** and the manatees and dugongs of order **Sirenia.**

Each of these orders arose independently from land ancestors. They exhibit the mammalian traits of being endo-thermic, breathing air, giving birth to living young that they suckle with milk from mammary glands, and having hair at some time in their lives. Unlike other mammals, however, these extraordinary creatures have become adapted to life in the ocean.

All marine mammals share four common features:

- Their *streamlined body shape* with limbs adapted for swimming makes an aquatic lifestyle possible. Drag is reduced by a slippery skin or hair covering.

- They *generate internal body heat* from a high metabolic rate and *conserve* this heat with layers of insulating fat and, in some cases, fur. Their large size gives them a favorable surface-to-volume ratio; with less surface area per unit of volume they lose less heat through the skin.

- The *respiratory system is modified* to collect and retain large quantities of oxygen. The biochemistry of blood and muscle is optimized for the retention of oxygen during deep, prolonged dives.

- A number of *osmotic adaptations* free marine mammals from any requirement for fresh water. Minimal intake of seawater, coupled with their kidneys' abilities to excrete a concentrated and highly saline urine, permits them to meet their water needs with the metabolic water derived from the oxidation of food.

ORDER CETACEA   The 79 living species of cetaceans are thought to have evolved from an early line of ungulates—hooved land mammals related to today's horses and sheep—whose descendants spent more and more time in productive shallow waters searching for food. Modern whales range from 1.8 to 33 meters (6 to 110 feet) in length and weigh up to 100,000 kilograms (110 tons). Their paddle-shaped fore-limbs are used primarily for steering, and their hind limbs are reduced to vestigial bones that do not protrude from the body. They are propelled mainly by horizontal tail flukes moved up and down by powerful muscles at the animal's posterior end. A thick layer of oily blubber provides insulation, buoyancy, and energy storage. One or two nostrils are located at the top of the head and have special valves to prevent intake of water when submerged. Whales have large, deeply convoluted brains and are thought to form complex family and social groupings.

Modern cetaceans are further divided into two suborders. **Figure 13.20** shows representative whales in each division. Whales in the suborder *Odontoceti*, the toothed whales, are active predators and possess teeth to subdue their prey. Toothed whales have a high brain-weight-to-body-weight ratio, and though much of their brain tissue is involved in formulating and receiving the sounds on which they depend for feeding and socializing, many researchers believe them to be quite intelligent. Smaller whales in this group include the killer whale and the familiar dolphins and porpoises of oceanarium shows. The largest toothed whale is the 18-meter (60-foot) sperm whale, which can dive to at least 1,140 meters (3,740 feet) in search of the large squids that provide much of its diet.

Toothed whales search for prey using **echolocation,** the biological equivalent of sonar. They generate sharp clicks and other sounds that bounce off prey species and return to be recognized. Odontocete whales are now thought to use sound offensively as well. Recent research indicates that some odontocetes can generate sounds loud enough to stun, debilitate, or even kill their prey. In one experiment dolphins produced clicks as loud as 229 decibels, equivalent to a blasting cap exploding close to the target organism. Sperm whales, it has been calculated, may generate sounds exceeding 260 decibels! (The decibel scale is logarithmic; compare this figure to the 130-decibel noise of a military jet engine at full power 20 feet away!) How this prodigious noise is generated is not yet known, nor do we know how such energy is radiated from the whale without damaging the organs that produce and focus it.

Whales in the suborder *Mysticeti*, the whalebone or baleen whales, have no teeth. Filter feeders rather than active predators, these whales subsist primarily on krill, a shrimp-like crustacean zooplankter obtained in productive polar or subpolar waters. They do not dive deep, but commonly feed a few meters below the surface. Their mouths contain inter-leaving triangular plates of bristly hornlike **baleen** used to filter the zooplankton from great mouthfuls of water (**Figure 13.21**). The plankton is concentrated as water is expelled, swept from the baleen plates by the whale's tongue, compressed to wring out as much seawater as possible, and swallowed through a throat not much larger in diameter than a grapefruit. A great blue whale, largest of all animals, requires about 3 metric tons (6,600 pounds) of krill each day during the feeding season. The short, efficient food chain from phytoplankton to zooplankton to whale provides the huge quantity of food required for their survival.

Mysticeti is an excellent name for these odd and wonderful animals. We know comparatively little of their social structure, intelligence, sound-producing abilities, navigational skills, or physiology. We do know that humpback whales use complex songs in group communication, and that blue whales may use very low frequency sound to communicate over tremendous distances. Some species migrate annually from polar to tropical waters and back. Until recently our primary response to all whales has been to slaughter them in countless numbers for meat and oil with little thought for their extraordinary abilities and assets. More of this depressing history may be found in the discussion of marine resources in Chapter 15.

ORDER CARNIVORA   The order Carnivora includes land predators ranging from dogs and cats to bears and weasels, but the members of the carnivoran suborder *Pinnipedia*—the seals, sea lions, and walruses (**Figure 13.22**)—are almost

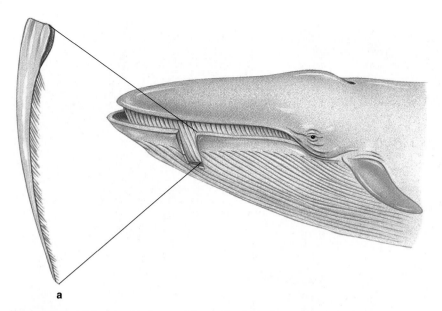

Figure 13.21 (**a**) A plate of baleen and its position in the jaw of a baleen whale. For clarity, the illustration shows part of the mouth cut away. (**b**) A student and a whale, up close and personal. This gray whale uses its stiff, coarse baleen plates to sieve crustaceans from shallow bottom mud.

exclusively marine. Unlike the cetaceans, the gregarious pinnipeds leave the ocean for varying periods of time to mate and raise their young.

True seals have a smooth head with no external ear flaps, the external part of the ear having been sacrificed to further streamline the body. They are covered with a short coarse hair without soft underfur. Seals are graceful swimmers that pursue small fish, their usual prey, with powerful side-to-side strokes of their hind limbs. These rear appendages are partially fused and always point back from the hind end of the body, so they are of very little use for locomotion on land.

The elephant seal, named for its long snout and large size, holds the diving depth record for all air-breathing vertebrates: 1,560 meters (5,120 feet). Sea lions, familiar to many as the performers in "seal" shows, have hind limbs with a greater range of motion and thus are more mobile on land. They have a streamlined head with small external ears and a pelt with soft underfur; unlike seals, they use their front flippers for propulsion. Walruses are much larger than either seals or sea lions and may reach weights of 1,800 kilograms (2 tons).

The suborder *Fissipedia* has many members (including cats, dogs, raccoons, and bears) but only one truly marine

Figure 13.22 Some representatives of the suborder Pinnipedia. (**a**) Three juvenile harbor seals eagerly await lunch in a marine mammal care facility. (**b**) A California sea lion, the "seal" of seal shows.

Figure 13.23 A female sea otter and her nearly mature offspring.

representative, the sea otter (**Figure 13.23**), a relative new-comer to the marine environment. Human demand for the fur, the densest and warmest of any animal, caused its near-extermination. The modern population of the Pacific sea otter descends from a very few individuals accidentally overlooked by fur hunters of the late nineteenth and early twentieth centuries. Playful and intelligent, otters rarely exceed 120 centimeters (4 feet) in length; they eat voraciously, consuming up to 20% of their body weight in mollusks, crustaceans, and echinoderms each day. Sometimes they lie on their back in the water, balance a rock on their chest, and hammer the shell of the prey against it until it cracks. They extract morsels of food with small nimble fingers, and rolling over in the water cleans away the debris. A morning spent watching sea otters in their coastal habitat is a morning well spent!

ORDER SIRENIA   The bulky, lethargic, small-brained dugongs and manatees, collectively called *sirenians*, are the only herbivorous marine mammals (**Figure 13.24**). Like the cetaceans, they appear to have evolved from the same ancestors as modern ungulates. They make their living grazing on sea grasses, marine algae, and estuarine plants in coastal temperate and tropical waters of North America, Asia, and Africa. Some species live in fresh water. The largest sirenians reach 4.5 meters (15 feet) in length and weigh 680 kilograms (1,500 pounds). They were first compared to mermaids by early Greeks who noted the manatee's habit of resting in an upright position in the water and holding a suckling calf to her breast. Sirenians have been hunted extensively—only about 10,000 individuals are thought to exist worldwide. Even though protected now, many are killed or wounded each year in Florida by the propellers of powerboats.

Marine mammals, all of which evolved from land ancestors, share common adaptations for life in the ocean. The three major groups of marine mammals are carnivorans, cetaceans, and sirenians.

Figure 13.24 A manatee, or sea cow.

http://info.brookscole.com/garrison

QUESTIONS FROM STUDENTS

1. Phytoplanktonic organisms account for a significant percentage of world primary productivity. Why are phytoplankton so successful? Is there something about the planktonic lifestyle that lends itself to high efficiency?

Phytoplankton are successful because of their size. Because water supports them, small marine autotrophs don't require the elaborate support systems of large terrestrial plants. Their small size allows easy diffusion of required nutrients into their single cells and prompt transport of wastes out; vessels and sap are not needed. Nearly all of their tissue is photosynthetic.

With transparent siliceous valves to protect them, diatoms are especially successful. The relatively recent evolution of diatoms was a high point in the development of marine autotrophs, and the additional oxygen these organisms contributed to the atmosphere has greatly influenced the success of life on land. More plants have led to more animals. After a few hundred million years of evolution, the planktonic lifestyle is elegantly tuned and interlocked into the complex and productive system we now observe in the world as a whole.

2. How do birds like the wandering albatross navigate across the trackless ocean?

No one is certain. Experiments done at Cornell University with homing pigeons suggest that homing birds use a combination of magnetic and optical cues to return to their starting points. Even polarized light and the positions of certain stars might be involved. However it works, the behavior is not learned but is instinctive to the bird. Study continues.

3. I've heard that sharks don't get cancer. Could we find out what prevents cancer in sharks and synthesize it for human use?

In fact, sharks do develop cancer. The mistaken impression that shark cartilage may prevent or cure cancer in humans has contributed to sales of shark-derived supplements (derived mainly from hammerheads and spiny dogfish) worth more than $25 million in 1998. A large percentage of the 100 million sharks killed each year are taken for these ineffective medicines.

4. How is a porpoise different from a dolphin?

Dolphin and *porpoise* are common names of two subtly different groups of odontocetes. *Porpoise*, as a term, refers to the smaller members of the group, which have spade-shaped teeth, a triangular dorsal fin, and a smooth front end tapering to a point. *Dolphins* are usually larger and have an extended bottlelike jaw filled with sharp, round teeth. The small jumping whales in most oceanarium shows are dolphins, but killer whales are a species of large porpoise. To make matters even more complicated, the common dolphin seen in ocean-themed amusement parks, *Tursiops truncatus*, is often referred to as a porpoise, even by show announcers. This confusion between common names points out how useful scientific names can be: The real name of the animal, *Tursiops*, is clear and unambiguous.

5. If some seals and whales dive to great depths, why don't they get the bends?

The bends is a painful and occasionally fatal condition brought on by nitrogen gas—the most plentiful component of air—leaving solution and forming bubbles in the blood. The condition is analogous to what happens in a newly opened soda bottle. Human divers take a source of air with them to depth, breathe the air under pressure, and dissolve excess nitrogen gas in their blood. If they have been down long enough, or deep enough, the release of pressure upon surfacing will be like taking the cap off the soda bottle. Whales don't use scuba tanks—they have no source of supplemental air at depth. They "tank up" at the surface by oxygenating their blood and tissues, but they have no excess gas to bubble from the blood at the end of their dives.

CHAPTER SUMMARY

The organisms of the pelagic world drift or swim in the ocean. (Animals and plants associated with the bottom are known as benthic organisms.)

The organisms that drift in the ocean are known collectively as plankton. The plantlike organisms that comprise phytoplankton are responsible for most of the ocean's primary productivity. Phytoplankton—and zooplankton, the small drifting or weakly swimming animals that consume them—are usually the first links in oceanic food webs. Plankton are most common along the coasts, in the upper sunlit layers of the temperate zone, in areas of equatorial upwelling, and in the southern subpolar ocean. Marine scientists have been inspired by the beauty and variety of plankton since first observing them under the microscope in the nineteenth century.

Actively swimming animals comprise the nekton. Nektonic organisms include invertebrates (such as squids, nautiluses, and shrimps) and vertebrates (such as fishes, reptiles, birds, and mammals). Each organism has a continuously evolving suite of adaptations that has brought it through the rigors of food finding, predator avoidance, salt balance, and temperature regulation—all the challenges of the marine environment—time and time again.

ON-LINE STUDY RESOURCES

The Web site for this book contains a wealth of helpful study aids. Log on to:

http://info.brookscole.com/garrison

and select *Essentials of Oceanography*, 3rd edition. Select Chapter 13 from the drop-down menu or click on one of these resource areas:

- For study and review, **Chapter at a Glance** gives you an outline of the chapter, **Chapter Summary** allows you to review the chapter's main ideas, and **Glossary** lists concepts and terms for the chapter along with their definitions.

- To test your mastery of the Terms and Concepts to Remember for this chapter, you can use the electronic **Flash Cards,** play the **Concentration** game, or work the **Crossword Puzzle.**
- For practice quizzing, try the multiple choice **Tutorial Quiz,** the ten-question **True/False Quiz,** or the **Image Analysis Quiz,** which poses questions based on art and photos from the chapter.

TERMS AND CONCEPTS TO REMEMBER

Arthropoda	krill
baleen	Mammalia
Carnivora	meroplankton
cartilage	mollusk
cephalopod	nanoplankton
Cetacea	nekton
Chondrichthyes	Osteichthyes
coccolithophore	pelagic (adj.)
diatom	phytoplankton
dinoflagellate	plankton
drag	plankton net
echolocation	salt gland
exoskeleton	schooling
flagella	Sirenia
foraminiferan	swim bladder
frustule	Teleostei
gas exchange	valve
gill membrane	vertebrate
holoplankton	zooplankton
invertebrate	

STUDY QUESTIONS

1. What are plankton? How are plankton collected? How are zooplankton different from phytoplankton?

2. Describe the most important phytoplankters. Which are most efficient in converting solar energy to energy in chemical bonds? By what means is this conversion achieved?

3. Where in the ocean is plankton productivity the greatest? Why?

4. Describe five nektonic organisms that are *not* fish.

5. What are the major categories of fishes? What problems have fishes overcome to be successful in the pelagic world?

6. How does a marine bird differ from a terrestrial bird—a pigeon, for example?

7. What characteristics are shared by all marine mammals?

8. Compare and contrast the groups of living marine mammals.

9. How are odontocete (toothed) whales different from mysticete (baleen) whales? Which are the better known and studied? Why?

10. Without looking ahead to Chapter 15, can you list some of the influences human activities have had on pelagic communities?

RESOURCES FOR FURTHER READING AND RESEARCH

The Web site for this book contains many ideas for further reading and research. Log on to:

http://info.brookscole.com/garrison

and select *Essentials of Oceanography,* 3rd edition. Select Chapter 13 from the drop-down menu or click on one of these resource areas:

- **Additional Readings** lists the major books and articles consulted in writing this chapter, along with comments from the author about their content and reading level.

- **Hypercontents** takes you to an extensive list of current links to Internet sites with news, research, and images related to individual subjects in the chapter. Just click on the icon that corresponds to a numbered section to see the links for that subject.

- **Internet Exercises** are critical thinking questions that involve research on the Internet with starter URLs provided.

- **InfoTrac College Edition Exercises** leads you to Critical Thinking Projects that use InfoTrac College Edition as a research tool.

- **Regional InfoTrac College Edition** articles are organized into East Coast, West Coast, and Gulf Coast regions, allowing you to study oceanography on a more local level.

 For more readings, go to InfoTrac College Edition, your on-line research library, at:

http://infotrac.thomsonlearning.com

a

Figure 14.9 A Pacific coast tide pool and intertidal shore. (**a**) A diagrammatic view. (**b**) Key.

Darwen Hennings

b

1 bushy red algae, *Endocladia*
2 sea lettuce, green algae, *Ulva*
3 rockweed, brown algae, *Fucus*
4 iridescent red algae, *Iridea*
5 encrusting green algae, *Codium*
6 bladderlike red algae, *Halosaccion*
7 kelp, brown algae, *Laminaria*
8 Western gull, *Larus*
9 intrepid marine biologist, *Homo*
10 California mussels, *Mytilus*
11 acorn barnacles, *Balanus*
12 red barnacles, *Tetraclita*
13 goose barnacles, *Pollicipes*
14 fixed snails, *Aletes*
15 periwinkles, *Littorina*
16 black turban snails, *Tegula*
17 lined chiton, *Tonicella*
18 shield limpets, *Collisella pelta*
19 ribbed limpet, *Collisella scabra*
20 volcano shell limpet, *Fissurella*
21 black abalone, *Haliotis*
22 nudibranch, *Diaulula*
23 solitary coral, *Balanophyllia*
24 giant green anemones, *Anthopleura*
25 coralline algae, *Corallina*
26 red encrusting sponges, *Plocamia*
27 brittle star, *Amphiodia*
28 commmon starfish, *Pisaster*
29 purple sea urchins, *Strongylocentrotus*
30 purple shore crab, *Hemigrapsus*
31 isopod or pil bug, *Ligia*
32 transparent shrimp, *Spirontocaris*
33 hermit crab, *Pagurus,* in turban snail shell
34 tide pool sculpin, *Clinocottus*

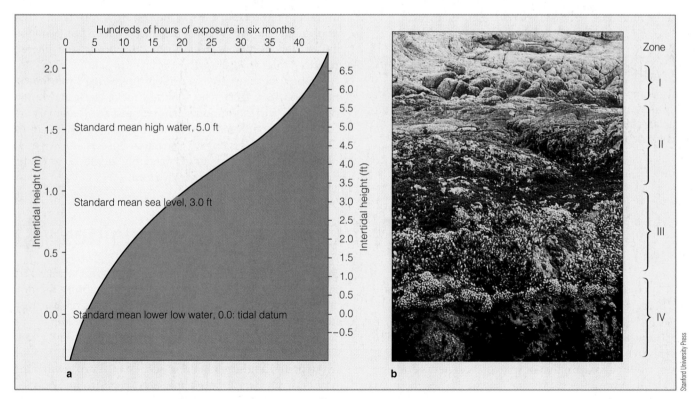

Hundreds of hours of exposure in six months

Intertidal height (m)

Standard mean high water, 5.0 ft

Standard mean sea level, 3.0 ft

Standard mean lower low water, 0.0: tidal datum

Intertidal height (ft)

Zone

I
II
III
IV

a

b

Stanford University Press

Figure 14.10 The relationship between amount of exposure and vertical zonation in a rocky intertidal community. (**a**) A graph showing intertidal height versus hours of exposure. The 0.0 point on the graph, the tidal datum, is the height of mean lower low water. (**b**) Vertical zonation, showing four distinct zones. The uppermost zone (I) is darkened by lichens and cyanobacteria; the middle zone (II) is dominated by a dark band of the red alga *Endocladia*; the low zone (III) contains mussels and gooseneck barnacles; and the bottom zone (IV) is home to sea stars (*Pisaster*) and anemones (*Anthopleura*). The bands in the photograph correspond approximately to the heights shown in the graph.

 http://info.brookscole.com/garrison

hours per week or month, the animals and plants sort themselves into three or four horizontal bands, or subzones, within the intertidal zone. Each distinct zone is an aggregation of animals and plants best adapted to the conditions within that particular narrow habitat. The zones are often strikingly different in appearance, even to a person unfamiliar with shoreline characteristics. This zonation is clearly evident in the rocky shore of **Figure 14.10b.**

For intertidal areas exposed to the open sea, wave shock is a challenging physical factor. **Motile** animals, like crabs, move to protective overhangs and crevices, where they cower during intense wave activity. Attached, or **sessile,** animals hang on tightly, often gaining assistance from rounded or very low-profile shells, which deflect the violent forces of rushing water around their bodies. Some sessile animals have a flexible foot that wedges into small cracks to provide a good hold; others, like mussels, form shock-absorbing cables that attach to something solid.

Desiccation (drying) by exposure to air and sunlight is another source of intertidal stress. Again, motile organisms have an advantage because they can move toward water left in tidal pools or muddy depressions by the retreating ocean. Producers and consumers must await the water's return, huddled in low spots, moist pockets, or cracks in the rocks, or within tightly closed shells. Water trapped within a shell can keep gills moist for the needed exchange of gases. A protective mucous coating can retard evaporative water loss from exposed soft animal body parts or blades of seaweed.

Though among Earth's most rigorous habitats, high-energy rocky and sandy shores are heavily populated by diverse organisms able to take advantage of the abundant feeding opportunities found there.

Sand Beach and Cobble Beach Communities

Not all intertidal areas are composed of firm rock; some are sandy, some are muddy, and others consist of gravel or cobbles. (A few shores combine these elements within a small area.) The usual rigors of the intertidal zone are intensified for organisms that survive on loose substrates. Indeed, it may surprise you to learn that in spite of its generally benign conditions, the ocean contains what may well be Earth's most hostile, rigorous, and dangerous environments for small living things: high-energy sand and cobble beaches.

As environments go, sand beaches don't seem particularly nasty places to us humans; most of my students consider the beach to be about the finest habitat around. Seals and sea lions spend a lot of time at the beach and seem to

enjoy the experience as much as people do. In short, for organisms of about our size, the problems of living on a beach are manageable.

For smaller organisms, however, a beach is a forbidding place. Sand itself is the key problem. Many sand grains have sharp pointed edges, so rushing water turns the beach surface into a blizzard of abrasive particles. Jagged grit works its way into soft tissues and wears away protective shells. A small organism's only real protection is to burrow below the surface, but burrowing is difficult without a firm footing. When the grain size of the beach is small, capillary forces can pin down small animals and prevent them from moving at all. If these organisms are trapped near the sand surface, they may be exposed to predation, to overheating or freezing, to osmotic shock from rain, or to crushing as heavy animals walk or slide on the beach.

As if this weren't enough, those that survive must contend with the difficulty of separating food from swirling sand and the dangers of leaving telltale signs of their position for predators or being excavated by crashing waves. A few can run for their lives: Some larger beach-dwelling crabs depend on their good eyesight and sprinting ability to outrace onrushing waves.

To these horrors must be added the usual problems of intertidal life discussed earlier. Not surprisingly, very few species have adapted to wave-swept sandy beaches! The few that have done so—mostly small, fast-burrowing clams and sand crabs (**Figure 14.11**) and sturdy polychaetes and other minute worms—consume a rich harvest of plankton and organic particles washed onto the beach and filtered from the water by the uppermost layer of sand.

Cobble beaches are even more uninviting (and they're murder on bare feet). The rounded rocks clack and bump together as waves pound the shore; most small animals are crushed. Except for nimble insectlike "beach hoppers" and a few species of scavenging terrestrial insects, most loose rock-strewn shores are understandably sterile of anything much larger than microscopic organisms.

Coral Reef Communities

The tropical ocean is blue, brilliantly transparent, relatively high in salinity, and notable for its deep and abrupt thermocline and relatively low surface concentrations of dissolved nutrients and gases. It supports surprisingly little life.

But what of the travel-poster view of the tropical ocean? What about the thousands of brightly colored fish, strange invertebrates, and breathtaking scenes of divers swimming through living reef formations? Such scenes, photographed on reefs at the edges of islands and continents, can be found in less than 2% of the tropical ocean. The key to the difference between the open tropical ocean and the tropical reefs lies in the productivity of the **reefs** themselves, wave-resistant structures dominated by strong and rigid masses of living (or once-living) organisms. Not all reefs are built of coral—other

a

b

Figure 14.11 Sand beach organisms. (**a**) Dime-sized coquina clams (*Donax*) lie at the surface awaiting a ride up the beach on an incoming wave. They will bury themselves in the loose sediment, push up their siphons, and filter the water for food. When the tide retreats, they will again pop to the surface and allow the waves to take them back down the beach. (**b**) A sand crab (*Emerita*), beloved of all beach-going children and beginning lab students, attempts to bury itself in anticipation of an onrushing wave. It gleans food from passing water with its feathery antennae.

reef builders include red and green algae, cyanobacteria, worms, even oysters—but we think first of coral reefs when the words *reef* and *tropics* are mentioned together.

Coral Animals

Although they look like flowers, **corals** are related to sea anemones and jellyfish. Some corals are solitary animals with bodies up to 30 centimeters (12 inches) in diameter, but most of the more than 500 species are ant-sized organisms crowded into colonies called **coral reefs** (**Figure 14.12**). The coral animals themselves construct the reefs by secreting hard skeletons of a crystalline form of calcium carbonate. The matrix of cup-shaped individual skeletons secreted by coral animals gives the colony its characteristic shape.

Figure 14.12 Close-up of hermatypic (reef-building) coral, showing expanded polyps.

An individual coral animal, or **polyp,** feeds by capturing and eating plankton that drift within reach of its tentacles. Victims are entrapped by stinging cells on each tentacle, transported to a central gastric cavity, and rapidly digested. Tropical corals feed at night; at dawn the polyps retract into their skeletal cups to withstand drying should the colony be exposed to air at low tide. The anatomy of a coral polyp is shown in **Figure 14.13.**

Tropical reef-building corals are **hermatypic,** a term derived from the Greek word for mound-builder. Their bodies contain masses of tiny symbiotic dinoflagellates. Coral's success in the nutrient-poor water of the tropics depends on its intimate biological partnership with specialized dinoflagellates. The microscopic dinoflagellates carry on photosynthesis, absorb waste products, grow, and divide within their coral host. The coral animals provide a safe and stable environment and a source of carbon dioxide and nutrients; the dinoflagellates reciprocate by providing oxygen, carbohydrates, and the alkaline pH necessary to enhance the rate of calcium carbonate deposition. The coral occasionally

Figure 14.17 A tripod fish, an abyssal benthic species. The long, curved projections are thought to aid in sensing the distant vibrations of prospective prey.

of prey many meters away (**Figure 14.17**). Some organisms whose mouths blend with the natural contours of the ooze act as living caves into which small creatures crawl for protection. The predator need not even swallow to get the prey into its gut—back-pointing spines direct the victim along a one-way path to the stomach! Other species are capable of smelling sunken dead organisms for miles downcurrent, and then spending weeks or months slowly following the scent to its source. The metabolic rate of organisms in cold water tends to be slow, so most deep-ocean animals require relatively little food, move slowly, and live very long lives. Some may feed less than once in a year and may live to be hundreds of years old. Deep benthic representatives are seen in **Figure 14.18.**

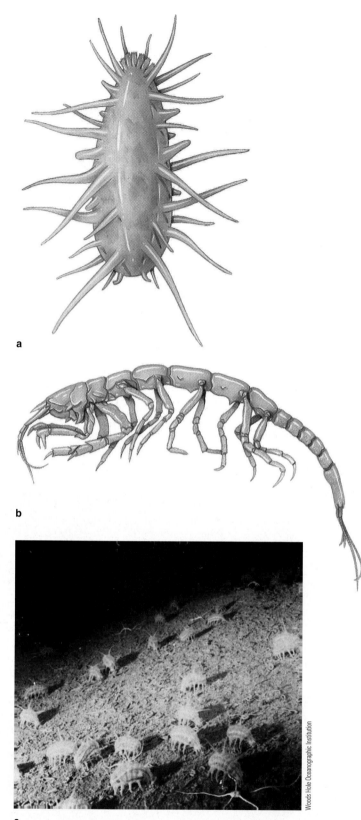

Figure 14.18 Abyssal benthic animals. (**a**) *Oneirophanta*, a 10-centimeter (4-inch) holothurian found on the abyssal plains of the North Atlantic. (**b**) *Apseudes*, a blind, thumb-sized crustacean found in the Kermadec Trench, north of New Zealand. (**c**) *Oneirophanta*—as in (**a**)—and brittle stars (foreground) search for food on a continental slope at a depth of about 1,000 meters (3,300 feet). Brittle stars like these are among the world's most cosmopolitan organisms, found on nearly all deep sediments.

The organisms within deep pelagic and benthic communities share some curious adaptations. Gigantism is a common characteristic; individuals of representative families in deep water often tend to be much larger than related individuals in the shallow ocean. Fragility is also common in the depths. Not only are heavy support structures unnecessary in the calm deep environment, but the low water pH and deficiency of dissolved calcium discourage skeletal development. Some animals have slender legs or stalks to raise them above the sediment, and some come apart like warm gelatin at the slightest touch. Except for its influence on enzyme activity, hydrostatic pressure is not a problem for these animals. Because they lack gas-filled internal spaces, their internal pressure is precisely the same as the pressure outside their bodies.

> The deep-sea floor is the ocean's most uniform habitat. It is populated by a large variety of highly specialized species.

Vent Communities

The oceanographic world was excited in 1976 when scientists from Scripps Institution of Oceanography discovered an entirely new type of marine community more than 3,000 meters (10,000 feet) below the surface. Using a towed camera platform, they were searching the seafloor along a spreading center 350 kilometers (220 miles) north and east of the Galápagos Islands. What they found were jets of superheated water (to 350°C, 650°F) blasting from rift vents in the young oceanic ridge. Clustered around the vents were dense aggregations of large, previously unknown animals. Bottom water in the area was laden with hydrogen sulfide, carbon dioxide, and oxygen, upon which specialized bacteria were found to live. These bacteria form the base of a food chain that extends to the unique animals. Large crabs, clams, sea anemones, shrimp, and unusual worms were found in this warm oasis.

Some of the "tube worms," contained in their own long parchmentlike tubes, were 3 to 4 meters (10 to 13 feet) long and the diameter of a human arm. These strange animals are pogonophorans, members of a small phylum of invertebrates also found in fairly shallow water. Three species of the appropriately named genus *Riftia* have been identified so far (**Figure 14.19a**). The tubes of these pogonophorans are flexible and capable of housing the length of the animal when it retracts. The animals extend tufts of tentacles from the openings of their tubes. Feeding was something of a puzzle because these animals have no mouth, digestive tract, or anus. The trunks of the worms were found to contain large "feeding bodies" tightly packed with bacteria similar to those seen in the water and on the bottom near the geothermal vents. The worms' tentacles absorb hydrogen sulfide from the water and transport it to the bacteria, which then use the hydrogen sulfide as an energy source to convert carbon dioxide to organic molecules. The ultimate source of the worms' energy (and the energy of most other residents in this community) is this energy-binding process, called chemosynthesis, which replaces photosynthesis in the world of darkness. Puzzle solved.

a

b

Figure 14.19 Some large organisms of hydrothermal vents. (**a**) *Riftia*, large tube worms (pogonophorans) that contain masses of chemosynthetic bacteria in special interior pouches. (**b**) A vent field dominated by the giant white clam *Calyptogena magnifica*. Each clam is about the size of a man's shoe and contains chemosynthetic bacteria within its gill filaments.

The clams and shrimp of the vent communities are equally unusual. For example, the large white clam *Calyptogena* grows among uneven basaltic mounds (**Figure 14.19b**). Each the size of a shoe, the clams shelter the same kinds of bacteria as *Riftia*. Though the clam retains its filter-feeding structures, it too derives nutrition from the bacteria. Small shrimp discovered at the vents in 1985 have been found to possess special organs that may allow them to see heat from the vents. Such an adaptation would permit them to range away from the vents for food, yet return to the warmth and richness of the home community.

Similar vent communities have now been found off Florida, California, and Oregon, and in several other locations along oceanic ridges. Cold seeps—and their attendant communities—have also been located, and these areas usually support a larger number of species than the hot-vent areas. Could vent communities occupy the active central rift valleys of a significant percentage of the 65,000 kilometers (41,000 miles) of Earth's oceanic ridges? Perhaps the deep-vent communities will prove to be more important in marine biology than has been previously supposed. Marine biologists are eager to continue their explorations.

Deep-vent communities, near black smokers and atop cold seeps, depend on chemosynthesis rather than photosynthesis for energy.

QUESTIONS FROM STUDENTS

1. If the rocky, sandy, or muddy intertidal zones represent such a challenging mix of environmental factors, why do so many organisms live there?

Difficulty in biology is a relative term. It may seem a circular argument, but wherever organisms live, conditions for life at that place are biologically tolerable, food is available, and environmental conditions are not so extreme as to preclude success. Organisms live in abundance where energy is available. Where there is food, or sunlight, or biodegradable compounds, there is life. Natural selection has sorted out the ways that work in this zone from the ways that do not, and the adaptations that work give the organisms living in the intertidal zone's many niches access to a rich harvest of nutrients.

2. Could there be any huge undiscovered Godzilla-type sea monsters in the deep ocean?

Probably not, unless they can extract energy directly from water molecules! The deep pelagic feeding situation is simply not rich enough to support the energy needs of an active population of violent, aggressive, city-eating (metrophagous?) reptiles. Scientists never say never, but classic science fiction films aside, it doesn't look promising.

3. Does the great diversity of marine species seen at the surface of some tropical ocean areas occur in the deep sea as well?

No. The cold, unchanging regions of the deep ocean are populated by the same kinds of specialized organisms all over the world. Down there, below the pycnocline, water is cold, food is scarce, and only a few species have adapted. A deep-bottom sample off Tahiti would yield the same sorts of brittle stars, sea cucumbers, and unusual cnidarians as a sample from similar depth in the Arctic Ocean. Conditions—and species—below about 3,500 meters (12,000 feet) are similar anywhere in the world ocean.

4. Can the specialized dinoflagellates that live within hermatypic coral polyps exist outside of a coral animal?

In laboratory cultures, yes. They change from the spherical shape seen within the coral to the typical biflagellate form characteristic of motile dinoflagellates. Researchers are uncertain whether all corals are host to the same species of dinoflagellates or whether several species exist. As far as we know, they do not normally live free in the ocean.

CHAPTER SUMMARY

Benthic organisms live on or in the ocean bottom. They may be distributed throughout their habitats randomly, uniformly, or (most commonly) in clumped distributions.

Nearshore benthic habitats in the temperate zones often contain multicellular algae, large nonvascular plants known as seaweeds. Carbohydrates produced by these highly productive large algae (and the vascular plants) can provide much of the energy needed by animals of the benthic communities.

Salt marshes and estuaries are among the ocean's most productive habitats, and estuaries shelter a remarkable variety of juvenile forms, some of which are refugees from the forbidding and competitive open-ocean environment. Rocky intertidal communities are among the ocean's richest and most diverse. Although the problems of rocky shore living are formidable, hundreds of organisms have adapted to its rigors because of the wealth of food available there. Sand and cobble beaches may seem more benign, but the difficulty of maintaining a dependable foothold and separating food from inedible particles severely limits the number of organisms able to live there. Except for vent communities associated with mid-ocean ridges, the deep seabed is the most sparsely populated benthic habitat due largely to the limited food supply. It stands in remarkable contrast to the world of tropical coral reefs, places of overwhelming productivity, diversity, and beauty.

ON-LINE STUDY RESOURCES

The Web site for this book contains a wealth of helpful study aids. Log on to:

http://info.brookscole.com/garrison

and select *Essentials of Oceanography*, 3rd edition. Select Chapter 14 from the drop-down menu or click on one of these resource areas:

- For study and review, **Chapter at a Glance** gives you an outline of the chapter, **Chapter Summary** allows you to review the chapter's main ideas, and **Glossary** lists concepts and terms for the chapter along with their definitions.

- To test your mastery of the Terms and Concepts to Remember for this chapter, you can use the electronic **Flash Cards,** play the **Concentration** game, or work the **Crossword Puzzle.**

- For practice quizzing, try the multiple-choice **Tutorial Quiz,** the ten-question **True/False Quiz,** or the **Image Analysis Quiz,** which poses questions based on art and photos from the chapter.

TERMS AND CONCEPTS TO REMEMBER

accessory pigments	intertidal zone
ahermatypic	kelps
atoll	mangroves
barrier reefs	motile
benthic	multicellular algae
blades	Phaeophyta
clumped distribution	polyp
coral reef	random distribution
corals	reef
desiccation	Rhodophyta
estuary	sessile
fringing reefs	stipes
gas bladders	thallus
hermatypic	uniform distribution
holdfast	wave shock

STUDY QUESTIONS

1. What factors influence the distribution of organisms within a benthic community? How are these distributions described? Why is random distribution so rare?

2. What are algae, and how are they different from plants? Are all algae seaweeds? How are seaweeds classified? Which seaweeds live at the greatest depths? Why?

3. What problems confront the inhabitants of the intertidal zone? How do you explain the richness of the intertidal zone in spite of these rigors? Which intertidal area has larger numbers of species and individuals: sand beach or rocky beach? Why?

4. Which benthic marine habitat is the most sparsely populated? Why?

5. If tropical oceans generally support very little life, why do coral reefs contain such astonishing biological diversity and density?

6. Explain Charles Darwin's classification scheme for coral reefs. Is the classification still in use?

7. What is the primary source of biological energy in rift vent and cold-seep communities?

RESOURCES FOR FURTHER READING AND RESEARCH

The Web site for this book contains many ideas for further reading and research. Log on to:

http://info.brookscole.com/garrison

and select *Essentials of Oceanography*, 3rd edition. Select Chapter 14 from the drop-down menu or click on one of these resource areas:

- **Additional Readings** lists the major books and articles consulted in writing this chapter, along with comments from the author about their content and reading level.

- **Hypercontents** takes you to an extensive list of current links to Internet sites with news, research, and images related to individual subjects in the chapter. Just click on the icon that corresponds to a numbered section to see the links for that subject.

- **Internet Exercises** are critical thinking questions that involve research on the Internet with starter URLs provided.

- **InfoTrac College Edition Exercises** leads you to Critical Thinking Projects that use InfoTrac College Edition as a research tool.

- **Regional InfoTrac College Edition** articles are organized into East Coast, West Coast, and Gulf Coast regions, allowing you to study oceanography on a more local level.

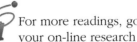 For more readings, go to InfoTrac College Edition, your on-line research library, at:

http://infotrac.thomsonlearning.com

Physical Resources

Petroleum and Natural Gas

Offshore petroleum and natural gas generated about $300 billion in worldwide revenues in 2001. The ocean makes a significant contribution to present world needs: About 35% of the crude oil and 26% of the natural gas produced in 2000 came from the seabed. About a third of known world reserves of oil and natural gas lie along the continental margins. Major U.S. marine reserves are located on the continental shelf of southern California, off the Texas and Louisiana Gulf coast, and along the North Slope of Alaska. The deep-sea floors probably contain little or no oil or natural gas.

Oil is a complex chemical soup containing perhaps a thousand compounds, mostly hydrocarbons. Petroleum is almost always associated with marine sediments, suggesting that the organic substances from which it was formed were once marine. Planktonic organisms or soft-bodied benthic marine animals are the most likely candidates. Their bodies apparently accumulated in quiet basins where the supply of oxygen was low and there were few bottom scavengers. The action of anaerobic bacteria converted the original tissues into simpler, relatively insoluble organic compounds that were probably buried—possibly first by turbidity currents and then later by the continuous fall of sediments from the ocean above. Further conversion of the hydrocarbons by high temperatures and pressures must have taken place at considerable depth, probably 2 kilometers (1.2 miles) or more beneath the surface of the ocean floor. Slow cooking under this thick sedimentary blanket for millions of years completed the chemical changes that produce oil.

If the organic material cooked too long, or at too high a temperature, the mixture turned to methane, the dominant component of natural gas. Deep sedimentary layers are older and hotter than shallow ones and have higher proportions of natural gas to oil. Very few oil deposits have been found below a depth of 3 kilometers (1.8 miles). Below about 7 kilometers (4.4 miles), only natural gas is found.

Oil is less dense than the surrounding sediments, so it can migrate from its source rock through porous overlying formations. It collects in the pore spaces of reservoir rocks when an impermeable overlying layer prevents further upward migration of the oil (**Figure 15.1**). When searching for oil, geologists use sound reflected off subsurface structures to look for the signature combination of layered sediments, depth, and reservoir structure before they drill.

Drilling for oil offshore is far more costly than drilling on land because special drilling equipment and transport systems are required. Most marine oil deposits are tapped from offshore platforms resting in water less than 100 meters (330 feet) deep. As oil demand (and therefore price) continues to rise, however, deeper deposits farther offshore will be exploited from larger platforms. The largest and heaviest platform is *Statfjord-B*, in position since 1981 northeast of the Shetland Islands in the North Sea. The tallest drilling platform is *Ursa*, a tension-leg structure deployed by Shell Oil

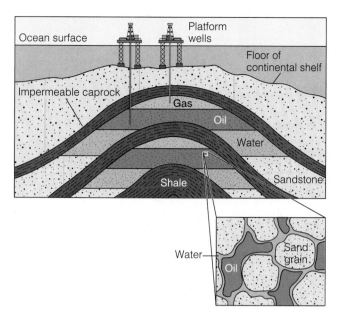

Figure 15.1 Oil and natural gas are often found together beneath a dome of impermeable caprock. Oil and gas are not found in great hollow reservoirs, but within pore spaces in rock (inset).

Company in 1998 (**Figure 15.2a**). *Ursa* floats in water more than 1,160 meters (3,800 feet) deep and is held in position off the Louisiana coast by 16 lengths of 81-centimeter (32-inch) steel cables anchored into the seabed and pulling against the semisubmerged platform's buoyancy (**Figure 15.2b**). At full production the 14 wells on platform *Ursa* are projected to produce 30,000 barrels of oil and 80 million cubic feet of natural gas per day. Total cost of platform *Ursa* exceeded $1.45 billion. Even larger tension-leg platforms are being planned.

The fastest-growing alternative to oil as an energy source is wind power (**Figure 15.3**). The world's largest "wind farm" stretches across 130 square kilometers (50 square miles) of high desert ridges in eastern Oregon and Washington. Its 460 turbines will power 70,000 homes and businesses. Unlike oil and natural gas, wind can't be used up. If the present rate of development continues, wind could provide 12% of Earth's electricity by 2025.

Though expensive to obtain, offshore oil and natural gas presently provide nearly a third and a quarter (respectively) of world needs. Wind power is the most promising energy alternative to oil and natural gas.

Sand and Gravel

Sand and gravel are not very glamorous physical resources, but they are second in dollar value to oil and natural gas. More than 1 billion metric tons (1.1 billion tons) of sand and gravel, valued at more than half a billion dollars, were mined offshore in 1998. Only about 1% of the world's total sand and

a

Figure 15.2 (**a**) Shell Oil Company's tension-leg platform *Ursa*, deployed in 1,160 meters (3,800 feet) of water 108 kilometers (130 miles) southeast of New Orleans. (**b**) Platform *Mars*, deployed off Louisiana in 912 meters (2,933 feet) of water in 1996, is pictured here in this imaginary view in relation to the Houston skyline. Tension-leg platforms like *Ursa* and *Mars* are held in position by steel cables anchored into the seabed, pulling against the semisubmerged platform's buoyancy. Twelve lateral cables anchored to the seabed (not shown) prevent sideways movement. Both platforms are designed to withstand hurricane-force waves of 22 meters (71 feet) and winds of 225 kilometers (140 miles) per hour.

gravel production is scraped and dredged from continental shelves each year, but the seafloor supplies about 20% of the sand and gravel used in the island nations of Japan and the United Kingdom. The world's largest single mining operation is the extraction of aragonite sands at Ocean Cay in the Bahamas. Sand is suction-dredged onto an artificial island and then shipped on specially designed vessels. This sand, about 97% calcium carbonate, is used in Portland cement, glass, and animal feed supplements, and also in the reduction of soil acidity.

Most of the exploitable U.S. deposits of marine sand and gravel are found off the coasts of Alaska, California, Washington, and the East Coast states from Virginia to Maine. Nearshore deposits are widespread, easily accessible, and

b

Figure 15.3 This installation off the coast of Denmark is one of the world's largest "wind farms." On some windy days, Denmark is said to have a 100% supply of electricity from wind power. Generation of electricity from wind is the fastest growing source of energy in the world.

used extensively in buildings and highways. Offshore oil wells in Alaska are built on huge man-made gravel platforms; the large quantities of gravel available at those locations make offshore drilling practical there.

Salts

As you may remember from Chapter 6, the ocean's salinity varies from about 3.3% to 3.7% by weight. When seawater evaporates, the remaining major constituent ions combine to form various salts, including calcium carbonate ($CaCO_3$), gypsum ($CaSO_4$), table salt (NaCl), and a complex mixture of magnesium and potassium salts (see Figure 6.18). Table salt makes up slightly more than 78% of the total salt residue.

Seawater is evaporated in large salt ponds in arid parts of the world (**Figure 15.4**). Operators can segregate the various salts from one another by shifting the residual brine from pond to pond at just the right time during the evaporation process. The magnesium salts are used as a source of magnesium metal and magnesium compounds. The potassium salts are processed into chemicals and fertilizers. Bromine (a useful component of certain medicines, chemical processes, and antiknock gasoline) is also extracted from the residue. Gypsum is an important component of wallboard and other building materials. About a third of the world's table salt is currently produced from seawater by evaporation. In North America, some of this salt is used for snow and ice removal. Salt is also used in water softeners, agriculture, and food processing. In 1998 the United States

Figure 15.4 Salt evaporation ponds at the southern end of San Francisco Bay in California. Operators can segregate the various salts from one another by shifting the residual brine from pond to pond at just the right time during the evaporation process. The colors in the ponds are imparted by algae and other microorganisms that thrive at varying levels of salinity. In general, the ponds with the highest salinity have a reddish cast.

produced by evaporation about 4.7 million metric tons (5.2 million tons) of table salt with a value of about $145 million.

Fresh Water

Only 0.017% of Earth's water is liquid, fresh, and available at the surface for easy use by humans. Another 0.6% is available as groundwater within half a mile of the surface, unfortunately, much of this is polluted or otherwise unfit for human consumption. The fact that fresh, pure water often costs more per gallon than gasoline underscores its scarcity and importance. More than any other factor in nature, the availability of **potable water** (water suitable for drinking) determines the number of people who can inhabit any geographic area, their use of other natural resources, and their lifestyle.

Fresh water is becoming an important marine resource. Exploitation of that resource by **desalination,** the separation of pure water from seawater, is already under way, mainly in the Middle East, West Africa, Peru, Florida, and Texas. More than 2,000 desalination plants are currently operating worldwide, producing a total of about 18 billion liters (4.7 billion gallons) of fresh water per day. The largest desalination plant, in Saudi Arabia, produces about 114 million liters (30 million gallons) daily!

Several desalination methods are currently in use. *Distillation* by boiling is the most familiar; about three-fourths of the world's desalinated water is produced in this way. Distillation uses a great deal of energy, making it a very expensive process. *Freezing* is another effective but costly method of desalination; ice crystals exclude salt as they form, and the ice can be "harvested" and melted for use. Solar or geothermal power may bring down the cost of distillation or freezing, but more efficient, less energy-intensive mechanisms are being developed. Among these is *reverse osmosis desalination*. In this process seawater is forced against a semipermeable membrane at high pressure (**Figure 15.5**). Fresh water seeps through the membrane's pores while the salts stay behind. About a quarter of desalinated water is produced in this way. Reverse osmosis uses less energy per unit of fresh water produced than distillation or freezing, but the necessary membranes are fragile and costly.

Biological Resources

Ancient kitchen middens (garbage dumps of bones and shells) found in many coastal regions demonstrate that humans have used the sea for thousands of years as a source of food and medicines. Now the human population threatens to outgrow its food supply. Contemporary food production and distribution practices are unable to satisfy the nutritional needs of all the world's 6.1 billion people, and starvation and malnutrition are major problems in many nations. Can the ocean help?

Figure 15.5 Part of the $108-million reverse osmosis desalination plant in Tampa Bay, Florida. The nation's largest, this freshwater plant produces 25 million gallons of drinking water each day.

Since 1961, human demand on Earth's biological marine resources has doubled. Compared to the production from land-based agriculture, the contribution of marine animals and plants to the human intake of all protein is still small, probably around 4%. Although most of that protein comes from fish, marine sources account for only about 18% of the total *animal* protein consumed by humans. Fish, crustaceans, and mollusks contribute about 14.5% of the total; fish meal and by-products included in the diets of animals raised for food account for another 3.5%. About 85% of the annual catch of fish, crustaceans, and mollusks comes from the ocean, and the rest from fresh water.

The sea will probably not be able to provide substantially more food to help alleviate future problems of malnutrition and starvation caused by human overpopulation; indeed, population growth will likely absorb any resource increase. Nevertheless, these resources currently sustain a great many people.

Fish, Crustaceans, and Mollusks

Fish, crustaceans, and mollusks are the most valuable living marine resources. Commercial fishermen took more than 130 million metric tons (143 million tons) of these animals in 2000. The distribution of the 2000 catch is shown in **Figure 15.6.**

Table 15.1 World Commercial Catch and Aquaculture Yield, 2000	
Species Group	**Millions of Metric Tons, Live Weight**
Herrings, sardines, anchovies	24.7
Carps, barbels, cyprinids	16.6
Cods, hakes, haddocks	8.7
Tunas, bonitos, billfishes	5.7
Salmons, trouts, smelts	2.3
Tilapias	1.9
Other fishes	42.7
Shrimps, crabs, lobsters, krill	5.6
Molluscs (oysters, scallops, squid, etc.)	18.3
Sea urchins, other echinoderms	0.1
Miscellaneous	3.8
Total for all marine sources, excluding marine mammals, marine algae, and aquatic plants	**130.4**

Of the thousands of species of marine fishes, crustaceans, and mollusks, fewer than 500 species are regularly caught and processed. The major groups listed in **Table 15.1** supply nearly 95% of the commercial marine catch. Note that

Figure 15.6 The top five marine fish harvesting nations and the top marine fishing areas. (Source: *National Geographic,* January 2001. Reprinted with permission.)

—— Fishing area boundary

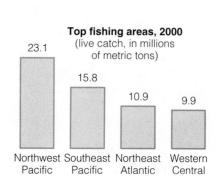

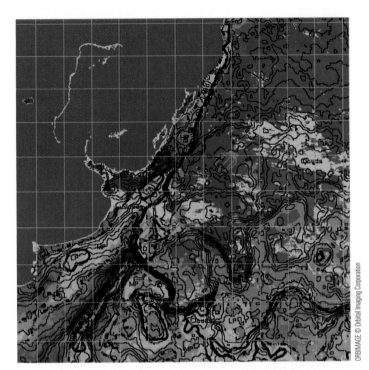

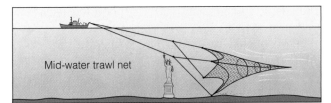
Mid-water trawl net

a

Figure 15.7 Orbimage's SeaStar Fisheries Information Service uses data from the Sea-Viewing Wide Field-of-View Sensor (SeaWiFS) on its OrbView-2 satellite to generate fish-finding charts like this one. This image, from information collected on 4 March 2003, shows high concentrations of plankton (green) off coastal Argentina, Uruguay, and Brazil. Computer models suggest the red hatch-marked areas contain high concentrations of desirable fish.

ORBIMAGE © Orbital Imaging Corporation

the largest commercial harvest is of the herring and its relatives, which account for about a fifth of the live weight of all living marine resources caught each year.

Fishing is a big business, employing more than 15 million people worldwide. As you read in this chapter's opener, it is also the most dangerous job in the United States: Commercial fishers suffer 155 deaths per 100,000 workers each year.[1] Though estimates vary widely, the value in 2000 for the worldwide marine catch was around $109 billion. Slightly more than half of the world's commercial marine catch is taken by the five countries listed in Figure 15.6. About 75% of the annual harvest is taken by commercial fishers who operate vast fleets working year-round, using satellite sensors, aerial photography, scouting vessels, and sonar to pinpoint the location of fish schools (**Figure 15.7**). Huge factory ships often follow the largest fleets to process, can, or freeze the animals on the run. Catching methods no longer depend on hooks and lines but rather on large trawl nets (**Figure 15.8**), purse seines, or gill nets. The living resources of the ocean are under furious assault: Between 1950 and 1997 the commercial marine fish catch increased more than fivefold.

© Shane Moore, Animals Animals/Earth Scenes

b

Figure 15.8 (**a**) Stern trawler fishing. After sonar on the trawler finds the fish, they're captured by a trawl net more than 122 meters (400 feet) wide. Boards angled to the water flow keep the net's mouth open. The largest nets extend about 0.8 kilometer (½ mile) behind the towing vessel and are large enough to hold a dozen 747 jetliners. (**b**) A haul of Alaskan pollock from the Bering Sea. (Pollock is a codlike fish popular in fish sticks and fast food.) About 60% of the fish landed in the United States are pulled from the Bering Sea, a resource worth $1.1 billion before processing. Once thought inexhaustible, pollock stocks are plunging—about 25% of the pollock in the Bering Sea is caught each year.

---

[1] Sebastian Junger's extraordinary 1997 book (and the 2000 movie) *The Perfect Storm* chillingly recalls these dangers.

The cost to obtain each unit of seafood has risen dramatically in spite of all this high-tech assistance. The increasing expense of fuel for the fishing fleets and processing plants, the rising cost of wages for the crews, and the greater distances that boats must cover to catch each ton of fish have all helped drive up the cost of seafood. In spite of greater efforts, the total marine catch leveled off in about 1970 and remained surprisingly stable until 1980, when greater demand and increasing prices began to drive the tonnage upward again. Another plateau occurred from 1989 to 1992. Then, under renewed attack, ocean yields rose again.

Despite new investments, world fisheries have almost certainly peaked, and we may now look forward to a period of declining harvests. Since 1970 the world human population has grown, so the average *per capita* world fish catch has fallen sharply.

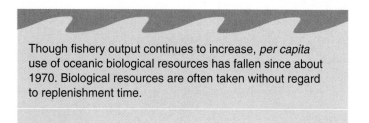

Though fishery output continues to increase, *per capita* use of oceanic biological resources has fallen since about 1970. Biological resources are often taken without regard to replenishment time.

Fishery Mismanagement   15.9

Can we continue to take huge amounts of food from the ocean year after year? The **maximum sustainable yield,** the maximum amount of each type of fish, crustaceans, and mollusks that can be caught without jeopardizing future populations, probably lies between 100 and 135 million metric tons (110–150 million tons) annually. As can be seen in Table 15.1, current yield is approaching the upper figure. Fleets are obtaining fewer tons per unit of effort and are ranging farther afield in their urgent search for food. The world's commercial catch peaked in 1989. We may be perilously close to the catastrophic collapse of more fisheries. As of 1997, 30% of recognized marine fisheries were over-exploited or already depleted, and 50% more were at their presumed limit of exploitation. The U.S. National Marine Fisheries Service estimates that 45% of the fish stocks whose status is known are now **overfished**—so many fish have been harvested that there is not enough breeding stock left to replenish the species. Recent trends can be inferred from **Figure 15.9.**

Even when faced with evidence of overfishing, the fishing industry seldom follows a rational course. The industry's dominant motivating force is quick financial return, even if it means depleting a stock and disrupting the equilibrium of a fragile ecosystem. Long-term stability is forsaken for short-term profit. When the catch begins to drop, the industry increases the number of boats and develops more efficient techniques for capturing the remaining animals. When the impending catastrophe is obvious, governments will sometimes intervene to set limits or close a fishery altogether. In 1999 New England officials approved a plan to close a large

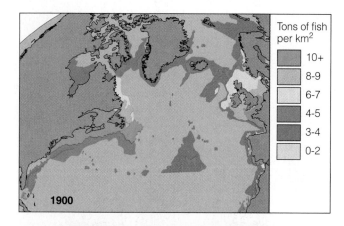

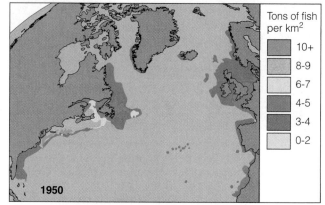

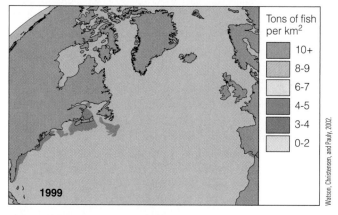

Figure 15.9 A century of dramatic decline in the fisheries of the North Atlantic. These data are for table fish (to be eaten directly by people), not for fish collected for oil or animal feed.

section of the Gulf of Maine to fishing in an attempt to replenish once-abundant cod stocks. As many as 2,500 fishers and 700 boats were idled, and losses may exceed $21 million a year. Given the choice between immediate profit and a long-term sustainable fishery, fishers made their priorities clear by initiating a recall campaign against officials who had approved the ban!

The intended organism is not the only victim. In some fisheries, **bykill**—animals unintentionally killed while desirable organisms are collected—exceeds the target catch. In the shrimp fishery, discarded creatures outnumber targeted shrimp by 125% to 830%. In 1993 in the Gulf of Mexico, fishing for shrimp resulted in a bykill of 34 million red snappers

Figure 15.10 An unintentional but unavoidable consequence of drift net fishing.

and 2,800 metric tons (3,090 tons) of sharks. Albatrosses are endangered in the southern ocean because they take baited longline hooks cast for southern bluefin tuna. Bottom trawling is, if anything, even more devastating. The habitat itself is disturbed; slow-growing organisms and complex communities are ransacked. Again, bykill exceeds target organisms by a wide margin. In 1995 the governor of Alaska said, "Last year's [Alaska bottom fishing] discards would have provided about 50 million meals." Worldwide bykill exceeded 27 million metric tons (30 million tons) in 1995, *a quantity nearly one-third of total landings!*

There are other disruptive, mistargeted fishing techniques. One employs **drift nets**—fine, vertically suspended nylon nets as large as 7 meters (25 feet) high and 80 kilometers (50 miles) long. Drift net technology was pioneered by a United Nations agency to help impoverished Asian nations turn a profit from what had been subsistence fishing. Until 1993, Taiwanese, Korean, and Japanese vessels deployed some 48,000 kilometers (30,000 miles) of these "walls of death" each night—more than enough to encircle Earth! Drift nets caught the fish and squid for which they were designed, but they also entangled everything else that touched them, including turtles, birds, and marine mammals (**Figure 15.10**). The process has been compared to stripmining—the ocean is literally sieved of its contents. An estimated 30 kilometers (18 miles) of these fine, nearly invisible nets were lost each night—about 1,600 kilometers (1,000 miles) per season. These remnants, made of nonbiodegradable plastic, become *ghost nets*, continuing to entangle fish and other animals perhaps for decades.

Though large-scale drift net fishing is now banned, pirate netting continues in the Pacific and has begun in the Atlantic. Completely unrestrained by regulation, these out-

law fishermen operate where their activities cause maximum damage to valuable reproductive stock.

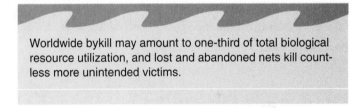

Worldwide bykill may amount to one-third of total biological resource utilization, and lost and abandoned nets kill countless more unintended victims.

Figure 15.11 The ocean under furious assault. The Chinese and Korean fishing fleets seen here in June 2001 are fishing for bait and squid across the South China Sea. From horizon to horizon, their bright lights pierced the dusk.

Figure 15.12 Whales being slaughtered for food.

"Madhouse Economics"

The fishing industry spent $124 billion to catch $70 billion worth of fish and other marine life in 1995. Government subsidies made up most of the $54-billion deficit. Artificial supports included fuel-tax exemptions, price controls, low-interest loans underwritten by government or state agencies, and grants of equipment. More than 1,600 large fishing vessels have been reflagged to maneuver around legal requirements in their original countries or to take advantage of subsidies. In an effort to preserve well-paying jobs, governments encourage construction of underutilized ships and research in new and often-destructive technology. The efficiency of the new ships reduces the number of fish available, thereby reducing the value of hulls and equipment on the market. Unable to sell their vessels, owners are forced to continue fishing to repay their loans. In an important 1995 paper in *Scientific American*, Carl Safina understandably described the situation as "madhouse economics."

Whaling

Since the 1880s, whales have been hunted to provide meat for human and animal consumption; oil for lubrication, illumination, industrial products, cosmetics, and margarine; bones for fertilizers and food supplements; and baleen for corset stays. **Figure 15.12** shows whales being butchered for food and oil. An estimated 4.4 million large whales existed in 1900; today slightly more than 1 million remain. Eight of the 11 species of large whales once hunted by the whaling industry are commercially extinct. The industry pursued immediate profits despite obvious signs that most of the "fishery" was exhausted.

Substitutes exist for all whale products, but the harvest of most commercial species did not stop until whaling became uneconomical. In 1986 the International Whaling Commission, an organization of whaling countries established to manage whale stocks, placed a moratorium on the slaughter of large whales. Except for a suspiciously large harvest of whales taken by the Japanese for "scientific purposes," commercial whaling ceased in 1987. Fewer than 700 large whales were taken in 1988.

Now, however, the number being taken is rising. What protection there is may have come too late to save some species from extinction—and protection may be only temporary. Under intense pressure from its major fishing industry, Norway resumed whaling in 1993. Japan never stopped. Their target, the minke whale, is the smallest and most numerous of the great whale species (see Figure 13.20). The meat and blubber of this whale are prized in Japan as expensive delicacies. The minke whale population has been estimated at 1.2 million, a number that may withstand the present level of harvest. A decision to increase minke whaling, however, may doom the minke to the same fate as most other whale species. Between them, the Norwegians and the Japanese caught 1,078 minke whales in 1999. In 2000, Japan extended "scientific" whaling to sperm whales and Bryde's whales.

There is a glimmer of hope. In 1994 the International Whaling Commission voted overwhelmingly to ban whaling in about 21 million square kilometers (8 million square miles) around Antarctica to protect most of the remaining large whales, which feed in those waters. The sanctuary is often ignored, however. In their quest for minke (and other whales), Japanese and Norwegian whalers have entered the area and harpooned many animals. Chile, Peru, and North Korea are now considering joining the hunt. Pirate whalers based in other countries can also catch whales in the Antarctic and sell the flesh on the Japanese market. Still, conservation efforts can do some good. Although it was hunted nearly to extinction, the California gray whale has long been off limits to all but aboriginal hunters. Its numbers have grown, and it was removed from the endangered list in 1993.

Many more small whales have been killed than large ones, but not for food or raw materials. For reasons that are not well understood, dolphins (which are small whales) gather above schools of yellowfin tuna in the open sea. Fishermen have learned to find the tuna by spotting the dolphins. Nets cast to catch the tuna also entangle the air-breathing dolphins, and the mammals drown. The dead dolphins are simply pitched over the side as waste. More than 6 million small whales have been killed in association with tuna fishing since 1971.

Passage of the U.S. Marine Mammals Protection Act in 1972 brought a drop in the number of dolphin deaths. However, from 1977 to 1987 the number of tuna boats in the U.S. fleet declined from more than 100 boats to 34 boats, while the foreign fleet rose from fewer than 10 to more than 70 boats. Foreign fishermen do not always abide by the dolphin-saving provisions of the act, but the U.S. Congress voted in 1988 to ban importation of tuna not caught in accordance with new methods designed to reduce the kill. In 1990, Amer-

ican tuna processing companies agreed to buy tuna only from fishermen whose methods do not result in the deaths of dolphins. American commercial fishermen have agreed to comply, and conservationists hope that the foreign fleets will follow their lead.

Whaling is legal and encouraged in some nations. Thousands of whales are still killed each year for food and oil, or by accident in tuna nets.

Botanical Resources

Marine plants are also commercially exploited. The most important commercial product is **algin,** made from the mucus that slickens seaweeds. When separated and purified, algin's long, intertwining molecules are used to stiffen fabrics; to form emulsions such as salad dressings, paint, and printer's ink; to prevent the formation of large crystals in ice cream; to clarify beer and wine; and to suspend abrasives. The U.S. seaweed gel industry produces more than $220 million worth of algin each year, and the annual worldwide value of products containing algin (and other seaweed substances) was estimated to be more than $42 billion in 2000.

Seaweeds are also grown to be eaten directly (**Figure 15.13**). The Japanese consume 150,000 metric tons (165,000 tons) of *nori* each year; seaweed and seaweed extracts are also eaten in the United States, Britain, Ireland, New Zealand, and Australia. Their mineral content and fiber are useful in human nutrition.

Aquaculture and Mariculture

Aquaculture is the growing or farming of plants and animals in any water environment under controlled conditions. Aquaculture production currently accounts for more than one-quarter of all fish consumed by humans. Most aquaculture production

Figure 15.13 Orderly plots of seaweed grown for human food off islands east of Bali.

occurs in China and the other countries of Asia. In 1998, more than 28 million metric tons (31 million tons) of fish—mostly freshwater fish—were produced worldwide. By 2010, fish aquaculture could overtake cattle ranching as a food source.

Mariculture is the farming of *marine* organisms, usually in estuaries, bays, or nearshore environments, or in specially designed structures using circulated seawater. Mariculture facilities are sometimes placed near power plants to take advantage of the warm seawater that flows from their cooling condensers.

Worldwide mariculture production is thought to be about one-eighth that of freshwater aquaculture. Several species of fish, including plaice and salmon, have been grown commercially, and marine and brackish-water fish account for 67% of the total production. Shrimp mariculture is the fastest growing and most profitable segment, with an annual global value exceeding $8 billion in 2000.

In the United States, annual revenue from mariculture exceeds $175 million, with most of the revenue generated from salmon and oyster mariculture (**Figure 15.14**). About half of the oysters consumed in North America are cultured. Not all mariculture produces food, however; cultured pearls are an important industry in Japan, China, and northern Australia. Japan leads the world in mariculture.

Ranching, or open-ocean mariculture, is an interesting variation. Juveniles are grown in a certain area, released into the ocean, and expected to return when mature. The natural homing instinct of salmon makes salmon ranching quite successful; about 20% of the world supply of salmon is ranched. Yet only about 1 in 50 ranched juveniles return to the area of release. In Japan, recent attempts to extend ranching techniques to yellowtail and tuna have met with only limited success.

Mariculture is an expanding industry; it is growing at about 8% annually in contrast to a growing decline for the world marine fishery as a whole. Mariculture produces mostly luxury seafoods such as oysters and abalone, and it uses fish meal from so-called *trash fish* as feed. (Trash fish are edible but considered unappealing because they are bony or bad tasting.) As the world population increases and the demand for protein grows among the world's millions of undernourished people, however, today's trash fish may become more acceptable human food.

Aquaculture accounts for more than one-quarter of all fish consumed by humans. Shrimp mariculture is the most profitable and fastest growing sector.

Drugs

The earliest recorded use of medicines derived from marine organisms appears in the *Materia Medica* of the emperor Shen Nung of China in 2700 B.C. Modern medical researchers estimate that perhaps 10% of all marine organisms are likely

Courtesy of the British Columbia Salmon Farmers' Association

Figure 15.14 Oyster mariculture in Puget Sound. Oyster larvae attached to old oyster shells are put in screen bags, which are placed on intertidal mud flats. The bags must be lifted monthly to clean out excess mud. Oysters are ready for harvesting in about three years.

to yield clinically useful compounds. One such medicine is derived from a Caribbean sponge and is already in use: Acyclovir, the first antiviral compound approved for humans, has been fighting herpes infections of the skin and nervous system since 1982. A class of anti-inflammatory drugs known as pseudopterosins, developed by researchers at the University of California, has also been successful.

Newly discovered compounds are also being tested. A common bryozoan—a small encrusting invertebrate—has been found to produce a potent anticancer chemical that is now being tested in human volunteers. Extracts from 30% of all tunicate species investigated show antiviral and antitumor activity; one of these extracts, Didemnin-B, shows promise as a treatment for malignant melanoma, the deadliest form of skin cancer. Another tunicate derivative, Esteinascidin 743, has been found to be useful in the treatment of skin, breast, and lung cancers. A related compound found in the same organism, Aplidine, shows promise in shrinking tumors in cases of pancreatic, stomach, bladder, and prostate cancer.

Cancer is not the only target. A compound derived from cyanobacteria stimulated the immune system of test animals by 225% and cells in culture by 2,000%; the drug may be useful in treating AIDS. Vidabarine, another antiviral drug developed from sponges, may attack the AIDS virus directly.

Nonextractive Resources

Transportation and recreation are the main nonextractive resources the oceans provide. People have been using the ocean for transportation for thousands of years. Through most of this time the transport of cargo has produced far more revenue than the movement of passengers. At present, oil tankers ship the greatest gross tonnage of any type of cargo (more than 275 million metric tons—303 million tons—annually). Oil accounts for about 53% of the total value of world trade transported by sea; iron, coal, and grain make up 24% of the rest.

Nearly half of the world's crude oil production is transported to market by ships. Tankers are needed because very few of the major oil-drilling sites are close to areas where the demand for refined oil products is highest. The largest tankers are more than 430 meters (1,300 feet) long and 66 meters (206 feet) wide, and they carry more than 500,000 metric tons (3.5 million barrels) of oil.

Modern harbors are essential to transportation. Cargoes are no longer loaded and off-loaded piece by piece by teams of longshoremen. Today's harbors bristle with automated bulk terminals, high-volume tanker terminals (both offshore and dockside), containership facilities (**Figure 15.15**), roll-on–roll-off ports for automobiles and trucks, and passenger facilities required by the growing popularity of cruising. Most of this specialized construction has occurred since 1960. New Orleans is now the busiest North American port;

Figure 15.15 The use of standardized containers has revolutionized the shipping industry. Containers can be loaded and sealed at a factory, transported by truck and rail to a port, shipped on specialized vessels to a distant harbor, moved to warehouses, and then opened. Efficiency is greatly improved. Damage and pilferage are minimized.

nearly 210 million tons of cargo—most of it grain—passed through its docks in 1999. The world's largest container terminals are in Long Beach/Los Angeles Harbor in California and Hong Kong (Kowloon) in China. In 1995 the Hong Kong facility was the first to move more than 1 million containers in a single month!

Transportation is sometimes combined with recreation (**Figure 15.16**). In the last decade, the cruise industry has experienced spectacular growth. Passengers on luxurious liners can enjoy a few relaxing days on the ocean crossing the North Atlantic, visiting tropical islands, or touring places accessible to the public only by ship. (Indeed, tourism is now

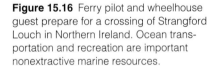

Figure 15.16 Ferry pilot and wheelhouse guest prepare for a crossing of Strangford Louch in Northern Ireland. Ocean transportation and recreation are important nonextractive marine resources.

the world's largest industry.) Ocean-related leisure pursuits, including sport fishing, surfing, diving, day cruising, sunbathing, dining in seaside restaurants, and just plain relaxing, contribute to the economy. In addition to being important producers of revenue, public aquariums and marine parks (like Sea World) are centers of education, research, and captive breeding programs. Even public interest and curiosity about whales are a source of recreational revenue. In the United States in 1994, whale-watching trips generated annual direct revenue of $122 million; indirect revenues including dolphin displays, whale artwork and books, and conservation group donations amounted to another $382 million.

And don't forget about real estate. As coastal land values will attest, people enjoy living near the ocean. By 2015 about 165 million Americans will be living in coastal areas on property valued in excess of $7 trillion.

> Nonextractive resources, which include transportation and recreation, are among the fastest growing and potentially most valuable ocean resources.

Resources and Pollution

The ocean's great volume and relentless motion dissipate and distribute natural and synthetic substances, but its ability to absorb is not inexhaustible. The utilization of ocean resources can result in the accidental (or intentional) release of harmful substances. We define **marine pollution** as the introduction into the ocean by humans of substances or energy that changes the quality of the water or affects the physical, chemical, or biological environment. Sources of marine pollution are shown in **Figure 15.17.**

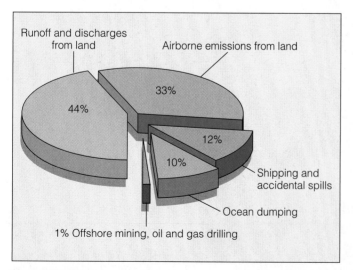

Figure 15.17 Sources of ocean oil pollution. (Source: *The State of the Marine Environment*, UNEP Regional Seas Reports and Studies, No. 115, Nairobi: United Nations Environmental Programme, 1990.)

No one knows to what extent we have contaminated the ocean. By the time the first oceanographers began widespread testing, the Industrial Revolution was well under way and changes had already occurred. Traces of synthetic compounds have now found their way into every oceanic corner. It is sad to consider that we will never know what the natural ocean was like or what remarkable plants and animals may have vanished as a result of human activity. Our limited knowledge of pristine conditions is gleaned from small seawater samples recovered from deep within the polar ice pack and from tiny air bubbles trapped in glaciers. There are few undisturbed habitats left to study, and few marine organisms are completely free of the effects of ocean pollutants.

> Pollution can be a side-effect of resource extraction and use.

Characteristics of a Pollutant

A **pollutant** causes damage by interfering directly or indirectly with the biochemical processes of an organism. Some pollution-induced changes may be instantly lethal; other changes may weaken an organism over weeks or months, or alter the dynamics of the population of which it is a part, or gradually unbalance the entire community.

In most cases an organism's response to a particular pollutant will depend on its sensitivity to the combination of *quantity* and *toxicity* of that pollutant. Some pollutants are toxic to organisms in tiny concentrations. For example, the photosynthetic ability of some species of diatoms is diminished when chlorinated hydrocarbon compounds are present in parts-per-trillion quantities. Other pollutants may seem harmless, as when fertilizers flowing from agricultural land stimulate plant growth in estuaries. Still other pollutants may be hazardous to some organisms but not to others. For example, crude oil interferes with the delicate feeding structures of zooplankton and coats the feathers of birds, but it simultaneously serves as a feast for certain bacteria.

Pollutants also vary in their *persistence*; some reside in the environment for thousands of years, while others last for only a few minutes. Some pollutants break down into harmless substances spontaneously or through physical processes (like the shattering of large molecules by sunlight). Sometimes pollutants are removed from the environment through biological activity. For example, some marine organisms escape permanent damage by metabolizing hazardous substances to harmless ones. Indeed, many pollutants are ultimately **biodegradable**—that is, able to be broken down by natural processes into simpler compounds. Many pollutants resist attack by water, air, sunlight, or living organisms, however, because the synthetic compounds of which they are composed resemble nothing in nature.

The ways in which pollutants are changing the ocean and the atmosphere are often difficult for researchers to determine. Environmental impact cannot always be predicted or explained. As a result, marine scientists vary widely in their opinions about what pollutants are doing to the ocean and atmosphere and what to do about it. Environmental issues are frequently emotional, and media reports tend to sensationalize short-term incidents (like oil spills) rather than more serious, long-term problems (like atmospheric changes or the effects of long-lived chlorinated hydrocarbon compounds).

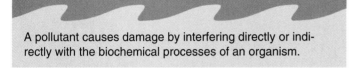

A pollutant causes damage by interfering directly or indirectly with the biochemical processes of an organism.

Oil

Oil is a natural part of the marine environment. Oil seeps have been leaking large quantities of oil into the sea for millions of years. The amount of oil entering the ocean has increased greatly in recent years, however, because of our growing dependence on marine transportation for petroleum products, offshore drilling, nearshore refining, and street runoff carrying waste oil from automobiles.

In the late 1990s people were using nearly 6.8 billion metric tons (7.5 billion tons) of crude oil each year, slightly more than half of which was transported to market in large tankers. In 1998 about 6 million metric tons (6.6 million tons) of oil entered the world ocean. Natural seeps accounted for only about 10% of this annual input—about 600,000 metric tons. About a third of the total was associated with marine transportation. Some of this oil was not spilled in well-publicized tanker accidents but was released during the loading, discharging, and flushing of tanker ships. Between 150,000 and 450,000 marine birds are killed each year by oil released from tankers.

Much more oil reaches the ocean in runoff from city streets, or as waste oil dumped down drains, poured into dirt, or hidden in trash destined for a landfill. Every year more than 900 million liters (about 240 million gallons) of used motor oil—about 22 times the volume of the *Exxon Valdez* spill—finds its way to the sea. This oil is much more toxic than crude or newly refined oil because it has developed carcinogenic and metallic components from the heat and pressure within internal combustion engines. It's no wonder that a sea surface completely free of an oil film is quite rare (**Figure 15.18**).

It is difficult to generalize about the effects that a concentrated release of oil—an oil spill from a tanker, coastal storage facility, or drilling platform—will have in the marine environment. The consequences of a spill vary according to several factors: its location and proximity to shore; the quantity and composition of the oil; the season of the year, currents, and weather conditions at the time of release; and the composition and diversity of the affected communities. Intertidal and shallow-water subtidal communities are most sensitive to the effects of an oil spill.

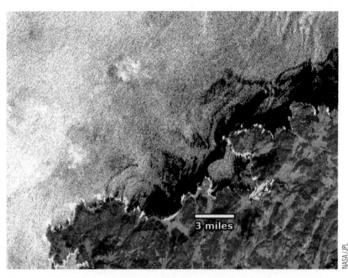

Figure 15.18 On 19 November 2002 a tanker carrying 20 million gallons of oil split open off the northwestern coast of Spain. Three days later the Canadian *RADARSAT* satellite imaged the resulting oil slick. The coastal city of La Coruña is located just above the 3-mile mark overlay.

Spills of *crude* oil are generally larger in volume and more frequent than spills of refined oil. Most components of crude oil do not dissolve easily in water, but those that do can harm the delicate juvenile forms of marine organisms even in minute concentrations. The remaining insoluble components form sticky layers on the surface that prevent free diffusion of gases, clog adult organisms' feeding structures, kill larvae, and decrease the sunlight available for photosynthesis. Even so, crude oil is not highly toxic, and it is biodegradable. Though crude oil spills look terrible and generate great media attention, most forms of marine life in an area recover from the effects of a moderate spill within about five years. For example, the 907 million liters (240 million gallons) of light crude oil released into the Persian Gulf during the 1991 Gulf War (**Table 15.2**) dissipated relatively quickly and will probably cause little long-term biological damage.

Spills of *refined* oil, especially near shore where marine life is abundant, can be more disruptive for longer periods of time. The refining process removes and breaks up the heavier components of crude oil and concentrates the remaining lighter, more biologically active ones. Components added to oil during the refining process also make it more deadly. Spills of refined oil are of growing concern because the amount of refined oil transported to the United States rose dramatically through the 1990s.

The volatile components of any oil spill eventually evaporate into the air, leaving the heavier tars behind. Wave action causes the tar to form into balls of varying sizes. Some of the tar balls fall to the bottom, where they may be assimilated by bottom organisms or incorporated into sediments. Bacteria will eventually decompose these spheres, but the process may take years to complete, especially in cold polar waters. This oil residue—especially if derived from refined oil—can have long-lasting effects on seafloor communities. The fate of spilled oil is summarized in **Figure 15.19.**

http://info.brookscole.com/garrison

Table 15.2 Largest Marine Oil Spills Since 1960

Rank	Date	Incident and Location	Size of Spill (millions of gallons)
1	26 Jan 1991	Discharged into Persian Gulf during Gulf War	240.0
2	3 June 1979	Well spill into the Bay of Campeche Mexico	140.0
3	19 July 1979	Tanker *Atlantic Empress* wreck and subsequent tow, Caribbean NE of Trinidad and Tobago	83.8
4	4 Feb 1983	Well spill at Nowruz, Iran, Persian Gulf	80.0
5	6 Aug 1983	Fire aboard tanker *Castillo de Bellver*, South Africa	78.5
6	16 March 1978	Tanker *Amoco Cadiz* aground off Portsall, Brittany, France	68.7
7	10 Nov 1988	Tanker *Odyssey*, breakup NE of St. Johns, Newfoundland, Canada	43.1
8	11 April 1991	Tanker *Haven*, Mediterranean Sea off Genoa, Italy	42.3
9	18 Mar 1967	Tanker *Torrey Canyon* aground at Land's End, southwestern England	38.2
10	19 Dec 1972	Tanker *Sea Star*, Gulf of Oman, Oman	37.9
•			
•			
•			
46	24 March 1989[a]	Tanker *Exxon Valdez*, Prince William Sound, Alaska	11.0

[a]The *Exxon Valdez* spill of March 1989 was the third worst in U.S. history. But when ranked with other massive marine oil spills involving both shipping and well accidents worldwide, it falls to number 46.
Source: International Oil Spill Database, Cutter Information Corporation

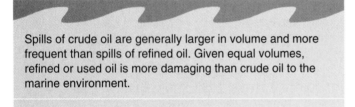

Spills of crude oil are generally larger in volume and more frequent than spills of refined oil. Given equal volumes, refined or used oil is more damaging than crude oil to the marine environment.

The methods used to contain and clean up an oil spill sometimes cause more damage than the oil itself. When the supertanker *Exxon Valdez* ran aground in Alaska's Prince William Sound on 24 March 1989, more than 40 million liters (almost 11 million gallons; 29,000 metric tons) of Alaskan crude oil—about 22% of her cargo—escaped from the crippled hull (see again Table 15.2). Only about 17% of this oil was recovered by a work crew of more than 10,000 people using containment booms, skimmer ships, bottom scrapers, and absorbent sheets. About 35% of the oil evaporated, 8% was burned, 5% was dispersed by strong detergents, and 5%

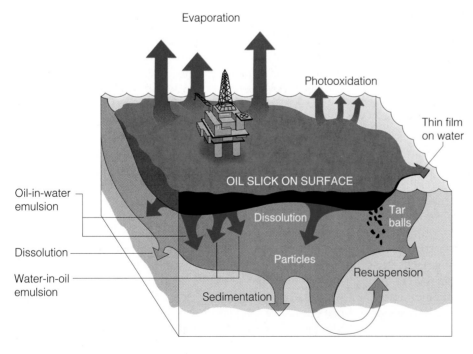

Figure 15.19 The fate of oil spilled at sea. Smaller molecules evaporate or dissolve into the water beneath the slick. Sunlight and atmospheric oxygen cause oxidation of the surface layer. Within a few days, water motion coalesces the oil into tar balls and semisolid emulsions of water-in-oil and oil-in-water. Tar balls and emulsions may persist for months after formation. Though bonding with sediment particles may sink oil droplets, suspended oil does not enter benthic subtidal sediments in large quantities. If crude oil is left undisturbed, bacterial activity will eventually consume it. Refined oil, however, can be more toxic, and natural cleaning processes take a proportionally longer time to complete.

biodegraded in the first five months. The rest of the oil, some 30% of the spill, formed oil slicks on Prince William Sound and fouled more than 450 kilometers (300 miles) of coastline.

Though it generated much media attention, the *Exxon Valdez* oil spill was only the 46th worst spill since 1960. Americans flush more oil down drains every year than was spilled by *Exxon Valdez*.

Recent analysis of the affected parts of Prince William Sound shows the cleaned areas to be in generally worse shape than areas left alone. Most of the small animals that make up the base of the food chain in these areas were cooked by the 65°C (150°F) water used to blast oil from between the rocks (**Figure 15.20**). Others were smothered when the high-pressure jets rearranged sand and mud. It appears that an overambitious cleanup program can be counterproductive. The biological cost of the spill will not be known for many

years. The cost of the cleanup has exceeded $3.5 billion. The legal costs could be even higher: In 1994 a federal jury awarded $5 billion in punitive damages to 14,000 plaintiffs including fishermen, native villagers, and fish processors. Exxon appealed the verdict; litigation continues.

Of course the best way to deal with oil pollution is to prevent it from happening in the first place. Tanker design is being modified to limit the amount of oil intentionally released in transport. Legislation is being considered that would limit new tanker construction to stronger, double-hull designs (although shipping experts are uncertain whether even a double-hull design could have prevented the *Exxon Valdez* spill). Perhaps most important, crew testing and training are being upgraded.

Heavy Metals and Synthetic Organic Chemicals

Among the dangerous heavy metals being introduced into the ocean are mercury and lead. Human activity releases about five times as much mercury and 17 times as much lead as is

Figure 15.20 Cleaning up the 1989 *Exxon Valdez* oil spill in Prince William Sound, Alaska. Ironically, nearly as much environmental damage was done in the cleanup process as in the original spill.

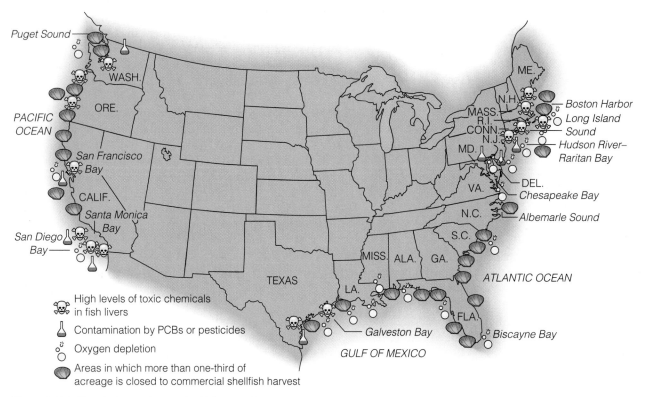

Figure 15.21 Chemical assault on major U.S. coastal areas.

Map labels:

Puget Sound
WASH.
ORE.
PACIFIC OCEAN
San Francisco Bay
CALIF.
Santa Monica Bay
San Diego Bay
TEXAS
Galveston Bay
GULF OF MEXICO
MISS. ALA. GA.
LA.
FLA.
Biscayne Bay
ME.
N.H.
MASS.
R.I.
CONN.
N.J.
MD.
VA.
N.C.
S.C.
DEL.
Boston Harbor
Long Island Sound
Hudson River–Raritan Bay
Chesapeake Bay
Albemarle Sound
ATLANTIC OCEAN

Legend:

☠ High levels of toxic chemicals in fish livers

⚗ Contamination by PCBs or pesticides

°∘○ Oxygen depletion

🦪 Areas in which more than one-third of acreage is closed to commercial shellfish harvest

derived from natural sources, and the incidence of mercury and lead poisoning, major causes of brain damage and behavioral disturbances in children, has increased dramatically over the last two decades. Lead particles from industrial wastes, landfills, and gasoline residue reach the ocean through runoff from land during rains, and the lead concentration in some shallow-water bottom-feeding species is increasing at an alarming rate.

Wellness-conscious consumers see fish as a safe and healthful food, but with the ocean receiving heavy-metal-contaminated runoff from the land, a rain of pollutants from the air, and the fallout from shipwrecks, we can only wonder how much longer seafood will be safe to eat. Consumers should be wary of seafoods taken near shore in industrialized regions.

Halogenated hydrocarbons—a class of synthetic hydrocarbon compounds that contain chlorine, bromine, or iodine—are used in pesticides, flame retardants, industrial solvents, and cleaning fluids. The concentration of **chlorinated hydrocarbons**—the most abundant and dangerous halogenated hydrocarbons—is so high in the water off New York State that officials have warned women of childbearing age and children under 15 not to consume more than half a pound of local bluefish a week. (They are told *never* to eat striped bass caught in the area.) One administrator of the U.S. Environmental Protection Agency has written, "Anyone who eats the liver from a lobster taken from an urban area is living dangerously."

The level of synthetic organic chemicals in seawater is usually very low, but some organisms at higher levels in the food chain can concentrate these toxic substances in their flesh. This **biological amplification** is especially hazardous to top carnivores in a food web. Biologists were alarmed by the recent discovery that many of the near-shore dolphins off U.S. coasts are intensely contaminated. The concentration of chlorinated hydrocarbons in these animals exceeds 6,900 parts per million (ppm), concentrations high enough to disrupt the dolphins' immune systems, hormone production, reproductive success, neural function, and ability to stave off cancers.[2] These levels vastly exceed the 50-ppm limit the U.S. government considers hazardous, and the 5 ppm considered the maximum acceptable level for humans! Investigations are continuing, as are probes into the effects of dioxin and other synthetic organic poisons accumulating in the oceanic sink.

Figure 15.21 indicates the locations of U.S. coastal areas degraded by synthetic organic chemicals. Production of such chemicals subject to biological amplification in food chains presently exceeds 91 million metric tons (100 million tons) each year. Even vaster volumes of ocean will be affected if the predicted 1% of that production reaches the sea.

Synthetic organic chemicals and heavy metals are persistent environmental poisons and can sometimes be concentrated by biological amplification through food chains.

[2] If you drag a beached porpoise into the ocean, you could theoretically receive a $10,000 fine for improper disposal of polluted materials!

Figure 15.22 Rachel Carson, author of the influential 1962 book *Silent Spring* on the misuse of synthetic pesticides. This work is generally credited with beginning the environmental movement in the United States.

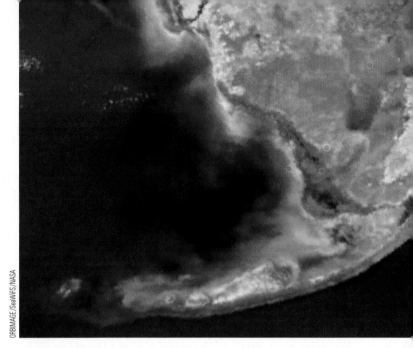

Figure 15.23 "Black water" in a Florida Bay, February 2002. Visibility in the usually clear water dropped to less than 3 meters (10 feet). A plankton bloom caused by an abundance of sewage-related nutrients in the water is the likely cause. Such blooms have economic consequences; normally abundant fish are almost completely absent in this area.

Eutrophication

Not all pollutants kill organisms. Some dissolved organic substances act as nutrients or fertilizers that speed the growth of marine autotrophs, causing eutrophication. **Eutrophication** is a set of physical, chemical, and biological changes that take place when excessive nutrients are released into the water. Too much fertility can be as destructive as too little. Eutrophication stimulates the growth of some species to the detriment of others, destroying the natural biological balance of an ocean area. The extra nutrients come from wastewater treatment plants, factory effluent, accelerated soil erosion, or fertilizers spread on land. They usually enter the ocean from river runoff and are particularly prevalent in estuaries. Eutrophication is occurring at the mouths of almost all the world's rivers.

The most visible manifestations of eutrophication are the red tides, yellow foams, and thick green slimes of vigorous plankton blooms. These blooms typically consist of one dominant phytoplankter that grows explosively and overwhelms other organisms. Huge numbers of algal cells can choke the gills of some animals and (at night, when sunlight is unavailable for photosynthetic oxygen production) deplete the free oxygen content of surface water (note the symbols for oxygen depletion in Figure 15.21).

In nearshore waters, low oxygen content is now thought to cause more mass fish deaths than any other single agent, including oil spills. It is a leading threat to commercial shellfisheries.

These exceptional algae blooms appear to be increasing in number and intensity (**Figure 15.23**). There is little mention of foam events before about 1930, but since 1978 there has been at least one every year. A similar pattern has been reported for red tides. The threat is especially great in countries that are experiencing rapid economic growth. After two decades of breathless development, severe harmful algal blooms (HABs) have begun to plague China's southeastern coastline. In 1998 a massive bloom killed up to 75% of Hong Kong's fish farms, wiping out many local fishers. The Chinese government estimates $240 million in direct economic losses from 45 HAB incidents from 1997 to 1999.

Uncontrolled algal growth through artificially high nutrient levels—eutrophication—can be dangerous to organisms and community diversity.

Solid Waste

Not all pollutants enter the ocean in a dissolved state; much of the burden arrives in solid form. Some solid waste is ultimately biodegradable, but plastic, which now makes up almost 7% of the solid waste stream, is not. Scientists estimate that some kinds of synthetic materials—plastic six-pack holders, for example—will not decompose for about 400 years!

Americans generate 120 million metric tons of plastic waste, about 500 kilograms (1,100 pounds) per person, each year. Plastics now account for more than 10% of all solid waste. Since the ocean is treated as the ultimate dump, some of this waste plastic finds its way to the sea. In 1997 more than 4,500 volunteers scoured the New Jersey–New York coast to collect debris. More than 75% of the 209 tons recovered was plastic. (Paper and glass accounted for another 15%.) In isolated areas the dunes of plastic debris can build to surreal proportions (**Figure 15.24**).

A 1987 survey by the Woods Hole Oceanographic Institution found that each square mile of ocean surface off the northeast coast of the United States has more than 46,000 pieces of plastic floating on the surface. This material includes ropes, fishing line and nets, plastic sheeting and bags, and granules of broken plastic cups. A staggering 100,000 marine

Figure 15.24 Plastic mounds on an isolated shore of Niihau, one of the Hawaiian Islands.

Figure 15.25 Sea lions (seen here) and seals die by the thousands each year after becoming entangled in plastic debris, especially discarded and broken fishing nets.

mammals and 2 million seabirds die *each year* after ingesting or being caught in plastic debris! Sea turtles mistake plastic bags for their jellyfish prey and die from intestinal blockages. Seals and sea lions starve after becoming entangled in nets or muzzled by six-pack rings (**Figure 15.25**). The same kinds of rings strangle fish and seabirds. Adding ingredients to plastics that would hasten their decomposition would add only 5% to 7% to their cost, but this price increase is currently unacceptable to industry.

What should we do with plastic and other solid wastes such as glass and paper, disposable diapers, scrap metal, building debris, and all the rest? Dumping it into the ocean is clearly unacceptable, yet places to deposit this material are becoming scarce. In 2000 the average New Yorker threw away more than a ton of waste annually. California's Los Angeles and Orange Counties generate enough solid waste to fill Dodger Stadium every eight days. Transportation of waste to sanitary landfills becomes more expensive as nearby landfills reach capacity.

Is recycling the answer? The Japanese currently recycle about 50% of their solid waste and are importing even more; scrap metal and waste paper headed for Japan are the two biggest exports from the Port of New York. Americans are buying back their own refuse in the form of appliances, automobiles, and the cardboard boxes that hold their televisions and compact disc players. Massachusetts and California have set a goal of recycling 25% of their waste; the city of Seattle is now approaching 30%. The direct savings to consumers, as well as the environmental rewards to ocean and air, will be significant.

Sewage 15.22

About 98% of sewage is water. Sewage treatment separates the fluid component from the solids, treats the water to kill disease organisms and reduce the levels of nutrients, and releases it into a river or the ocean. The remaining **sewage sludge** is a semisolid mixture of organic matter containing bacteria and viruses, toxic metal compounds, synthetic organic chemicals, and other debris. It is digested, thickened, dried, and shipped to landfills, burned to generate electricity, or dumped into the ocean. The liquid effluent wanders from its outlet and circulates with currents, but sludge and other insoluble residues may stay near the outfall or dump site for years. The amount of wastewater and sewage sludge increased by 60% in the 1990s.

Until 1992 almost 2 billion liters (half a billion gallons) of partially treated sewage poured into Boston Harbor every day. A new sewage treatment plant opened in that year, 22 years after the deadline mandated by the U.S. Clean Water Act. About the same amount of treated sewage pours from outfall pipes off Los Angeles. Treatment plants in southern California are sometimes overwhelmed after heavy rainstorms, and raw sewage enters the ocean in large quantity. Rain or shine, areas around the entrance to San Diego Harbor are often so contaminated with sewage from Tijuana, Mexico, that anyone who enters the water runs the risk of bacterial or viral infection.

Not only human sewage is involved. In June 1995, 26 million gallons of hog feces and urine spilled into a narrow stream feeding North Carolina's New River estuary. Thousands of fish died from direct effects, and hundreds of thousands more perished as algae grew and consumed oxygen.

Sludge may be an even greater problem than raw sewage. The water just south of Long Island, New York, has been one of the most intensely used ocean dumping sites in the world. The area is covered with sludge, which creates an oxygen-poor environment in which few animals can survive. During storms some of this material washes up on local beaches, contaminates shellfish beds, and routinely causes disease outbreaks among people consuming raw oysters and clams from the area. After a public outcry, a new dumping site was selected beyond the edge of the continental shelf. Ten million tons of wet sludge produced by treatment plants in New York and New Jersey since March 1986 has been dropped there by huge barges. The plume from this sludge now contaminates the Gulf Stream.

Liquid effluent and sludge also contribute to eutrophication. As we've seen, some forms of algae thrive in nutrient-rich wastewater, where they multiply to prodigious numbers, secrete toxins, or deplete oxygen in the surrounding water and cause the death of fish and crustaceans.

The Costs of Pollution

In 1998 government and industry in the United States spent about $220 billion—an average of $800 for each American—on the control of atmospheric, terrestrial, and marine pollution. This figure was equivalent to about 1.6% of the gross national product, or 2.7% of capital expenditures by U.S. business. That same year the United States lost 4% of its gross national product through environmental damage. Clearly the financial costs of pollution will continue to increase.

But there are other costs. Failure to control pollution will eventually threaten our food supply (marine and terrestrial), destroy whole industries, produce a greater disparity between have and have-not nations, and cause a decline in the health of all the planet's inhabitants. To these costs must be added the aesthetic costs of an ocean despoiled by pollution; few of us look forward to sharing the beach with oiled birds, jettisoned diapers, or clumps of medical waste.

Atmospheric Changes

The ocean and the atmosphere are extensions of each other, and human activity has changed the atmosphere as it has changed the ocean. Pollutants injected into the air can have global consequences for the ocean and for all of Earth's inhabitants. Potentially two of the most destructive atmospheric problems are depletion of the ozone layer and global warming.

Ozone Layer Depletion

Ozone is a molecule formed of three atoms of oxygen. Ozone occurs naturally in the atmosphere. A diffuse layer of ozone mixed with other gases—the **ozone layer**—surrounds the world at a height of about 20 to 40 kilometers (12 to 25 miles).

Seemingly harmless synthetic chemicals released into the atmosphere—primarily **chlorofluorocarbons** (**CFCs**) used as cleaning agents, refrigerants, fire-extinguishing fluids, spray-can propellants, and insulating foams—are converted by the energy of sunlight into compounds that attack and partially deplete Earth's atmospheric ozone. Ozone levels in the stratosphere have decreased by at least 3% over most of the United States since 1969. A 4% drop has been noted over Australia and New Zealand, and a 50% decrease has been observed near the North and South Poles. The amount of depletion varies with latitude (and with the seasons) because of variations in the intensity of sunlight (**Figure 15.26**)

This decline in ozone alarms scientists because stratospheric ozone intercepts some of the high-energy ultraviolet radiation coming from the sun. Ultraviolet radiation injures living things by breaking strands of DNA and unfolding protein molecules. Species normally exposed to sunlight have evolved defenses against average amounts of ultraviolet radiation, but increased amounts could overwhelm those defenses. Land plants such as soybeans and rice would be subjected to sunburn that decreases their yields. Even plankton in the uppermost 2 meters (6.5 feet) of ocean would be affected: Recent research indicates an alarming decrease in

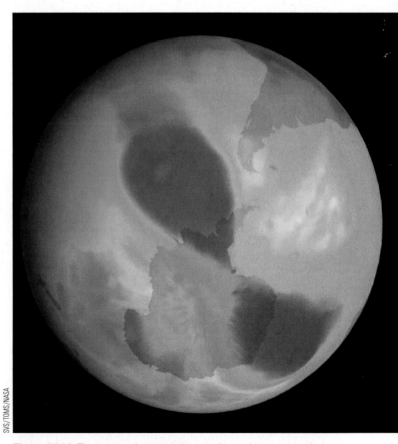

SVS/TOMS/NASA

Figure 15.26 The seasonal ozone hole over Antarctica, as recorded by the *Earth Probe* satellite in September 2002. The lowest ozone values (the "hole") are indicated by magenta and purple. This hole is smaller than in recent years, in part because of international efforts to reduce the use of damaging chemicals and in part because of warmer than normal air in the surrounding stratosphere.

phytoplankton primary productivity of 6% to 12% in the coastal waters around Antarctica.

Our own species would not escape: A 1% decrease in atmospheric ozone would probably be accompanied by a 5% to 7% increase in the incidence of human skin cancer. According to medical researchers' estimates, the 4% ozone depletion over Australia and New Zealand will cause at least a 20% increase in cases of human skin cancers over the next two decades. Strong ultraviolet light can also suppress the immune system and cause eye cataracts.

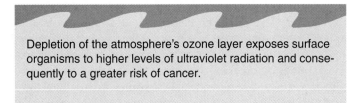

Depletion of the atmosphere's ozone layer exposes surface organisms to higher levels of ultraviolet radiation and consequently to a greater risk of cancer.

In June 1990, representatives of 53 nations agreed to ban major production and use of ozone-destroying chemicals by the year 2000. A sense of great urgency surrounds ongoing research to find safe substitutes for these substances. Recent data indicate that these measures are having an effect. CFC concentrations peaked near the beginning of 1994. Researchers believe we could see a consistent decrease in the sizes of polar ozone holes by 2005 or 2010.

Global Warming

The surface temperature of Earth fluctuates slowly over time. The global temperature trend has been generally upward in the 18,000 years since the last ice age, but the *rate* of increase has recently accelerated. This rapid warming is prob-ably the result of an enhanced **greenhouse effect,** the trapping of heat by the atmosphere. Glass in a greenhouse is transparent to light but not to heat. The light is absorbed by objects inside the greenhouse, and its energy is converted into heat. The temperature inside a greenhouse rises because the heat is unable to escape. On Earth **greenhouse gases**—carbon dioxide, water vapor, methane, CFCs, and others—take the place of glass. Heat that would otherwise radiate away from the planet is absorbed and trapped by these gases, causing a rise in surface temperature. **Figure 15.27** shows this mechanism.

The greenhouse effect is necessary for life; without it, Earth's average atmospheric temperature would be about $-18°C$ (0°F). Earth has been kept warm by natural greenhouse gases. The sources of these gases are volcanic and geothermal processes, the decay and burning of organic matter, and respiration and other biological sources. The removal of these gases by photosynthesis and absorption by seawater appears to prevent the planet from overheating. But the human demand for quick energy to fuel industrial growth, especially since the beginning of the Industrial Revolution, has injected unnatural amounts of new carbon dioxide into the atmosphere from the combustion of fossil fuels. Carbon dioxide is now being produced at a greater rate than it can be absorbed by the ocean. **Figure 15.28** shows how much carbon dioxide concentrations have increased in the atmosphere since about 1750. The atmosphere's carbon dioxide content now rises at the rate of 0.4% each year. Though CFC levels are now declining, methane levels continue to increase. (The primary sources of methane are intestinal gases from cows and termites and the decay of refuse and vegetation cleared from land.)

There has been a 5°C (9°F) rise in global temperature from the end of the last ice age until today. Carbon dioxide and other human-generated greenhouse gases produced

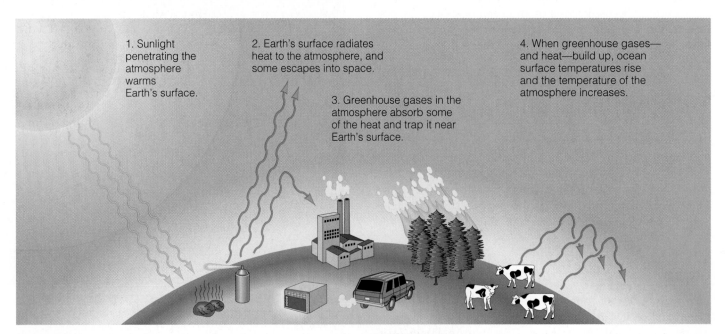

1. Sunlight penetrating the atmosphere warms Earth's surface.

2. Earth's surface radiates heat to the atmosphere, and some escapes into space.

3. Greenhouse gases in the atmosphere absorb some of the heat and trap it near Earth's surface.

4. When greenhouse gases—and heat—build up, ocean surface temperatures rise and the temperature of the atmosphere increases.

Figure 15.27 How the greenhouse effect works. Since the beginning of the Industrial Revolution, emissions of CO_2, methane, and other greenhouse gases have increased. Most researchers believe these gases have contributed to a general warming of Earth's atmosphere and ocean.

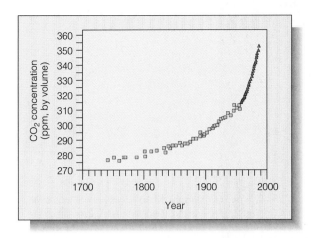

Figure 15.28 The rise in atmospheric carbon dioxide concentrations since 1750. Measurements up to about 1960 (■) came from samples of air trapped in the Antarctic ice; later measurements (▲) were made directly near the crest of Mauna Loa, Hawaii.

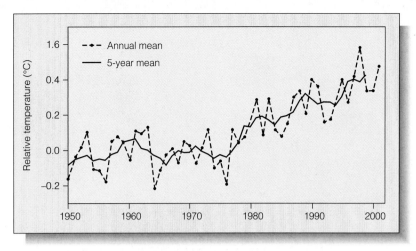

Figure 15.29 Global annual surface temperature relative to the 1951–1980 mean based on surface air measurements at meteorological stations and satellite measurements of sea surface temperature. (NASA-Goddard, after Hansen et al., 2002.)

since 1880 are thought to be responsible for about half of that increase. A new survey of 300 bore-hole temperatures on four continents has confirmed that Earth is getting warmer and that the rate of warming has been increasing since 1900. Atmosphere and surface temperatures confirm that the average global temperature has increased 0.4° to 0.8°C (0.7° to 1.4°F) over the last century. The year 2001 was the warmest on record (**Figure 15.29**). A consensus of experts who met in Shanghai in January 2001 confirmed that if greenhouse emissions continue to rise, the average temperature should increase another 5.8°C (10.5°F) over the next century.[3]

The south polar ice sheets are already shrinking in the warmth. Because of the influx of water, and because the water itself is becoming warmer and expanding, sea level is rising today faster than at any time in the past. Imagine the effect of a significant rise in sea level on the harbors, coastal cities, river deltas, and wetlands where one-third of the world's people now live. As **Figure 15.30** suggests, costs to society would be enormous.

Unfortunately, it will be exceedingly difficult to curtail our production of carbon dioxide and methane. In the last hundred years, industrial production has increased fifty-fold; we have burned roughly 1 billion barrels of oil, 1 billion metric tons of coal, and 10 billion cubic meters of natural gas. Carbon dioxide is a major product of combustion for these hydrocarbon compounds. The world's energy demand is projected to increase 3.5 times between now and 2025, with carbon dioxide emissions 65% higher than today.

At a meeting in Kyoto, Japan, in 1997, leaders and representatives of 160 countries established carbon dioxide emission targets for each developed country. The United States, for example, would reduce its carbon dioxide emissions to 7% less than 1990 levels by 2012. This goal is thought to be economically untenable, and the Kyoto Protocol has not been ratified by the U.S. Senate. In any case, those levels would only *slow* the eventual effects on Earth's climate.

Changes in the composition of the atmosphere can influence global climate, ocean temperatures, and sea level. Experts estimate that at least half of recent global warming is anthropogenic (human-caused).

Alternatives to fossil fuel must be found if we are to maintain world economies and prevent an increase in global temperature with all its uncertainties. Other than wind power, the only alternative source of energy that currently produces significant amounts of power is nuclear energy, which now generates about 17% of the electricity produced in the United States. Despite much publicity to the contrary, these pressurized water reactors have good records of dependable power production and safety.[4] The problem with nuclear power lies not so much in the everyday operation of the reactors but in disposing of the nuclear wastes they produce. By 1992 about 55,000 highly radioactive spent-fuel assemblies were in temporary storage in deep pools of cooled water; they must be stored for 10,000 years before their levels of radioactivity will be low enough to pose no environmental hazard. Radioactive substances emit ionizing radiation, a form of energy that is able to penetrate and permanently damage cells. It is mandatory that artificial sources of ionizing radiation be isolated from the environment.

Will citizens of industrial countries (and countries that wish to become industrialized) agree to lessen the danger of

---

[3] This estimate is based on projections of data, not on models.

[4] The Soviet reactor at Chernobyl that exploded in 1986 was of a much different design.

Reprinted by permission of the *Los Angeles Times*

Figure 15.30 For island nations such as the Maldives, even a small rise in sea level could spell disaster. Strung out across 880 kilometers (550 miles) of the Indian Ocean, 80% of this island nation of 263,000 people lies less than 1 meter (3.3 feet) above sea level. Of the country's 1,180 islands, only a handful would survive the median estimate of sea level rise by 2100. Most of the population lives in fishing villages on low islands like the one shown here, where the effects of this century's sea level rise of 10 to 25 centimeters (4 to 10 inches) have already been felt.

increased global warming by slowing their economic growth, decreasing their dependence on fossil fuel combustion, and developing safe alternative sources of energy? Some insight may be gained from the behavior of ranchers and industrialists in the rain forests of New Guinea, the Philippines, and Brazil. The Amazon rain forest of Brazil is being burned at a rate of about 12 square kilometers (almost 5 square miles) *per hour*, acreage equivalent to the area of West Virginia every year. Huge stands of trees that should be nurtured to absorb excess carbon dioxide are being destroyed. The cleared land is used for farms, cattle ranches, roads, and cities. The priorities are clear.

A Cautionary Tale

Let me tell you a story.

Easter Island was home to a culture that rose to greatness amidst abundant resources, attained extraordinary levels of achievement, and then died suddenly and alone, terrified, in the empty vastness of the Pacific. The inhabitants had destroyed their own world.

Europeans first saw Easter Island (Rapa Nui) in 1722. The Dutch explorer Jacob Roggeveen encountered the small volcanic speck on Easter morning while on a scouting voyage. The island, dotted with withered grasses and scorched vegetation, was populated by a few hundred skittish, hungry, ill-clad Polynesians who lived in caves. During his one-day visit, Roggeveen was amazed to see more than 200 massive stone statues standing on platforms along the coast (**Figure**

15.31). At least 700 more statues were later found partially completed, lying in the quarries as if they had just been abandoned by workers. Roggeveen immediately recognized a problem: "We could not comprehend how it was possible that these people, who are devoid of thick heavy timber for making machines, as well as strong ropes, nevertheless had been able to erect such images." The islanders had no wheels, no powerful animals, and no resources to accomplish this artistic, technical, and organizational feat. And there were too few of them—600 to 700 men and about 30 women.

When Captain James Cook visited in 1774, the islanders paddled to his ships in canoes "put together with manifold small planks . . . cleverly stitched together with very fine twisted threads." Cook noted the natives "lacked the knowledge and materials for making tight the seams of the canoes, they are accordingly very leaky, for which reason they are compelled to spend half the time in bailing." The canoes held only one or two people each, and they were less than 3 meters (10 feet) long. Only three or four canoes were seen on the entire island, and Cook estimated the human population at less than 200. By the time of Cook's visit, nearly all the statues had been overthrown, tipped into pits dug for them, often onto a spike placed to shatter their faces upon impact.

Archaeological research on Easter Island has revealed a chilling story. Only 165 square kilometers (64 square miles) in extent, Easter Island is one of the most isolated places on Earth, the easternmost outpost of Polynesia. Voyagers from the Marquesas first reached Easter Island about A.D. 350, possibly after being blown off course by storms. The place was a miniature paradise. In its fertile volcanic soils grew dense forests of palms, daisies, grasses, hauhau trees, and toromino

Figure 15.31 Giant statues look to the horizon on Easter Island (Rapa Nui). The civilization that built the statues had all but destroyed its world when the first Europeans arrived in 1722.

shrubs. Large ocean-going canoes could be built from the long, straight, buoyant Easter Island palms, and strengthened with rope made from the hauhau trees. Toromino firewood cooked the fish and dolphins caught by the newly arrived fishermen, and forest tracts were cleared to plant crops of taro, bananas, sugarcane, and sweet potatoes. The human population thrived.

By the year 1400 the population had blossomed to between 10,000 and 15,000 people. Gathering, cultivating, and distributing the rich bounty for so many inhabitants required complex political control. As numbers grew, however, stresses began to be felt: The overuse and erosion of agricultural land caused crop yields to decline, and the nearby ocean was stripped of benthic organisms so the fishermen had to sail greater distances in larger canoes. As resources became inadequate to support the growing population, those in power appealed to the gods. They redirected community resources to carve worshipful images of unprecedented size and power. The people cut down more large trees for ropes and rolling logs to place the heads on huge carved platforms.

By this time the island's seabirds had been consumed, and no new birds came to nest. The rats that had hitchhiked to the island on the first canoes were raised for food. Palm seeds were prized as a delicacy. All available land was under cultivation. As resources continued to shrink, wars broke out

over dwindling food and space. By about 1550 no one could venture offshore to harpoon dolphins or fishes because the palms needed to construct seagoing canoes had all been cut down. The trees used to provide rope lashings were extinct, their wood burned to cook what food remained. The once-lush forest was gone. Soon the only remaining ready source of animal protein was being utilized: The people began to hunt and eat each other.

Central authority was lost and gangs arose. As tribal wars raged, the remaining grasses were burned to destroy hiding places. The rapidly shrinking population retreated into caves from which raids were launched against enemy forces. The vanquished were consumed, their statues tipped into pits and destroyed. Around 1700 the human population crashed to less than a tenth of its peak numbers. No statues stood upright when Cook arrived.

Even if the survivors had wanted to leave the island, they could not have done so. No suitable canoes existed; none could be made. What might the people have been thinking as they chopped down the last palm? Generations later, their successors died in silence, wondering what the huge stone statues had been looking for.

As you read this chapter, you may have been struck by similarities between Easter Island and Earth. But the Easter Islanders had no books and no histories of other doomed societies. Might we be able to learn from their mistakes?

What Can Be Done?

We think of ourselves as more advanced, more intelligent, more foresighted than the Easter Islanders. Yet our births exceed deaths by about three people per second. The six-billionth human was born in October 1999. *Each year* there are 90 million more of us—a total equal to the combined populations of Mexico City, Tokyo and Yokohama, São Paulo, New York, Shanghai, and Calcutta. The number of people has tripled in this century and is expected to double again before reaching a plateau sometime in the next century. Another billion humans will join the world population in the next ten years, 92% of them in third-world countries.[5]

This exploding population will not be content with using the same proportion of resources used today. Citizens of the world's least-developed countries are influenced by education and advertising to demand a developed-world standard of living. They look with justifiable envy at the United States, a country with 4.5% of the planet's population that consumes 55% of its raw material resources and 25% of its energy, while generating 30% of industry-related carbon dioxide. Can the world support a population whose expectations are rising as rapidly as their numbers? The burgeoning human population is the greatest environmental problem of all (**Figure 15.32**).

We cannot expect science to solve the problem for us. Most of the decisions and necessary actions fall outside pure science in the areas of values, ethics, morality, and philosophy. The solution to environmental problems, if one exists, *lies in education and action.* Each of us is obliged to become informed on issues that affect Earth, its ocean, and its air—to learn the arguments and weigh the evidence. Once informed, we must act in rational ways. Chaining yourself to an oil tanker is not rational, but selecting well-designed, long-lasting and recyclable products made by responsible companies with minimal impact on the environment (and encouraging others to do so) certainly is.

The present trade-off between financial and ecological considerations is often strongly tilted in the direction of immediate gain, of short-term profit, of immediate convenience. Education may be the only way to modify these destructive behaviors.

Humanity is part of the natural world, not its master. We may be able to learn to live in harmony with this small, beautiful, blue world. True convenience and true progress depend on the preservation of open space, serious and sustained attention to population control, conversion to a steady-state economy instead of one that must grow to stay alive, business incentives for preservation, the use of renewable resources, and, above all, *public education in environmental issues.* We must ask ourselves difficult questions: What is the optimal quality of life? How can I achieve balance between my mate-

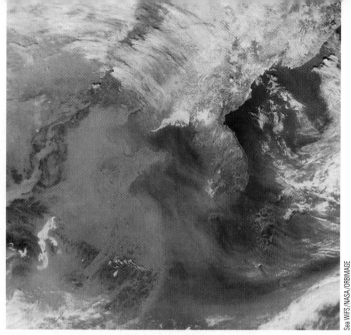

Figure 15.32 A thick haze lingers over the eastern reaches of the world's most populous country. Factory floor to the world, China is home to nearly 1.3 billion people. China's gross national product has grown by an average of 9.8% per year in real terms since 1980, and China will exceed the United States as the world's largest economy in about 2020. If present trends continue, China will surpass the United States as the number one emitter of greenhouse gases about five years later. The new entrepreneurial spirit of the Chinese people and the successes of their enterprises are clearly evident on the labels of your clothing, your shoes, your appliances and housewares, and high-technology hardware. This true-color image was taken by the *OrbView* satellite in January 2002. India will surpass China in population in about 2030.

rial needs and the needs of Earth? What do I want to leave for my children? How can I reserve the quiet, renewing ocean for myself, for all species, and for the future? Our cities are crowded and our tempers are short. Times of turbulent change lie before us. *The trials ahead will be severe.*

Each of us, individually, needs to take a stand. We must preserve the sunsets and fog, the waves to ride, the cold clean windblown spray on our faces. Margaret Mead summarized our potential for making a difference: "Never doubt that a small group of thoughtful, committed citizens can change the world. Indeed it is the only thing that ever has." We need to start now.

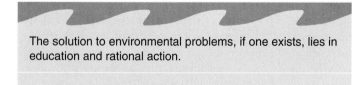

The solution to environmental problems, if one exists, lies in education and rational action.

QUESTIONS FROM STUDENTS

1. What can an individual do to minimize his or her impact on the ocean and atmosphere?

Remember that Earth and all its life-forms are interconnected. There are no true consumers, only users: Nothing can truly be thrown away (there is no "away"). We must

---

[5] In the United States alone, the population is growing by the equivalent of four Washington D.C.s every year, another New Jersey every three years, another California every 12.

abandon the pollute-and-move-on ethic that has guided the actions of most humans for thousands of years, and we must work toward a society more in harmony with the fundamental rhythms of life that sustain us. *Our task is not to multiply and subdue Earth*. It may not be too late to change our ways. We need to act individually to effect change. We should think globally and act locally.

2. Is pollution always a bad thing?

Some forms of pollution bring temporary benefits. For example, some of Florida's once-endangered manatees are thriving at the warm outfalls of coastal power stations. On a larger scale, if increased greenhouse warming does develop, some computer models indicate increased rainfall, longer growing seasons, and increased crop yields over broad latitude bands in the temperate zones of both hemispheres. On the whole, though, the less human intervention in complex natural systems, the better.

3. What are the most dangerous threats to the environment overall?

The underlying causes of the problems discussed in this chapter are human population growth and a growth-dependent economy. Stanford professor Paul Ehrlich said, "Arresting global population growth should be second in importance on humanity's agenda only to avoiding nuclear war." The present world population, now more than 6.1 billion, seems doomed to reach at least 10 billion before leveling off. And what if everybody wants to live in first-world comfort?

4. What role does public perception play in pollution issues?

A large role, indeed! Until recently, relatively small-scale but highly visible insults have claimed most of the public's attention and have driven us to action. The messy breakup of the oil tanker *Torrey Canyon* off the southern coast of England in 1969 galvanized world opinion and set the stage for the present environmental movement. In the summer of 1988, beach-goers were horrified to discover that more than 80 kilometers (50 miles) of northern New Jersey and Long Island beaches had been temporarily closed because of medical debris littering the shore. Some of the dozens of vials of blood, syringes, stained bandages, and surgical sutures tested positive for the viruses that cause AIDS and hepatitis B. Similar incidents occurred in Rhode Island and Massachusetts.

Appalling and visible though such incidents are, their long-term effects on the ocean as a whole are negligible. Public attention has recently turned to issues of larger consequence. The drought and heat of the past few summers brought terms like *greenhouse effect* and *ozone layer* to local newspapers and dinner table conversation. Weekend fishermen worry about eating their catch. They wonder about the unseen threats as much as the obvious ones. Perceptions are changing.

5. When will we run out of oil?

By the time the world's crude oil supplies have been substantially depleted, the total amount of oil extracted is expected to range from 1.6 to 2.4 trillion barrels. If this estimate is correct, consumption could continue at the present level until some time between 2025 and 2040, at which time it would drop quite rapidly. But, in fact, we will never run *completely* out of oil—there will always be some oil within Earth to reward great effort at extraction. The days of unlimited burning of so valuable a commodity are nearing an end, however. Future civilizations will surely look back in horror at their ancestors actually burning something as valuable as lubricating oil.

6. The problems of overfishing and bykill concern me. What seafood can I eat, and what should I avoid?

This list prepared by the Monterey Bay Aquarium is an excellent guide:

BEST CHOICES
Abalone *(farmed)*
Catfish *(U.S. farmed)*
Caviar *(farmed)*
Clams *(farmed)*
Crab, Dungeness
Halibut *(Pacific)*
Hoki
Lobster, Rock *(CA, Australia)*
Mussels *(farmed)*
Oysters *(farmed)*
Sablefish/Black Cod *(AK, BC)*
Salmon *(CA, AK; wild-caught)*
Salmon, canned
Sand Dabs
Sardines
Shrimp/Prawns *(trap-caught)*
Squid/Calamari *(CA market squid)*
Striped Bass *(farmed)*
Sturgeon *(farmed)*
Tilapia *(farmed)*
Tuna, Albacore
Tuna, canned white *(albacore)*
Tuna, Yellowfin/Ahi *(troll/pole-caught)*

PROCEED WITH CAUTION
Clams *(wild-caught)*
Cod, Pacific
Crab, Imitation/Surimi
Crab, King
Crab, Snow
Lobster, American
Mahi-Mahi
Mussels *(wild-caught)*
Oysters *(wild-caught)*
Pollock
Sablefish/Black Cod *(CA, WA, OR)*
Salmon *(OR, WA; wild-caught)*
Scallops, Bay
Shark, Thresher *(U.S. West Coast)*
Shrimp *(U.S. wild-caught)*
Shrimp, Bay
Sole, English/Petrale/Dover
Swordfish *(U.S. West Coast)*
Trout, Rainbow *(farmed)*
Tuna, Yellowfin/Ahi
Tuna, canned chunk light

AVOID
Caviar, Beluga/Osetra/Sevruga
Chilean Sea Bass
Cod, Atlantic/Icelandic
Lingcod
Monkfish
Orange Roughy
Rockfish/Rock Cod/Pacific Snapper
Salmon *(farmed/Atlantic)*
Scallops, Sea
Sharks *(except U.S. West Coast Thresher)*
Shrimp *(wild-caught or farmed)*
Sturgeon *(wild-caught)*
Swordfish *(Atlantic)*
Tuna, Bluefin

AK = Alaska
BC = British Columbia
CA = California
OR = Oregon
U.S. = United States
WA = Washington

Figure 15.33 Recommended use and avoid list for seafood. (*Source:* Monterey Bay Aquarium. For updates, go to: http://www.mbayaq.org/cr/cr-seafood/download.asp. Reprinted with permission.)

© 2002 Monterey Bay Aquarium Foundation.

CHAPTER SUMMARY

Marine resources include physical resources such as oil, natural gas, building materials, and chemicals; biological resources such as seafood and kelp; and nonextractive resources like transportation and recreation. The contribution of marine resources to the world economy has become so large that international laws now govern their allocation.

In spite of their abundance, marine resources provide only a fraction of the worldwide demand for raw materials, human food, and energy. Similar resources on land can usually be obtained more safely and at lower cost.

Our species has always exercised its capacity to consume resources and pollute its surroundings, but only in the last few generations have our efforts affected the ocean and atmosphere on a planetary scale. The introduction into the biosphere of unnatural compounds (or natural compounds in unnatural quantities) has had, and will continue to have, unexpected detrimental effects. The destruction of marine habitats and the uncontrolled harvesting of the ocean's living resources have also disturbed delicate ecological balances. We find ourselves in difficult situations for which solutions do not come easily.

ON-LINE STUDY RESOURCES

The Web site for this book contains a wealth of helpful study aids. Log on to:

http://info.brookscole.com/garrison

and select *Essentials of Oceanography*, 3rd edition. Select Chapter 15 from the drop-down menu or click on one of these resource areas:

- For study and review, **Chapter at a Glance** gives you an outline of the chapter, **Chapter Summary** allows you to review the chapter's main ideas, and **Glossary** lists concepts and terms for the chapter along with their definitions.

- To test your mastery of the Terms and Concepts to Remember for this chapter, you can use the electronic **Flash Cards,** play the **Concentration** game, or work the **Crossword Puzzle.**

- For practice quizzing, try the multiple-choice **Tutorial Quiz,** the ten-question **True/False Quiz,** or the **Image Analysis Quiz,** which poses questions based on art and photos from the chapter.

TERMS AND CONCEPTS TO REMEMBER

algin	chlorinated hydrocarbons
aquaculture	chlorofluorocarbons (CFCs)
biodegradable	desalination
biological amplification	drift nets
biological resources	eutrophication
bykill	greenhouse effect

greenhouse gases	ozone
mariculture	ozone layer
marine pollution	physical resources
maximum sustainable yield	pollutant
nonextractive resources	potable water
nonrenewable resources	renewable resources
overfishing	sewage sludge

STUDY QUESTIONS

1. Distinguish between physical and biological resources, and between renewable and nonrenewable resources.

2. What are the three most valuable physical resources? How does the contribution of each to the world economy compare to the contribution of that resource derived from land?

3. Does the ocean provide a substantial percentage of all protein needed in human nutrition? Of all animal protein? What is the most valuable biological resource? The fastest growing fishery?

4. What is a nonextractive resource? Give some examples.

5. What is pollution? What factors determine how dangerous a pollutant is?

6. Why is refined oil more hazardous to the marine environment than crude oil? Which is spilled more often?

7. What heavy metals are most toxic? How do these substances enter the ocean? How do they move from the ocean to marine organisms and people?

8. Few synthetic organic chemicals are dangerous in the very low concentrations in which they enter the ocean. How are these concentrations increased? What can be the outcome when these substances are ingested by organisms in a marine food chain?

9. What are the signs of overfishing? How does the fishing industry often respond to these signs? What is the result?

10. What synthetic chemicals appear to be causing depletion of Earth's protective ozone layer? What is the likely result?

11. What is the greenhouse effect? Is it always detrimental? What gases contribute to the greenhouse effect? Why do most scientists believe Earth's average surface temperature will increase over the next few decades? What may result?

12. What, if anything, can be done to minimize the environmental disturbances that result from the exploitation of marine resources?

 Earth Systems Today CD Question

1. Go to "Weather and Climate" and then to "Ozone Hole." Watch the 1996 to 2000 ozone-hole animation showing the seasonal changes in the size of the ozone hole. What can you imply from past data for the future of ozone-hole size? Should mid-latitude dwellers be concerned?

RESOURCES FOR FURTHER READING AND RESEARCH

The Web site for this book contains many ideas for further reading and research. Log on to:

http://info.brookscole.com/garrison

and select *Essentials of Oceanography*, 3rd edition. Select Chapter 15 from the drop-down menu or click on one of these resource areas:

- **Additional Readings** lists the major books and articles consulted in writing this chapter, along with comments from the author about their content and reading level.

- **Hypercontents** takes you to an extensive list of current links to Internet sites with news, research, and images related to individual subjects in the chapter. Just click on the icon that corresponds to a numbered section to see the links for that subject.

- **Internet Exercises** are critical thinking questions that involve research on the Internet with starter URLs provided.

- **InfoTrac College Edition Exercises** leads you to Critical Thinking Projects that use InfoTrac College Edition as a research tool.

- **Regional InfoTrac College Edition** articles are organized into East Coast, West Coast, and Gulf Coast regions, allowing you to study oceanography on a more local level.

 For more readings, go to InfoTrac College Edition, your on-line research library, at:

http://infotrac.thomsonlearning.com

Afterword

The marine sciences are at the threshold of a new age. The recent revolutions in biology and geology are being assimilated, and the road ahead seems clearer. A revolution in the design of sampling devices, robot submersible vehicles, and data processing has brought new vigor to oceanography. Satellite-borne sensors can provide in an instant data that would have taken years to collect using surface ships. Shipboard technology has become so sophisticated that Wyville Thomson or Fridtjof Nansen would hardly recognize our sensors or sampling devices.

The tools may be different, but the spirits of those who use them remain the same. Today's marine scientists are like all the men and women who have gone before: *We want to know about the ocean.* We haunt our mailboxes for journals bearing the latest research news, search television listings for any new ocean shows, inspect new samples with the enthusiasm of little kids, and share our insights with anyone at the drop of a hat. I am personally delighted that you have traveled with me this far. Those of us who enjoy an oceanographic background (and this now includes you) look at Earth with greater understanding than we did before we began. The whole concept of an ocean world appeals to us, gives us profound pleasure, and sobers us with a deep sense of responsibility. In no other field of science do so many ideas interweave to form so rich a tapestry.

Our journey together is over, but before we go our separate ways, I have four last ideas to share:

- *Change* has been a recurrent theme of this book. Earth's climate has changed with time, as has its atmospheric composition, its ocean chemistry, the size and positions of its continents, and its life-forms. Earth may seem a calm and stable home, but it is really a violent place for inhabitation by such seemingly delicate objects as living things. Even so, life and the ocean have grown old together. The story of Earth is the story of change and chance; its history is written in the rocks, the water, and the genes of the millions of organisms that have evolved here. We are survivors.

  That survival may now be in question. Change is now progressing at an unnatural rate, and these human-induced changes are imposing stress on natural systems. What we do *with* and *to* the ocean is literally of planetary consequence. In the last century we have developed the physical, chemical, and biological machinery to destroy or rejuvenate the world ocean and all of its life. A painful time of inadvertent global experimentation lies just ahead.

  All of us who love the colors and textures of this small wet world need to act to moderate the negative effects of the looming environmental crisis. In Chinese, the written character for the word *crisis* has two components: danger and opportunity. Informed citizens will express their concern, discuss this concern with others, and act whenever possible to minimize the threats and take advantage of new opportunities. Intelligence and beauty must triumph; we have no other rational alternative.

- Appreciation of the ocean doesn't come exclusively from the realm of science. Philosophers, artists, composers, and poets have had much to say about the sea. Read Homer's description of the ocean in *The Odyssey* (try Books IV, X, and XI). See how Lord Byron's feeling for the ocean colors his poetry (see, for instance, *Childe Harold's Pilgrimage,* stanzas 183 and 184). Read modern poet Robinson Jeffers's powerful *Continent's End.* Share Prospero's marine magic in Shakespeare's *The Tempest.* Find some of the evocative woodcuts of Rockwell Kent and the impressionistic ocean paintings of English artist J.M.W. Turner. Listen to Benjamin Britten's *Four Sea Interludes* from *Peter Grimes,* and Ralph Vaughan-Williams's *Sea Symphony* and *Sinfonia Antarctica* (but take care not to blow out your sound system). Read the ocean novels of Herman Melville and Jack London, and try reading the journals and accounts of the famous explorers and scientists you have met in this book. Sit on a quiet beach at night with the stars of the Milky Way shining softly overhead. The pervasive inspiration of the wave-breathing ocean is never far away.

- Don't let your involvement stop here. Lifelong learning is the truest joy, a pleasure that does not diminish with age, a source of wisdom and calm. We can learn much about patience, hope, and optimism from the ocean. We can learn much about the world—and about ourselves—by looking for the oceanic connections among things. I hope your interest in learning about the ocean has just been kindled. There is much good in the world. Go and add to it.

- Share your insight with family and friends. *Use* your new knowledge—make it your own. You don't need to be a college professor to talk to people about the beauty, history, and future of the ocean. My wife has always been patient with and receptive to my oceanic tilt. Our children are my tolerant built-in students. Our daughter, son, son-in-law, and soon-to-be granddaughter are participants in Figures 12.12 and 15.16. Along with your children, they will inherit the world.

APPENDIX I

Measurements and Conversions

Other than the United States, only two countries in the world—Liberia and Myanmar—do not use metric measurements. The metric system, a contribution of the French Revolution, conquered Europe along with Napoleon. It is based on a decimal system, a system familiar to Americans because of our decimal money system: 10 cents to a dime, 10 dimes to a dollar.

The first move toward a rational system of measurement was made in 1670 by Gabriel Mouton, the vicar of St. Paul's Church in Lyon, France. Instead of the then-prevalent measurement system based on the width of the king's hand, or the length of his outstretched right arm, or the weight of a particular basket of stones kept in the palace, Mouton suggested a length measure based on the arc of 1 minute of longitude, to be subdivided decimally. Other measurements would follow from this unit of length. His proposal contained the three major characteristics of the metric system: using Earth itself as a basis for measurement, subdividing decimally (by 10s), and using standard prefixes (*kilo, centi, milli,* and so on). These ideas were debated for 125 years before being implemented by a commission appointed by Louis XVI in one of his last official acts before being imprisoned during the French Revolution. One ten-millionth of the distance from the North Pole to the equator (on the line of longitude passing through Paris) was selected as the standard unit of length, the meter. A new unit of weight was derived from the weight of 1 cubic meter of pure water. Temperature was to be based on pure water's boiling and freezing points. A list of prefixes for decimal multiples and submultiples was proposed. In 1795, a firm decision was made to establish the system throughout France, and in 1799, the metric system was implemented "for all people, for all time."

At first, people objected to the changes, but the government insisted that old measurements be included side by side with the equivalent new (metric) ones. In everyday competition, the advantages of the metric system proved decisive; in 1840, it was declared a legal monopoly in France. The French public had been won over to the new, simple, rational system of measurement. All of Europe—and, eventually, virtually all other countries—followed.

Not the United States, however. Though Ben Franklin proposed that the country convert in the eighteenth century, the people of the United States have continued to insist that the metric system—now known as the Système International (SI)—is too difficult to learn and work with. The federal government has urged conversion to metric units to increase opportunities for international trade. In August 1988, President Ronald Reagan signed the Omnibus Trade and Competitiveness Act. This act amended the 1975 Metric Conversion Act, stating that by 1992 all federal agencies must, wherever feasible, use the metric (SI) system in their purchases, grants, and other business. (It should be noted that Canada began to convert in 1970 and has been metric since 1980.)

The government may be making the change, but the public clings tenaciously to inches, pints, and pounds. Why? Is it really simpler to add $\frac{1}{16}$ of an inch, $\frac{1}{32}$ of an inch, and $\frac{3}{8}$ of an inch to cut a bookshelf to length? Can you remember how many cups to a quart? How many pints to a gallon? How many miles to a league? The reason we continue to use the old English Imperial system (which, of course, the English have long since abandoned) is that it is familiar to us. We know how long 5 inches is, and how much a quart is, and what 72° Fahrenheit represents. Perhaps by following the French example—by having measurements expressed everywhere in English *and* metric measurements—we may be able to make a complete conversion within a generation or two. That's why American and metric measurements are used together throughout this book. The process has already begun, of course: You use 35mm film, 2-liter soft drink containers, 750-milliliter wine bottles, 100-watt light bulbs—and you might run a 10-K (10-kilometer) race on Saturday.

The conversion factors listed here will give you an idea of how American and metric (SI) units are equivalent. Don't panic—the system is as rational and logical as it has always been. Note that 1 meter equals 100 centimeters and that 1 centimeter equals 10 millimeters. Note that 2.54 centimeters equals 1 inch. (See the table of conversion factors if you wish to convert from one system to the other.) Some numerical oceanographic data are included in supplemental tables.

Conversion Factors

Area

1 square inch (in.^{2})	6.45 square centimeters
1 square foot (ft^2)	144 square inches 0.0929 square meter
1 square centimeter (cm^2)	0.155 square inch 100 square millimeters
1 square meter (m^2)	10^4 square centimeters 10.8 square feet
1 square kilometer (km^2)	247.1 acres 0.386 square mile 0.292 square nautical mile

Mass

1 kilogram (kg)	2.2 pounds 1,000 grams
1 metric ton	2,205 pounds 1,000 kilograms 1.1 tons
1 pound	16 ounces 454 grams 0.45 kilogram
1 ton	2,000 pounds 907.2 kilograms 0.91 metric ton

Volume

1 cubic inch (in.^{3})	16.4 cubic centimeters
1 cubic foot (ft^3)	1,728 cubic inches 28.32 liters 7.48 gallons
1 cubic centimeter (cc; cm^3) 0.61 cubic inch	1 milliliter
1 liter	1,000 cubic centimeters 61 cubic inches 1.06 quarts 0.264 gallon
1 cubic meter (m^3)	10^6 cubic centimeters 264.2 gallons 1,000 liters
1 cubic kilometer (km^3)	10^9 cubic meters 10^{15} cubic centimeters 0.24 cubic mile

Length

1 micrometer (μm)	0.001 millimeter 0.0000349 inch
1 millimeter (mm)	1,000 micrometers 0.1 centimeter 0.001 meter 0.03937 inch
1 centimeter (cm)	10 millimeters 0.394 inch 10,000 micrometers
1 meter (m)	100 centimeters 39.4 inches 3.28 feet 1.09 yards
1 kilometer (km)	1,000 meters 1,093 yards 3,281 feet 0.62 statute mile 0.54 nautical mile
1 inch (in.)	25.4 millimeters 2.54 centimeters
1 foot (ft)	30.5 centimeters 0.305 meter
1 yard	3 feet 0.91 meter
1 fathom	6 feet 2 yards 1.83 meters
1 statute mile	5,280 feet 1,760 yards 1,609 meters 1.609 kilometers 0.87 nautical mile
1 nautical mile	6,076 feet 2,025 yards 1,852 meters 1.15 statute miles
1 league	15,840 feet 5,280 yards 4,804.8 meters 3 statute miles 2.61 nautical miles

Speed

1 statute mile per hour	1.61 kilometers per hour 0.87 knot
1 knot (nautical mile per hour)	51.5 centimeters per second 1.15 miles per hour 1.85 kilometers per hour
1 kilometer per hour	27.8 centimeters per second 0.62 mile per hour 0.54 knot

Temperature

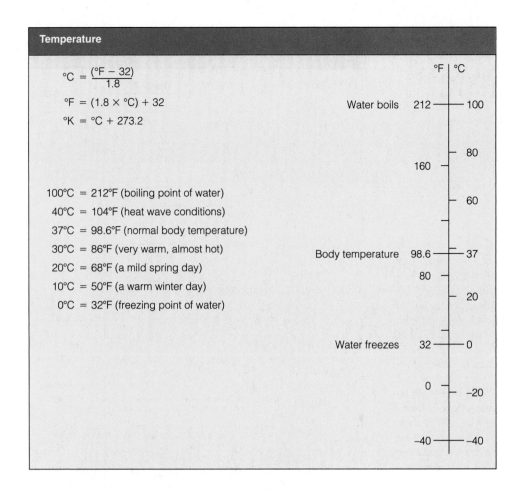

$$°C = \frac{(°F - 32)}{1.8}$$

$$°F = (1.8 \times °C) + 32$$

$$°K = °C + 273.2$$

100°C = 212°F (boiling point of water)
40°C = 104°F (heat wave conditions)
37°C = 98.6°F (normal body temperature)
30°C = 86°F (very warm, almost hot)
20°C = 68°F (a mild spring day)
10°C = 50°F (a warm winter day)
0°C = 32°F (freezing point of water)

	°F	°C
Water boils	212	100
		80
	160	
		60
Body temperature	98.6	37
	80	
		20
Water freezes	32	0
	0	–20
	–40	–40

Some Familiar Metric Approximations

Measurement	Metric Unit	Approximate Size of Unit
Length	millimeter	diameter of a paper clip wire
	centimeter	a little more than the width of a paper clip (about 0.4 inch)
	meter	a little longer than a yard (about 1.1 yards)
	kilometer	somewhat farther than ½ mile (about 0.6 mile)
Mass (Weight)	gram	a little more than the mass (weight) of a paper clip
	kilogram	a little more than 2 pounds (about 2.2 pounds)
	metric ton	a little more than a short ton (about 2,200 pounds)
Volume	milliliter	a fifth of a teaspoon
	liter	a little larger than a quart (about 1.06 quarts)
Pressure	kilopascal	atmospheric pressure is about 100 kilopascals

Source: U.S. Metric Board Report.

Geological Time

As we saw in Chapter 1, astronomers and geologists have determined that Earth originated about 4.6 billion years ago. They have divided Earth's age into eras, roughly corresponding to major geological and evolutionary changes that have taken place, as shown in the chart in **Figure 1.** Note that the time spans of the different eras are not shown to scale; if they were, the chart would run off the page.

Recall that life began fairly soon after Earth's crust, atmosphere, and ocean formed. One way to conceive of the time span over which life evolved is to imagine it as a 24-hour clock, with life originating at midnight (**Figure 2**). In this scheme, invertebrates with hard parts (which make good fossils) became abundant about 4:30 P.M., and animals began to leave the ocean for land about 8 P.M. Our closest human ancestors (*Homo sapiens*) apeared about 2 seconds before midnight, agriculture began only ¼ second before midnight, and the Industrial Revolution has been around for ⁷⁄₁₀₀₀ of a second.

Era	Period	Epoch	Millions of Years Ago (Ma)
CENOZOIC	Quaternary	Recent	0.01
		Pleistocene	1.65
	Tertiary	Pliocene	5
		Miocene	24
		Oligocene	37
		Eocene	58
		Paleocene	66
MESOZOIC	Cretaceous	Late	
		Early	98
	Jurassic		144
	Triassic		208
PALEOZOIC	Permian		245
	Carboniferous		286
	Devonian		360
	Silurian		408
	Ordovician		438
	Cambrian		505
			570
PROTEROZOIC			2,500
ARCHEAN			

Source: Geological Time Scale, Decade of North American Geology, Geological Society of America, 1983.

Figure 1 A geological time scale.

Times of Major Geological and Biological Events

1.65 Ma to present. Major glaciations. Modern humans emerge and begin what may be greatest mass extinction of all time on land, starting with ice age hunters.

65–1.65 Ma. Unprecedented mountain building as continents rupture, drift, collide. Major climatic shifts; vast grasslands emerge. Major radiations of flowering plants, insects, birds, mammals. Origin of earliest human forms.

65 Ma. Asteroid impact? Mass extinction of all dinosaurs and many marine organisms.

135–65 Ma. Pangaea breakup continues, broad inland seas form. Major radiations of marine invertebrates, fishes, insects, dinosaurs. Origin of flowering plants.

181–135 Ma. Pangaea breakup begins. Rich marine communities. Major radiations of dinosaurs.

205 Ma. Asteroid impact? Mass extinction of many organisms in seas, some on land; dinosaurs, mammals survive.

240–205 Ma. Recovery, radiations of marine invertebrates, fishes, dinosaurs. Gymnosperms the dominant land plants. Origin of mammals.

240 Ma. Mass extinction. Nearly all species in seas and on land perish.

280–240 Ma. Pangaea, worldwide ocean forms; shallow seas squeezed out. Major radiations of reptiles, gymnosperms.

360–280 Ma. Tethys Sea forms. Recurring glaciations. Major radiations of insects, amphibians. Spore-bearing plants dominate. Gymnosperms present. Origin of reptiles.

370 Ma. Mass extinction of many marine invertebrates, most fishes.

435–360 Ma. Laurasia forms, Gondwana moves north. Vast swamplands, early vascular plants. Radiations of fishes continue. Origin of amphibians.

435 Ma. Glaciations as Gondwana crosses South Pole. Mass extinction of many marine organisms.

500–435 Ma. Gondwana moves south. Major radiations of marine invertebrates, early fishes.

550–500 Ma. Landmasses dispersed near equator. Simple marine communities. Origin of animals with hard parts.

700–550 Ma. Supercontinent Laurentia breaks up; widespread glaciations.

2,500–570 Ma. Oxygen present in atmosphere. Origin of aerobic metabolism. Origin of protistans, algae, fungi, animals.

3,800–2,500 Ma. Origin of photosynthetic bacteria.

4,600–3,800 Ma. Formation of Earth's crust, early atmosphere, oceans. Chemical evolution leading to origin of life (anaerobic bacteria).

4,600 Ma. Origin of Earth.

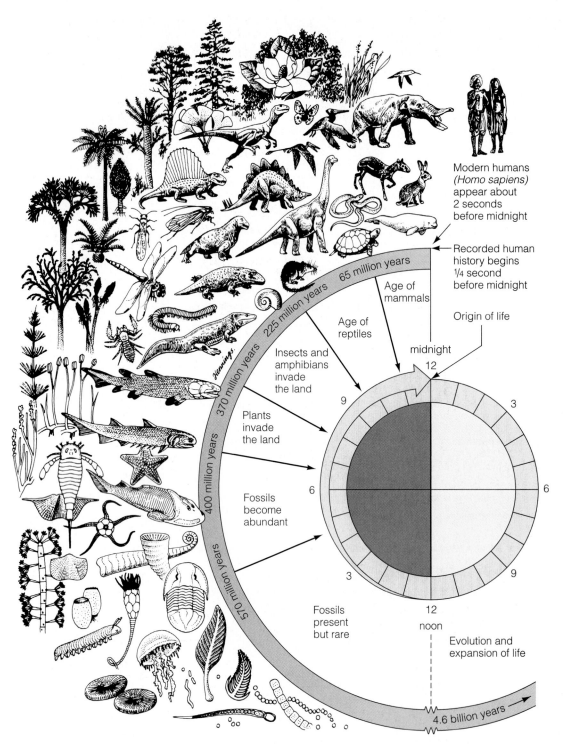

Modern humans
(Homo sapiens)
appear about
2 seconds
before midnight

Recorded human
history begins
¼ second
before midnight

Origin of life

65 million years

Age of
mammals

225 million years

Age of
reptiles

370 million years

Insects and
amphibians
invade
the land

midnight
12

400 million years

Plants
invade
the land

570 million years

Fossils
become
abundant

Fossils
present
but rare

noon

Evolution and
expansion of life

4.6 billion years

Figure 2 Greatly simplified history of the development of different forms of life on Earth through biological evolution, compressed to a 24-hour time scale.

Latitude and Longitude, Time, and Navigation

The ocean is large and easy to get lost in. A backyard, like that shown in **Figure 1,** is smaller, but we can still be lost in it if we don't have a frame of reference. Note that the yard is framed by a fence. We can refer to this frame to establish our position—in this case, at the intersection of perpendicular lines drawn from fence posts 2 and C. Many towns are arranged in this way: Fourth and D streets intersect at a precise spot based on the municipal frame of reference.

But the World Is Round: Spherical Coordinates

If the world were flat, a simple scheme of rectangular coordinates would serve all mapping purposes—a rectangle, like the yard in Figure 1, has four sides from which to measure. A sphere has no edges, no beginnings or ends, so what shall we use as a frame of reference for Earth? Since Earth turns, the poles, the axis of rotation, are the only absolute points of reference. We can draw an imaginary line equidistant from the North and South Poles, a line that *equates* the globe into northern and southern halves: the equator. Other lines, drawn parallel to the equator, further divide the sphere north and south of the equator. These lines, or parallels, are lines of **latitude (Figure 2)**.

We can further subdivide Earth by drawing lines at regular intervals through both poles. Note that unlike the parallels, these lines, called meridians, are all equally long. Meridians are lines of **longitude (Figure 3)**.

If you travel north from the equator, you can count the parallels (lines of latitude) that cross your path to find out how far you have gone. Likewise, if you travel east from a reference meridian, you can count the meridians (lines of longitude) that cross your path to find out how far you have gone. Just as a football player on the field knows his distance from the goal by the yard lines that cross his run, so you know how far north or east you have gone by the lines that have crossed your path.

Since there are no continuous lines of fence posts on the spherical Earth, our reference frame for latitude and longitude must be marked from the equator and poles by some other means. This is done by degrees.

Why Degrees?

Degrees measure fractions of a circle. We need to know what fraction of Earth's circumference separates us from the equator and from the reference meridian to have a definite idea of our location.

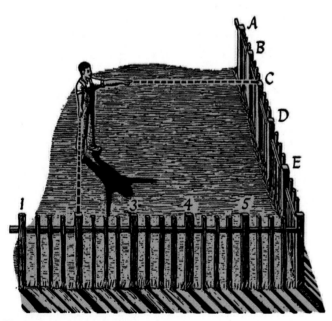

Figure 1

Figure 2

Figure 3

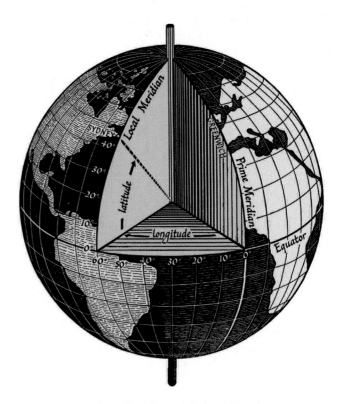

Figure 4 The latitude and longitude of Sydney, Canada.

Babylonian astronomers first divided the circle into 360 degrees (°). Why 360? The moon cycles around Earth every 30 days. It takes about 12 months ("moonths") to make a year. Thus, 30 × 12 = 360, the number of days they supposed was in a year. Circles were divided the same way. As we saw in Chapter 2, the Greek librarian Hipparchus applied this division to the surface of Earth.

In **Figure 4,** we have marked the position of Sydney, Canada. A line drawn from Sydney to the center of Earth intersects the plane of the equator at an angle of 46° to the north. That is its latitude.

The reference meridian, the meridian from which all others are marked, is known as the prime meridian. Unlike the equator, there is no earthly reason the prime meridian should pass through any particular place. It passes through Greenwich, England, because an international agreement signed in 1884 decreed it so. The meridian on which Sydney, Canada, lies intersects the plane of the prime meridian at an angle of 60°. The angular distance of Sydney from the prime meridian is 60° to the west. That is its longitude. So its position is 46°N 60°W.

We can do this for each hemisphere. A line drawn to the center of Earth from Sydney, *Australia,* intersects the plane of the equator at an angle of 34° south latitude. Sydney, Australia, lies 151° east of the prime meridian. Thus, its position is 34°S 151°E. (Note that the greatest possible longitude is 180°; once you pass 180°, the line opposite the prime meridian, you begin to come around the other side of Earth, and the angle to Greenwich decreases.)

What Does Time Have to Do with This?

Meridians are often numbered from prime meridian in 15° increments. Earth takes 24 hours to complete a 360° rotation. Divide 360° by 24 hours and you get 15, the number of degrees the sun moves across the sky in 1 hour. Meridians on a globe are often spaced to represent 1 hour's turning of Earth toward or away from the sun, toward or away from the moon.

You can use this fact to find your east-west position, your longitude. Imagine that you have a radio that can tell you the precise time of noon at Greenwich.[1] If your local noon comes *before* Greenwich noon, you are east of Greenwich. For instance, if the sun is highest in your sky at 10 A.M. Greenwich time, you are 2 hours before—30° east of—Greenwich. Earth must turn 2 more hours before the sun will shine directly above the Greenwich meridian. If your local noon is *after* Greenwich noon, you are west of Greenwich. Suppose that the sun is at high noon and your chronometer, set at Greenwich time, says 6 P.M. That means that Earth has been turning 6 hours since noon at Greenwich, and 6 hours times 15° per hour is 90°. That's your longitude relative to Greenwich: 90°W.

Navigation

Longitude is half the problem. To find latitude and obtain a position, we need to measure the angle north or south of the equator. But we can't use the time difference between local noon and Greenwich noon to determine longitude because the sun moves from east to west and we want to measure north-south position. Instead, we use the angle of the North Star above the horizon. Polaris, the current North Star, lies almost exactly above the North Pole. If we were standing at the North Pole, the North Star would

---

[1]Any shortwave radio will do. Tune it to 2.5, 5, 10, 15, or 20 mHz for radio stations WWV (Colorado) or WWVH (Hawaii). These stations broadcast time signals giving a measure of coordinated universal time, an international time standard based on the time at Greenwich. For a telephone report of coordinated universal time, call WWV at (303) 499-7111, or go to www.time.gov.

appear almost directly overhead; ideally, the angle from the horizon to the star would be 90°, the same as the latitude of the North Pole. At Sydney, Canada, the angle from the horizon to the star would be about 46°—again, the same as the latitude. What would the angle of Polaris be at the equator, 0° latitude? (If you enjoyed your high school geometry course, you might try to prove that the angle from the horizon to Polaris is equal to the latitude at any position in the Northern Hemisphere.)

Polaris is not visible in the Southern Hemisphere, so how can we find south latitude? By finding the angle above the horizon of other stars. In practice, navigators in both hemispheres use a sextant to measure angles from the horizon to selected stars, planets, the moon, and the sun. The time of the observation is carefully noted. The navigator takes these readings to his or her stateroom, consults a series of mathematical tables, does some relatively simple calculations to compensate for observational errors, and comes up to the pilothouse with the vessel's latitude and longitude, accurate (in the best of circumstances) to within ½ mile, marked on a small slip of paper. The daily results are always entered into the ship's log.

New Tricks

Discovering position by measuring the singular positions of heavenly bodies—celestial navigation—is a dying art. Global positioning satellites, loran-C, inertial platforms, radar, and other electronic wonders have largely replaced the romance of a navigator standing on the bridge squinting through a sextant. The slip of paper has been supplanted by the glow of back-lit liquid-crystal readouts or a chart with an X marking the ship's position, accurate to within 50 feet, feeding out of a slot. Still when the power fails, the human navigator becomes the most popular person on board.

Maps and Charts

It is easier to draw a diagram to show someone how to get to a place than to describe the process in words. For centuries, travelers have made special diagrams—maps and charts—to jog their own memories and to show others how to reach distant destinations. A **map** is a representation of some part of Earth's surface, showing political boundaries, physical features, cities and towns, and other geographical information. A **chart** is also a representation of Earth's surface, but it has been specially designed for convenient use in navigation. It is intended to be worked on, not merely looked at. A **nautical chart** is primarily concerned with navigable water areas. It includes information such as coastlines and harbors, channels, obstructions, currents, depths of water, and the positions of aids to navigation.

Any flat map or chart is necessarily a distortion of the spherical Earth. If we roll a flat sheet of paper around a globe to form a cylinder, the paper will contact the globe only along one curve. Let's assume that it's the equator. If the lines of latitude and longitude on the globe are covered with ink, only the equator will contact the paper and print an exact replica of itself. Unroll the cylinder, and

that part of the new map will be a perfect representation of Earth. To include areas north and south of the equator, we will have to "throw them forward" onto the paper; we need to *project* them in some way.

Now imagine our globe to be a translucent sphere. If we place a bright light at its center, we can project the lines of latitude and longitude onto the rolled paper cylinder (**Figure 1**). Careful tracing of these lines will result in a map, but the areas away from the equator will be distorted: The farther from the equator, the greater the distortion. A useful modification of this projection—one that does not distort high latitudes as dramatically—was devised by Gerhardus Mercator, a Flemish cartographer who published a map of the world in 1569. Though landmasses and ocean areas are not depicted as accurately in a Mercator projection as they would be on a globe, such a map is still useful because it enables mariners to steer a course over long distances by plotting straight lines.

The distortion in Mercator projections has led generations of schoolchildren to believe that Greenland is the same size as South America (**Figure 2**). Mercator charts can distort our perceptions of

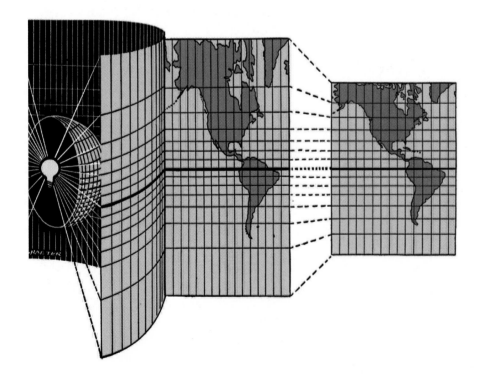

Figure 1 Central projection of a globe upon a cylinder, and a modified map structure, the Mercator, made to the same scale along the equator.

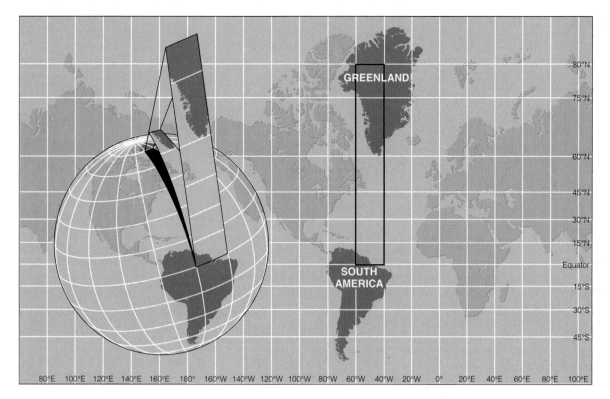

Figure 2 A gore of the globe peeled and protected according to the scheme devised by Gerhardus Mercator. This is the protection used on modern sailing charts. Note that this projection's distortion at high latitudes makes Greenland and South America appear about the same size. Next time you're near a globe, check their real sizes.

the ocean as well: The area of the continental shelves at high latitudes, the amount of primary productivity in the polar regions, and the importance of ocean currents at the northerly or southerly extremes of an ocean basin may be exaggerated if presented in Mercator projection. The projection used in this book—a further modification of the Mercator projection known as the Miller projection—was chosen for its more accurate representation of surface area at high latitudes.

Mapmakers have invented other projections, each with advantages and disadvantages for particular uses. Some are conical projections: a flat sheet of paper wrapped into a cone with its edge touching the globe at a line of latitude north (or south) of the equator and the point of the cone above the North (or South) Pole. Conical projections do not distort high-latitude areas in the same way a Mercator projection does, and if drawn for the ocean area in which a mariner is sailing, can be used to draw great circle routes as straight lines. However, the distortions inherent in a conical projection prevent it from being used to represent more than about one-third of the globe on a single sheet of paper. Other projections try to minimize distortion around a specific location. All map and

chart projections are distorted in some way; a sphere cannot be flattened onto a plane without deformation. Marine scientists necessarily become familiar with various chart projections and are careful to use the proper chart for its intended purpose.

FOR FURTHER STUDY

Herring, T. 1996. "The Global Positioning System." *Scientific American*, February, 44–50.

Krause, G., and M. Tomczak. 1995. "Do Marine Scientists Have a Scientific View of the Earth?" *Oceanography* 8 (no. 1): 11–16. Chart distortions often distort our interpretation of data, as this well-illustrated paper demonstrates.

Wilford, J. N. 1998. "Revolution in Mapping." *National Geographic*, February, 6–39. The usual thorough treatment of a rapidly changing topic.

APPENDIX V

The Law of the Sea

Prehistoric peoples living near the shore were the earliest users of marine resources. With the rise of nation-states and military establishments, the fight for control of marine resources began. Some nations assumed that the ocean belonged to all, and they endeavored to guarantee free right to passage and resources. Others decided that the ocean belonged to none and tried to control access to ports and resources by force. In 1604 Hugo Grotius, a learned Dutch jurist, wrote *De Jure Praedae* (*On the Law of Prize and Booty*), a treatise justifying the action of a Dutch admiral who had successfully defended Dutch trading rights in a dispute with Portugal. One chapter of this work, in which Grotius defended free ocean access for all nations, was reprinted in 1609 under the title *Mare Liberum* (*A Free Ocean*). *Mare Liberum* formed the basis for all modern international laws of the sea.

About a century later, in 1703, the concept of territorial seas adjacent to land was recognized. A country's seaward boundary was set at about 5 kilometers (3 miles)—the distance a cannonball could be fired from shore. This 5-kilometer (3-mile) limit stood until 1945.

The United Nations and the International Law of the Sea

After World War II the technology became available to search for oil and natural gas on continental shelves. After U.S. oil companies found rich deposits beyond the 5-kilometer (3-mile) limit off Louisiana, President Harry Truman issued a proclamation annexing the physical and biological resources of the continental shelf contiguous to the United States. Other nations rushed to make similar claims.

The United Nations then became involved. A committee of the General Assembly began to formulate policy, which was later presented at the First United Nations Conference on Law of the Sea in 1958 in New York. Twenty-four years of effort by delegates from many interested nations resulted in the 1982 Draft Convention on the Law of the Sea. In April 1982 the United Nations adopted the convention by a vote of 130 to 4, with 17 abstentions. (The United States, Turkey, Venezuela, and Israel voted against the convention.) By 1988 more than 140 countries had signed all or most parts of the treaty. It is now legally binding, but signatories have selectively chosen to respect or ignore its individual provisions.

Here are some important features of the 1982 Draft Convention:

- **Territorial waters** are defined as extending 12 miles (18.2 kilometers) from shore. A nation has the right to jurisdiction within its territorial waters. Straits used for international navigation are excluded from a nation's territorial waters in that any vessel has the right to innocent passage.

- The 200 nautical miles (370 kilometers) from a nation's shoreline constitute its **exclusive economic zone (EEZ)**. Nations hold sovereignty over resources, economic activity, and environmental protection within their EEZs.

- All ocean areas outside the EEZs are considered the **high seas.** Traditionally, the high seas are common property to be shared by the citizens of the world. An International Seabed Authority was established to oversee the extraction of mineral resources from the deep sea.

- The values of protecting the ocean and preventing marine pollution were endorsed.

- Subject to some conditions, the freedom of scientific research in the ocean was encouraged.

The convention places about 40% of the world ocean under the control of the coastal countries, within the EEZs. The resources of the remaining 60%, the high seas, are to be shared by the citizens of the world.

The United States Exclusive Economic Zone

The United States did not sign the 1982 Draft Convention for a variety of reasons. Among these was concern that private enterprise would be deprived of profits if it were made to share high-seas resources with other countries. Instead, the United States unilaterally claimed sovereign rights and jurisdiction over all marine resources within its own 200-nautical-mile region, which it called the **United States Exclusive Economic Zone.** The proclamation—similar in most ways to the 1982 United Nations Treaty but lacking the provision of shared high-seas resources—was signed by President Ronald Reagan on 10 March 1983. The U.S. EEZ brings within national domain more than 10.3 million square kilometers (4 million square miles) of continental margins, an area 30% larger than the land area of the United States and a region of diverse geological and oceanographic settings (**Figure 1**).

The first step in exploring this new region has been to map the surface of the seafloor, a project that is still going on. The first bottom surveys were conducted off the Pacific coast of the United States because of the energy and mineral resources known to exist there. Massive deposits of various metallic sulfides are being explored at the hydrothermal vents on the Gorda and Juan de Fuca Ridges. More than a hundred previously unknown volcanoes have been

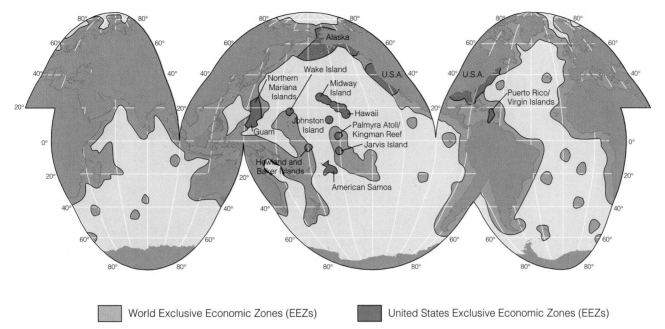

World Exclusive Economic Zones (EEZs)   United States Exclusive Economic Zones (EEZs)

Figure 1 The U.S. Exclusive Economic Zone (EEZ), shown here in red, covers a vast area of the continental margin. The edge of the continent as it extends out under the ocean contains a vast wealth of resources.

mapped within the U.S. EEZ off our Pacific coast. Huge faults, submarine landslides, seamounts, and details of the spreading crest of the oceanic ridges are a few of the features that have been discovered so far. Knowledge of the modern tectonic setting for ore formation has been valuable in locating new ore deposits on land.

Manganese nodules and crusts have been discovered within the U.S. EEZ off the Atlantic and Pacific coasts, Hawaii, and the Pacific Island territories. Cobalt, present in the manganese crusts, is being studied to determine how the deposit is formed and how it can be retrieved economically.

The EEZ project also includes research into meteorology, accurate weather forecasting, environmental studies, effects of plate movement on the ocean floor, effects of mining on seafloor organisms, and geohazards such as submarine landslides and earthquakes. It is hoped that this new era of marine exploration and research will lead to informed decisions about offshore activities that will benefit not only the United States but also other maritime nations. Maybe the treasure trove is really there, waiting for these new developments in marine technology and international cooperation to tap the riches of the sea.

Reference: Jacobson, J. L and A. Rieser. 1998. "The Evolution of Ocean Law." *Scientific American Presents*, vol. 9 #3, Fall 1998, pp. 100–105.

Working in Marine Science

Working in the marine sciences is wonderfully appealing to many people. They sometimes envision a life of diving in warm, clear water surrounded by tropical fish, or descending to the seabed in an exotic submersible outfitted like Captain Nemo's fictional submarine in *20,000 Leagues Under the Sea,* or living with intelligent dolphins in a marine life park. Then reality sets in. There are rewards from working in the marine sciences, but they tend to be less spectacular than the first dreams of students looking to the ocean for a life's work.

A marine science worker is paid to bring a specific skill to a problem. If that problem lies in warm, tropical water or in a marine park, fine. But more likely, the problem will yield only to prolonged study in an uncomfortable, cold, or dangerous environment. The intangible rewards can be great; the physical rewards are often slim. Having said that, let me add that no endeavor is more interesting or exciting, and few are more intellectually stimulating. Doing marine science is its own reward.

Training for a Job in Marine Science

Marine science is, of course, science. And science requires mathematics—you need math to do the chemistry, physics, measurements, and statistics that lie at the heart of science. Your first step in college should be to take a math placement test, enroll in an appropriate math class, and spend time doing math. *Math is the key to further progress in any area of marine science.*

With your math skills polished, start classes in chemistry, physics, and basic biology. Surprisingly, except for one or two introductory marine science classes, you probably won't take many marine science courses until your junior year. These introductory classes will be especially valuable because a balanced survey of the marine sciences can aid you in selecting an appealing specialty. Then, with a good foundation in basic science, you can begin to concentrate in that specialty.

Other skills are important too. The ability to write and speak well is crucial in any science job. Also critical is computer literacy. Expertise in photography or foreign languages or the ability to field-strip and rebuild a diesel engine or hydraulic winch will put you a step above the competition at hiring time. Certification as a scuba diver is almost mandatory; you can never have too much diving experience. (Remember, though, diving is only a tool, a way to deliver an informed set of eyes and an educated brain to a work site.) You should be in good health. Indeed, good aerobic fitness is essential in most marine science jobs; stamina is often a crucial factor in long experiments under difficult conditions at sea. It is also desirable to be physically strong—marine equipment is heavy and often bunglesome. And it helps greatly if you are not prone to seasickness.

Deciding what school to attend will depend on your skills. Readers of this book will probably be enrolled in a general oceanography course in a college or university. The first step would be to discuss your interests with your professor (or his or her teaching assistants). You'll need to attend a four-year college or university to complete the first phase of your training. If you're attending a two-year institution, picking a specific transfer institution can come later, but keep a few things in mind: No matter where you take your first two years of training, you need thorough preparation in basic science. You should attend an institution with strengths in the area of your specialty (such as geology, biology, and marine chemistry). And you should be reasonable in your expectations of acceptance if you're a transfer student (that is, don't try for Stanford or Yale with a B average).

Another thing: Most marine scientists have completed a graduate degree (a master's degree or doctorate). Most graduate students hold teaching or research assistantships (that is, they get paid for being grad students). In all, progress to a final degree is a long road, but the journey is itself a pleasure.

If the thought of four or more years of higher education does not appeal to you, does that mean there's no hope? Not at all. Many students begin a program with the goal of becoming a marine technician, animal trainer at a marine life park, marina or boatyard employee or manager, or crew member on a private yacht. Those jobs don't always require a bachelor's degree. Jobs at Sea World and other marine theme parks do require athletic ability, extreme patience, public speaking skills, a love of animals, and, usually, diving experience. Few positions are available, but there is some turnover in the ranks of junior trainers, and being hired is certainly possible.

Becoming a marine technician is an especially attractive alternative to the all-out chemistry-physics-math academic route. For every highly trained marine scientist, there are perhaps five technical assistants who actually do the experiments, maintain the equipment, work daily with organisms, and build special apparatus. Marine technicians tend to spend more time at hands-on tasks than marine scientists. Most of these folks (including the author of the letter that ends this appendix) have the equivalent of a two-year technical degree, usually from a community college.

Don't quit your job, burn your bridges, leave your family, sell your possessions, and dedicate yourself monklike to marine science. Do some investigation. Nothing is as valuable as *actually going out and talking to people who do things that you'd like to do.* Ask them if they enjoy their work. Is the pay OK? Would they start down the same road if they had it to do all over again? You may decide to expand your involvement in marine science in a more informal way, by becoming a volunteer; joining the Sierra Club, Audubon Society, Greenpeace, or other environmental group; working for your state's fish and game office as a seasonal aide; or attending lectures at local colleges and universities.

If you decide to continue your education, don't be discouraged by the time it will take. Have a general view of the big picture, but proceed one semester at a time. Again, remember that the educational journey is itself a great pleasure. Don Quixote reminds us of the joys of the road, not the inn.

The Job Market

Marine science is very attractive to the general public. People are naturally drawn to thoughts of working in the field. Unfortunately, there aren't a great many jobs in the marine sciences. But there will always be some jobs, and people will fill them. Those people will be the best prepared, most versatile, and most highly motivated of those who apply. Perhaps not surprisingly, marine biology is the most popular marine science specialty. Unfortunately, it is also the area with the smallest number of nonacademic jobs. Museums, aquariums, and marine theme parks employ biologists to care for animals and oversee interpretive programs for the public. A few marine biologists are employed as monitoring specialists by water management agencies like sanitation districts, which discharge waste into the ocean. Electrical utilities that use seawater to cool the condensers in power-generating plants almost always have a handful of marine biologists on staff to watch the effects of discharged heat on local marine life and to write the reports required by watchdog agencies. State and federal agencies employ marine biologists to read and interpret those documents and to set standards. Relatively small businesses, like private shipyards, agricultural concerns, and chemical plants, can't afford their own staff biologists, so private consulting firms staffed by marine biologists and other specialists have arisen to assist in the preparation of the environmental impact reports required of businesses under various legislation.

There are more jobs in physical oceanography: marine geology, ocean engineering, and marine chemistry and physics. Thousands of marine geologists work for oil and mineral companies; indeed, with the increasing emphasis on offshore resources, the market for these people may be increasing. Marine engineers are needed to design, construct, and maintain offshore oil rigs, ships, and harbor structures. Marine chemists are hard at work figuring ways to stop corrosion and to extract chemicals from seawater. Physicists are vitally interested in the transmission of underwater sound and light, in the movement of the ocean, and in the role the ocean plays in global weather and climate. Economists, lawyers, writers, and mathematicians also work in the marine science field.

Many biological and physical oceanographers are teachers and professors. Indeed, there are nearly as many marine scientists employed in the academic world as there are in private industry and government. If you like the idea of teaching, you might consider this avenue. The demand for science teachers at all educational levels is already great and is expected to increase.

Four factors will be significant in influencing your employability:

1. *Experience.* Employers are favorably impressed by experience, especially work experience related to the duties of the position for which you are applying. Volunteer work counts.

2. *Grades.* Good grades are important, especially for positions in government agencies. A grade point average of 3.0 or higher in all college work increases your chances of employment and should give you a higher starting salary.

3. *Geographical availability.* Don't restrict yourself geographically. Not everyone can work in Hawaii or California, but four out of ten marine scientists work in just three states: California, Maryland, and Virginia.

4. *Diversification.* Again, mastery of more than one specialty gives you an employment edge. Being a plankton connoisseur and also able to repair a balky computer while ordering in-port supplies over a radiotelephone in Spanish makes a lasting impression.

Report from a Student

Students in marine science programs graduate, get jobs, and move on. One of the pleasures of being a professor is hearing from them. One of our former students, an employee of the Marine Science Institute at the University of California, Santa Barbara, reported his activities as part of a team using the submersible *Alvin* to investigate plumes of warm water issuing from hydrothermal vents along the southern Juan de Fuca Ridge. The nature of his work—and his enthusiasm for it—is clearly evident in this excerpt. Dan Dion writes:

> The buoyant plume experiment wasn't going very well. The chemistry dives were pushed back because of technical difficulties and poor weather (rough seas cut two dives). The first two buoyant plume dives ended in failure. The first one because of mechanical/electrical problems, the second because of a computer crash. Everyone worked around the clock to get things in order for dive 2440. I was scheduled to go down with John Trefrey, from the Florida Institute of Technology. Cindy Van Dover was our pilot. We launched *Alvin* right on schedule at 0800, and descended from the glacier blue water into the bioluminescent snowstorm of the euphotic zone. During the hour and a half descent we listened to the music of Enya in the soft light of the sub as we busily prepared ourselves for the experiment: booting up the computers, loading film in the cameras, tapes in the recorders, etc. We had three laptop computers to deal with in the cramped spaces of the sub. I was in charge of two of them, one that plotted our in-sub navigation (from transponders), and the other that controlled and recorded data from the [continuous temperature-depth-conductivity probe]. The third laptop was connected to the chemical analyzer and John was in control of that. All of the instruments were operating perfectly. I periodically saved the computer file in the event of another crash. We reached the bottom right on target; Monolith Vent was in sight, 2261 meters below the surface. We did a video survey of the vent, especially a chimney that was rapidly growing back after the geologists had decapitated it just a few days earlier. We ascended to 55 meters-off-bottom and began our drive-throughs. To me, the navigator, it was the ultimate video game. From the computer screen I would guide *Alvin* through a dark abyss, calling out headings that would maneuver us into a "lawn mower" pattern crisscrossing the plume. It was quite visible, and even beautiful; wispy, intricate patterns of "smoke" which seemed to dance like graceful ghosts. We completed passes at 35, 20, 10, and 5 meters above the bottom, then one last one at 45 meters. Eight hours of sub time went by so quickly! Our dive was a huge success; in addition to all the samples we obtained, we generated over 25 megabytes of data. I used everything I learned . . . from computer skills to navigation and marlinspike seamanship (and, of course, chemistry!).

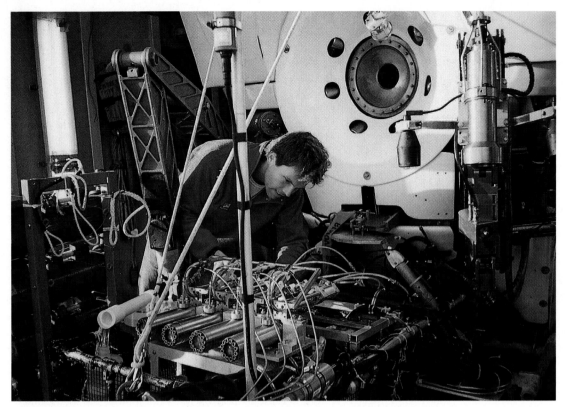

Figure 1 Dan Dion attaching hydraulic actuators to water-sampling bottles, in preparation for a dive by *Alvin*. The bottles were part of a sampling program that included measurements of water conductivity, transmissivity, temperature, and iron and manganese ion content near a hydrothermal vent. This information was later merged with data from transponder navigation to obtain a three-dimensional map of plume structure and chemistry.

Figure 2 Dan Dion enters the hatch of *Alvin* to visit hydrothermal vents 2,261 meters (7,416 feet) below the surface off the coast of Oregon.

Marine science is equipment training sessions, long cruises, seminars and lectures, visiting experts, hot sand volleyball games, and chilly labs with classical music. Marine science is a long and demanding road; but it is, quite honestly, great fun. Captain Nemo and his sub never had it this good!

FOR MORE INFORMATION

Two useful Internet sites

This Web site, hosted by Joe Wible, Hopkins Marine Station of Stanford University, contains links to many career-related sites:

www-marine.stanford.edu/HMSweb/careers.html

This Web site lists the member institutions of the Joint Oceanographic Institutions, universities at the forefront of oceanographic research:

www.joi-odp.org/JOI/Members.html

Organizations

American Society of Limnology and Oceanography A nonprofit, professional scientific society that seeks to promote the interests of limnology and oceanography and related sciences and to further the exchange of information across the range of aquatic science disciplines.

www.aslo.org/

Association for Women in Science A nonprofit association dedicated to achieving equity and full participation for women in science, mathematics, engineering, and technology.

www.awis.org/

Center for Marine Conservation A nonprofit membership organization dedicated to protecting marine wildlife and its habitats and to conserving coastal and ocean resources.

www.cmc-ocean.org/

Marine Advanced Technology Training Center For training of marine technicians—specialists in the deployment and maintenance of tools used in marine science.

www.marinetech.org/

Marine Technology Society An international, interdisciplinary society devoted to ocean and marine engineering science and policy.

www.mtsociety.org/

National Marine Educators Association An organization that brings together those interested in the study and enjoyment of the world of water—both fresh and salt.

www.marine-ed.org/

National Oceanic and Atmospheric Administration A government agency that guides the use and protection of our oceans and coastal resources, warns of dangerous weather, charts the seas and skies, and conducts research to improve our understanding and stewardship of the environment.

www.noaa.gov/

Oceanic Engineering Society An organization that promotes the use of electronic and electrical engineers for instrumentation and measurement work in the ocean environment and the ocean/atmosphere interface.

www.oceanicengineering.org/

The Oceanography Society A professional society for scientists in the field of oceanography.

www.tos.org/

The Society for Marine Mammalogy A professional organization that supports the conservation of marine mammals and the educational, scientific, and managerial advancement of marine mammal science.

www.marinemammalogy.org/

Joint Oceanographic Institutions The Joint Oceanographic Institutions (JOI) is a consortium of U.S. academic institutions that brings to bear the collective capabilities of the individual oceanographic institutions on research planning and management of the ocean sciences.

www.joi-odp.org/

Note: The organizations listed here are but a sampling of the resources that are available to help you learn more about careers in the marine sciences. Each contact you make in your search for information will lead you to more contacts and more information.

Glossary

absorption Conversion of sound or light energy into heat.

abyssal hill Small sediment-covered inactive volcano or intrusion of molten rock less than 200 meters (650 feet) high, thought to be associated with seafloor spreading. Abyssal hills punctuate the otherwise flat abyssal plain.

abyssal plain Flat, cold, sediment-covered ocean floor between the continental rise and the oceanic ridge at a depth of 3,700 to 5,500 meters (12,000 to 18,000 feet). Abyssal plains are more extensive in the Atlantic and Indian Oceans than in the Pacific.

abyssal zone The ocean between about 4,000 and 5,000 meters (13,000 and 16,500 feet) deep.

accessory pigment One of a class of pigments (such as fucoxanthin, phycobilin, and xanthophyll) present in various photosynthetic plants and that assist in the absorption of light and the transfer of its energy to chlorophyll. Also called *masking pigment*.

accretion An increase in the mass of a body by accumulation or clumping of smaller particles.

acid A substance that releases a hydrogen ion (H+) in solution.

active margin Continental margin near an area of lithospheric plate convergence. Also called *Pacific-type margin*.

adhesion Attachment of water molecules to other substances by hydrogen bonds. Wetting.

ahermatypic Describing coral species lacking symbiotic dinoflagellates and incapable of secreting calcium carbonate at a rate suitable for reef production.

air mass A large mass of air with nearly uniform temperature, humidity, and density throughout.

algin A mucilaginous commercial product of multicellular marine algae. Widely used as a thickening and emulsifying agent.

alkaline Basic. See *base*.

amphidromic point A "no-tide" point in an ocean caused by basin resonances, friction, and other factors around which tidal crests rotate. About a dozen amphidromic points exist in the world ocean. Sometimes called a *node*.

Antarctic Bottom Water The densest ocean water (1.0279 g/cm^3), formed primarily in Antarctica's Weddell Sea during Southern Hemisphere winters.

Antarctic Circumpolar Current See *West Wind Drift*.

aphotic zone The dark ocean below the depth to which light can penetrate.

aquaculture The growing or farming of plants and animals in a water environment under controlled conditions. Compare *mariculture*.

Arthropoda The phylum of animals that includes shrimp, lobsters, krill, barnacles, and insects. The phylum Arthropoda is the world's most successful.

artificial system of classification A method of classifying an object based on attributes other than its reason for existence, its ancestry, or its origin. Compare *natural system of classification*.

asthenosphere The hot, plastic layer of the upper mantle below the lithosphere, extending some 300 kilometers (187 miles) below the surface. Convection currents within the asthenosphere power plate tectonics.

astronomical tide A tide caused by the interaction of the gravitational attraction of the sun, moon, and other heavenly bodies. See also *meteorological tide*.

atmospheric circulation cell Large circuit of air driven by uneven solar heating and the Coriolis effect. Three circulation cells form in each hemisphere. See also *Hadley cell; Ferrel cell; polar cell*.

atoll A ring-shaped island of coral reefs and coral debris enclosing, or almost enclosing, a shallow lagoon from which no land protrudes. Atolls often form over sinking, inactive volcanoes.

authigenic sediment Sediment formed directly by precipitation from seawater. Also called *hydrogenous sediment*.

autotroph An organism that makes its own food by photosynthesis or chemosynthesis.

backshore Sand on the shoreward side of the berm crest, sloping away from the ocean.

backwash Water returning to the ocean from waves washing onto a beach.

baleen The interleaved, hard, fibrous, hornlike filters within the mouth of baleen whales.

barrier island A long, narrow, wave-built island lying parallel to the mainland and separated from it by a lagoon or bay. Compare *sea island*.

barrier reef A coral reef surrounding an island or lying parallel to the shore of a continent, separated from land by a deep lagoon. Coral debris islands may form along the reef.

basalt The relatively heavy crustal rock that forms the seabeds, composed mostly of oxygen, silicon, magnesium, and iron. Its density is about 2.9 g/cm^3.

base A substance that combines with a hydrogen ion (H+) in solution.

bathyal zone The ocean between about 200 and 4,000 meters (700 and 13,000 feet) deep.

bathymetry The discovery and study of ocean floor contours.

bay mouth bar An exposed sandbar attached to a headland adjacent to a bay and extending across the mouth of the bay.

beach A zone of unconsolidated (loose) particles extending from below water level to the edge of the coastal zone.

beach scarp Vertical wall of variable height marking the landward limit of the most recent high tides. Corresponds with the berm at extreme high tides.

benthic zone The zone of the ocean bottom. See also *pelagic zone.*

berm A nearly horizontal accumulation of sediment parallel to shore. Marks the normal limit of sand deposition by wave action.

berm crest The top of the berm; the highest point on most beaches. Corresponds to the shoreward limit of wave action during most high tides.

big bang The hypothetical event that started the expansion of the universe from a geometric point. The beginning of time.

biodegradable Able to be broken by natural processes into simpler compounds.

biogenous sediment Sediment of biological origin. Organisms can deposit calcareous (calcium-containing) or siliceous (silicon-containing) residue.

biological amplification Increase in concentration of certain fat-soluble chemicals such as DDT or heavy-metal compounds, in successively higher trophic levels within a food web.

biological resource A living animal or plant collected for human use. Also called a *living resource.*

biomass The mass of living material in a given area or volume of habitat.

biosynthesis The initial formation of life on Earth.

blade Algal equivalent of a vascular plant's leaf. Also called a *frond.*

breakwater An artificial structure of durable material that interrupts the progress of waves to shore. Harbors are often shielded by a breakwater.

buffer A group of substances that tend to work against change in the pH of a solution by combining with free ions.

buoyancy The ability of an object to float in a fluid by displacement of a volume of fluid equal to it in mass.

bykill Animals unintentionally killed when desirable organisms are collected.

calcareous ooze Ooze composed mostly of the hard remains of organisms containing calcium carbonate.

calcium carbonate compensation depth (CCD) The depth at which the rate of accumulation of calcareous sediments equals the rate of dissolution of those sediments. Below this depth, sediment contains little or no calcium carbonate.

calorie The amount of heat needed to raise the temperature of 1 gram (0.035 ounce) of pure water by 1°C (1.8°F).

capillary wave A tiny wave with a wavelength of less than 1.73 centimeters (0.68 inch), whose restoring force is surface tension; the first type of wave to form when the wind blows.

Carnivora The order of mammals that includes seals, sea lions, walruses, and sea otters.

cartilage A tough, elastic tissue that stiffens or supports.

cartographer A person who makes maps and charts.

cephalopod The group of marine predators that includes squid, octopuses, and nautiluses.

Cetacea The order of mammals that includes porpoises, dolphins, and whales.

Challenger expedition The first wholly scientific oceanographic expedition, 1872–76. Named for the steam corvette used in the voyage.

chart A map that depicts mostly water and the adjoining land areas.

chemical bond An energy relationship that holds two atoms together as a result of changes in their electron distribution.

chemical equilibrium In seawater, the condition in which the proportion and amounts of dissolved salts per unit volume of ocean are nearly constant.

chemosynthesis The synthesis of organic compounds from inorganic compounds using energy stored in inorganic substances such as sulfur, ammonia, and hydrogen. Energy is released when these substances are oxidized by certain organisms.

chlorinated hydrocarbons The most abundant and dangerous class of halogenated hydrocarbons, synthetic organic chemicals hazardous to the marine environment.

chlorinity A measure of the total weight of chloride, bromide, and iodide ions in seawater. We may derive salinity from chlorinity by multiplying by 1.80655.

chlorofluorocarbons (CFCs) A class of halogenated hydrocarbons thought to be depleting Earth's atmospheric ozone. CFCs are used as cleaning agents, refrigerants, fire-extinguishing fluids, spray-can propellants, and insulating foams.

Chondrichthyes The class of fishes with cartilaginous skeletons: the sharks, skates, rays, and chimaeras.

clamshell sampler Sampling device used to take shallow samples of the ocean bottom.

clay Sediment particle smaller than 0.004 millimeter in diameter; the smallest sediment size category.

climate The long-term average of weather in an area.

climax community A stable, long-established community of self-perpetuating organisms that tends not to change with time.

clumped distribution Distribution of organisms within a community in small, patchy aggregations, or clumps; the most common distribution pattern.

coast The zone extending from the ocean inland as far as the environment is immediately affected by marine processes.

coastal cell The natural sector of a coastline in which sand input and sand outflow are balanced.

coastal upwelling Upwelling adjacent to a coast, usually induced by wind.

coccolithophore A very small planktonic alga carrying discs of calcium carbonate, which contributes to biogenous sediments.

cohesion Attachment of water molecules to each other by hydrogen bonds.

Columbus, Christopher (1451–1506) Italian explorer in the service of Spain who discovered islands in the Caribbean in 1492. Although traditionally credited as the discoverer of America, he never actually sighted the North American continent.

community The populations of all species that occupy a particular habitat and interact within that habitat.

compass An instrument for showing direction by means of a magnetic needle swinging freely on a pivot and pointing to magnetic north.

conduction The transfer of heat through matter by the collision of one atom with another.

constructive interference The addition of wave energy as waves interact, producing larger waves.

continental crust The solid masses of the continents, composed primarily of granite.

continental drift The theory that the continents move slowly across the surface of Earth.

continental margin The submerged outer edge of a continent, made of granitic crust. Includes the continental shelf and the continental slope. Compare *ocean basin*.

continental rise The wedge of sediment forming the gentle transition from the outer (lower) edge of the continental slope to the abyssal plain. Usually associated with passive margins.

continental shelf Gradually sloping, submerged extension of a continent, composed of granitic rock overlain by sediments. Has features similar to the edge of the nearby continent.

continental slope The sloping transition between the granite of the continent and the basalt of the seabed. The true edge of a continent.

convection Movement within a fluid resulting from differential heating and cooling of the fluid. Convection produces mass transport or mixing of the fluid.

convection current A single closed-flow circuit of rising warm material and falling cool material.

convergence zone The line along which waters of different density converge. Convergence zones form the boundaries of tropical, subtropical, temperate, and polar areas.

convergent plate boundary A region where plates are pushing together and where a mountain range, island arc, and/or trench will eventually form. Often a site of much seismic and volcanic activity.

Cook, James (1728–1779) Officer in the British Royal Navy who led the first European voyages of scientific discovery.

coral Any of more than 6,000 species of small cnidarians, many of which are capable of generating hard calcareous (aragonite, $CaCO_3$) skeletons.

coral reef A linear mass of calcium carbonate (aragonite and calcite) assembled from coral organisms, algae, mollusks,

worms, and so on. Coral may contribute less than half of the reef material.

core The innermost layer of Earth, composed primarily of iron, with nickel and heavy elements. The inner core is thought to be solid, the outer core liquid. The average density of the outer core is about 11.8 g/cm^3, and that of the inner core is about 16 g/cm^3.

Coriolis, Gaspard Gustave de (1792–1843) The French scientist who in 1835 worked out the mathematics of the motion of bodies on a rotating surface. See *Coriolis effect*.

Coriolis effect The apparent deflection of a moving object from its initial course when its speed and direction are measured in reference to the surface of the rotating Earth. The object is deflected to the right of its anticipated course in the Northern Hemisphere and to the left in the Southern Hemisphere. The deflection occurs for any horizontal movement of objects with mass and has no effect at the equator.

cosmogenous sediment Sediment of extraterrestrial origin.

covalent bond A chemical bond formed between two atoms by sharing a pair of electrons.

crust The outermost solid layer of Earth, composed mostly of granite and basalt; the top of the lithosphere. The crust has a density of 2.7–2.9 g/cm^3 and accounts for 0.4% of Earth's mass.

current Mass flow of water. (The term is usually reserved for horizontal movement.)

cyclone A weather system with a low-pressure area in the center around which winds blow counterclockwise in the Northern Hemisphere and clockwise in the Southern Hemisphere. Not to be confused with a tornado, a much smaller weather phenomenon associated with severe thunderstorms. See also *extratropical cyclone; tropical cyclone*

deep-water wave A wave in water deeper than half its wavelength.

deep zone The zone of the ocean below the pycnocline, in which there is little additional change of density with increasing depth. Contains about 80% of the world's water.

degree An arbitrary measure of temperature. One degree Celsius (°C) = 1.8 degrees Fahrenheit (°F).

delta The deposit of sediments found at a river mouth, sometimes triangular in shape (like its name after the Greek letter).

density The mass per unit of volume of a substance, usually expressed in grams per cubic centimeter (g/cm^3).

density curve A graph showing the relationship between a fluid's temperature or salinity and its density.

density stratification The formation of layers in a material, with each deeper layer being denser (weighing more per unit of volume) than the layer above.

depositional coast A coast on which processes that deposit sediment exceed erosive processes.

desalination The process of removing salt from seawater or brackish water.

desiccation Drying.

destructive interference The subtraction of wave energy as waves interact, producing smaller waves.

diatom Earth's most abundant, successful, and efficient single-celled phytoplankton. Diatoms possess two interlocking valves made primarily of silica. The valves contribute to biogenous sediments.

dinoflagellate One of a class of microscopic single-celled flagellates, not all of which are autotrophic. The outer covering is often of stiff cellulose. Planktonic dinoflagellates are responsible for "red tides."

disphotic zone The lower part of the photic zone, where illumination is not sufficient for photosynthetic productivity.

disturbing force The energy that causes a wave to form.

diurnal tide A tidal cycle of one high tide and one low tide per day.

divergent plate boundary A region where plates are moving apart and where new ocean or rift valley will eventually form. A spreading center forms the junction.

doldrums The zone of rising air near the equator known for sultry air and variable breezes. See also *intertropical convergence zone (ITCZ)*.

downwelling Circulation pattern in which surface water moves vertically downward.

drag The resistance to movement of an organism induced by the fluid through which it swims.

drift net Fine, vertically suspended net that may be 7 meters (25 feet) high and 80 kilometers (50 miles) long.

dynamic theory of tides Model of tides that takes into account the effects of finite ocean depth, basin resonances, and the interference of continents on tide waves.

eastern boundary current Weak, cold, diffuse, slow-moving current at the eastern boundary of an ocean (off the west coast of a continent). Examples include the Canary Current and the Humboldt Current.

ebb current Water rushing out of an enclosed harbor or bay because of the fall in sea level as a tide trough approaches.

echolocation The use of reflected sound to detect environmental objects. Cetaceans use echolocation to detect prey and avoid obstacles.

echo sounder A device that reflects sound off the ocean bottom to sense water depth. Its accuracy is affected by the variability of the speed of sound through water.

ectotherm An organism incapable of generating and maintaining steady internal temperature from metabolic heat and therefore whose internal body temperature is approximately the same as that of the surrounding environment. A cold-blooded organism.

eddy A circular movement of water usually formed where currents pass obstructions, or between two adjacent currents flowing in opposite directions, or along the edge of a permanent current.

Ekman transport Net water transport, the sum of layer movement due to the Ekman spiral. Theoretical Ekman transport in the Northern Hemisphere is 90° to the right of the wind direction.

electron A tiny negatively charged particle in an atom responsible for chemical bonding.

El Niño A southward-flowing nutrient-poor current of warm water off the coast of western South America, caused by a breakdown of trade wind circulation.

endotherm An organism capable of generating and regulating metabolic heat to maintain a steady internal temperature. Birds and mammals are the only animals capable of true endothermy. A warm-blooded organism.

ENSO Acronym for the coupled phenomena of El Niño and the Southern Oscillation. See also *El Niño; Southern Oscillation.*

equatorial upwelling Upwelling in which water moving westward on either side of the geographical equator tends to be deflected slightly poleward and replaced by deep water often rich in nutrients. See also *upwelling.*

equilibrium theory of tides Idealized model of tides that considers Earth to be covered by an ocean of great and uniform depth capable of instantaneous response to the gravitational and inertial forces of the sun and the moon.

Eratosthenes of Cyrene (276–192 B.C.) Greek scholar and librarian at Alexandria who first calculated the circumference of Earth about 230 B.C.

erosion A process of being gradually worn away.

erosional coast A coast on which erosive processes exceed depositional ones.

estuary A body of water partially surrounded by land where fresh water from a river mixes with ocean water, creating an area of remarkable biological productivity.

euphotic zone The upper layer of the photic zone in which net photosynthetic gain occurs. Compare *photic zone.*

eustatic change A worldwide change in sea level, as distinct from local changes.

eutrophication A set of physical, chemical, and biological changes brought about when excessive nutrients are released into the water.

evaporite Deposit formed by the evaporation of ocean water.

excess volatiles Compounds found in the ocean and the atmosphere in quantities greater than can be accounted for by the weathering of surface rock. Such compounds probably entered the ocean and the atmosphere from deep crustal and upper mantle sources through volcanism.

exclusive economic zone (EEZ) The offshore zone claimed by signatories to the 1982 United Nations Draft Convention on the Law of the Sea. The EEZ extends 200 nautical miles (370 kilometers) from a contiguous shoreline. See also *United States Exclusive Economic Zone.*

exoskeleton A strong, lightweight, form-fitted external covering and support common to animals of the phylum Arthropoda. The exoskeleton is made partly of chitin and may be strengthened by calcium carbonate.

experiment A test that simplifies observation in nature or in the laboratory by manipulating or controlling the conditions under which observations are made.

extratropical cyclone A low-pressure mid-latitude weather system characterized by converging winds and ascending air rotating counterclockwise in the Northern Hemisphere and clockwise in the Southern Hemisphere. An extratropi-

cal cyclone forms at the front between the polar and Ferrel cells.

fault A fracture in a rock mass along which movement has occurred.

Ferrel cell The middle atmospheric circulation cell in each hemisphere. Air in these cells rises at 60° latitude and falls at 30° latitude. See also *westerlies.*

fetch The uninterrupted distance over which the wind blows without a significant change in direction, a factor in wind wave development.

fjord A deep, narrow estuary in a valley originally cut by a glacier.

flagellum (plural *flagella*) A whiplike structure used by some small organisms and gametes to move through the environment.

flood current Water rushing into an enclosed harbor or bay because of the rise in sea level as a tidal crest approaches.

food General term for organic molecules capable of providing energy to heterotrophs when combined with oxygen during biochemical respiration.

food web A group of organisms associated by a complex set of feeding relationships in which the flow of food energy can be followed from primary producers through consumers.

foraminiferan One of a group of planktonic amoeba-like animals with a calcareous shell, which contributes to biogenous sediments.

Forchhammer's principle See *principle of constant proportions.*

foreshore Sand on the seaward side of the berm, sloping toward the ocean, to the low-tide mark.

fracture zone Area of irregular, seismically inactive topography marking the position of a once-active transform fault.

fringing reef A reef attached to the shore of a continent or an island.

front The boundary between two air masses of different density. The density difference can be caused by differences in temperature and/or humidity.

frontal storm Precipitation and wind caused by the meeting of two air masses, associated with an extratropical cyclone. Generally, one air mass will slide over or under the other, and the resulting expansion of air will cause cooling and, consequently, rain or snow.

frustule Siliceous external cell wall of a diatom, consisting of two interlocking valves fitted together like the halves of a box.

fully developed sea The theoretical maximum height attainable by ocean waves given wind of a specific strength, duration, and fetch. Longer exposure to such wind will not increase the size of the waves.

galaxy A large rotating aggregation of stars, dust, gas, and other debris held together by gravity. There are perhaps 50 billion galaxies in the universe and 50 billion stars in each galaxy.

gas bladder In multicellular algae, an air-filled structure that assists in flotation.

gas exchange Simultaneous passage, through a semipermeable membrane, of oxygen into an animal and carbon dioxide out of it.

geostrophic gyre A gyre in balance between the Coriolis effect and gravity; literally, "turned by Earth."

gill membrane The thin boundary of living cells separating blood from water in a fish's (or other aquatic animal's) gills.

Global Positioning System (GPS) A satellite-based navigation system that provides very accurate location at low consumer cost.

granite The relatively light crustal rock—composed mainly of oxygen, silicon, and aluminum—that forms the continents. Its density is about 2.7 g/cm^3.

greenhouse effect Trapping of heat in the atmosphere. Incoming short-wavelength solar radiation penetrates the atmosphere, but the outgoing longer-wavelength radiation is absorbed by greenhouse gases and reradiated to Earth, causing a rise in surface temperature.

greenhouse gases Gases in Earth's atmosphere that cause the greenhouse effect; these include carbon dioxide, methane, and CFCs.

groin A short, artificial projection of durable material placed at a right angle to shore in an attempt to slow longshore transport of sand from a beach. Usually deployed in repeating units.

Gulf Stream The strong western boundary current of the North Atlantic, off the Atlantic coast of the United States.

gyre Circuit of mid-latitude currents around the periphery of an ocean basin. Most oceanographers recognize five gyres plus the West Wind Drift.

habitat The place where an individual or population of a given species lives. Its "mailing address."

hadal zone The deepest zone of the ocean, below a depth of 5,000 meters (16,500 feet).

Hadley cell The atmospheric circulation cell nearest the equator in each hemisphere. Air in these cells rises near the equator because of strong solar heating there and falls because of cooling at about 30° latitude. See also *trade winds*.

halocline The zone of the ocean in which salinity increases rapidly with depth. See also *pycnocline*.

heat A form of energy produced by the random vibration of atoms or molecules.

heat capacity The heat, measured in calories, required to raise 1 gram of a substance 1 degree Celsius. The input of 1 calorie of heat raises the temperature of 1 gram of pure water by 1 degree Celsius.

Henry the Navigator (1394–1460) Prince of Portugal who established a school for the study of geography, seamanship, shipbuilding, and navigation.

hermatypic Describing coral species that possess symbiotic dinoflagellates within their tissues and are capable of secreting calcium carbonate at a rate suitable for reef production.

heterotroph An organism that derives nourishment from other organisms because it is unable to synthesize its own food molecules.

hierarchy Grouping of objects by degrees of complexity, grade, or class. A hierarchical system of nomenclature is based on distinctions within groups and between groups.

high-energy coast A coast exposed to large waves.

high seas That part of the ocean past the exclusive economic zone that is considered common property to be shared by the citizens of the world. About 60% of the ocean area.

high tide The high-water position corresponding to a tidal crest.

holdfast A complex branching structure that anchors many kinds of multicellular algae to the substrate.

holoplankton Permanent members of the plankton community. Examples are diatoms and copepods. Compare *meroplankton*.

horse latitudes Zones of erratic horizontal surface air circulation near 30°N and 30°S latitudes. Over land, dry air falling from high altitudes produces deserts at these latitudes (for example, the Sahara).

hot spot A surface expression of a plume of magma rising from a relatively stationary source of heat in the mantle.

hurricane A large tropical cyclone in the North Atlantic or eastern Pacific, whose winds exceed 119 kilometers (74 miles) per hour.

hydrogen bond Relatively weak bond formed between a partially positive hydrogen atom and a partially negative oxygen, fluorine, or nitrogen atom of an adjacent molecule.

hydrogenous sediment Sediment formed directly by precipitation from seawater. Also called *authigenic sediment*.

hydrostatic pressure The constant pressure of water around a submerged organism.

hydrothermal vent Spring of hot, mineral- and gas-rich seawater found on some oceanic ridges in zones of active seafloor spreading.

hypothesis A speculation about the natural world that may be verified or disproved by observation and experiment.

ice age One of several periods (lasting several thousand years each) of low temperature during the last million years. Glaciers and polar ice were derived from ocean water, lowering sea level at least 100 meters (328 feet). (See Appendix II, "Geological Time.")

interference Addition or subtraction of wave energy as waves interact. Also called *resonance*. See also *constructive interference; destructive interference*.

intertidal zone The marine zone between the highest high-tide point on a shoreline and the lowest low-tide point. The intertidal zone is sometimes subdivided into four separate habitats by height above tidal datum, typically numbered 1 to 4, land to sea.

intertropical convergence zone (ITCZ) The equatorial area at which the trade winds converge. The ITCZ usually lies at or near the meteorological equator. Also called the *doldrums*.

invertebrate Animal lacking a backbone.

ion An atom (or small group of atoms) that becomes electrically charged by gaining or losing one or more electrons.

island arc Curving chain of volcanic islands and seamounts almost always found paralleling the concave edge of a trench.

isostatic equilibrium Balanced support of lighter material in a heavier, displaced supporting matrix. Analogous to buoyancy in a liquid.

Jason-1 Satellite launched by NASA in 2001 carrying a very accurate radar altimeter able to determine sea surface height to an accuracy within 2 centimeters (about 1 inch).

kelp Informal name for any species of large phaeophyte.

kingdom The largest category of biological classification. Five kingdoms are presently recognized.

krill *Euphausia superba,* a thumb-sized crustacean common in Antarctic waters.

lagoon A shallow body of seawater generally isolated from the ocean by a barrier island. Also the body of water enclosed within an atoll, or the water within a reverse estuary.

land breeze Movement of air offshore as marine air heats and rises.

La Niña An event during which normal tropical Pacific atmospheric and oceanic circulation strengthens, and the surface temperature of the eastern South Pacific drops below average values. Usually occurs at the end of an ENSO event. See *ENSO*.

latent heat of fusion Heat removed from a liquid during freezing (or added to a solid during thawing) that produces a change in state but not a change in temperature. For pure water, 80 calories per gram at 0°C (32°F).

latent heat of vaporization Heat added to a liquid during vaporization (or released from a gas during condensation) that produces a change in state but not a change in temperature. For pure water, 540 calories per gram.

latitude lines Regularly spaced imaginary lines on Earth's surface running parallel to the equator.

law A large construct explaining events in nature that have been observed to occur with unvarying uniformity under the same conditions.

Library of Alexandria The greatest collection of writings in the ancient world, founded in the third century B.C. by Alexander the Great. Could be considered the first university.

limiting factor A physical or biological environmental factor whose absence or presence in an inappropriate amount limits the normal actions of an organism.

Linnaeus, Carolus Carl von Linné (1707–1778). Swedish "father" of modern taxonomy.

lithification Conversion of sediment into sedimentary rock by pressure or by the introduction of a mineral cement.

lithosphere The brittle, relatively cool outer layer of Earth, consisting of the oceanic and continental crust and the outermost, rigid layer of mantle.

littoral zone The band of coast alternately covered and uncovered by tidal action; the intertidal zone.

longitude lines Regularly spaced imaginary lines on Earth's surface running north and south and converging at the poles.

longshore bar A submerged or exposed line of sand lying parallel to shore and accumulated by wave action.

longshore current A current running parallel to shore in the surf zone, caused by the incomplete refraction of waves approaching the beach at an angle.

longshore drift Movement of sediments parallel to shore, driven by wave energy.

longshore trough Submerged excavation parallel to shore adjacent to an exposed sandy beach. Caused by the turbulence of water returning to the ocean after each wave.

low-energy coast A coast only rarely exposed to large waves.

lower mantle The rigid portion of Earth's mantle below the asthenosphere.

low tide The low-water position corresponding to a tidal trough.

low tide terrace The smooth, hard-packed beach seaward of the beach scarp on which waves expend most of their energy. Site of the most vigorous onshore and offshore movement of sand.

lunar tide Tide caused by gravitational and inertial interaction of the moon and Earth.

macroplankton Animal plankters larger than 1 to 2 centimeters (½ to 1 inch). An example is the jellyfish.

Magellan, Ferdinand (c. 1480–1521) Portuguese navigator in the service of Spain who led the first expedition to circumnavigate Earth in 1519–22. He was killed in the Philippines during the expedition.

magma Molten rock capable of fluid flow. Called *lava* aboveground.

magnetometer A device that measures the amount and direction of residual magnetism in a rock sample.

Mammalia The class of mammals.

mangrove Large flowering shrub or tree that grows in dense thickets or forests along muddy or silty tropical coasts.

mantle The layer of Earth between the crust and the core, composed of silicates of iron and magnesium. The mantle has an average density of about 4.5 g/cm^3 and accounts for about 68% of Earth's mass.

map A representation of Earth's surface usually depicting mostly land areas. See also *chart*.

mariculture The farming of marine organisms, usually in estuaries, bays, or nearshore environments or in specially designed structures using circulating seawater. Compare *aquaculture*.

marine pollution The introduction by humans of substances or energy into the ocean that change the quality of the water or affect the physical and biological environment.

marine science The process (or result) of applying the scientific method to the ocean, its surroundings, and the life forms within it. Also called *oceanography* or *oceanology*.

masking pigment See *accessory pigment*.

mass extinction A catastrophic, global event in which major groups of species perish abruptly.

Maury, Matthew (1806–1873) Father of physical oceanography. Probably the first person to undertake the systematic study of the ocean as a full-time occupation, and probably the first to understand the global interlocking of currents, wind flow, and weather.

maximum sustainable yield The maximum amount of fish, crustaceans, and mollusks that can be caught without impairing future populations.

mean sea level The height of the ocean surface averaged over a few years' time.

meroplankton Temporary members of the plankton community. Examples are very young fishes and barnacle larvae. Compare *holoplankton*.

metabolic rate The rate at which energy-releasing reactions proceed within an organism.

Meteor expedition German Atlantic expedition begun in 1925; the first to use an echo sounder and other modern optical and electronic instrumentation.

meteorological tide A tide influenced by the weather. Arrival of a storm surge will alter the estimate of a tide's height or arrival time, as will a strong, steady onshore or offshore wind. See also *astronomical tide*.

metrophagy Tendency of very large marine reptiles to eat Asian cities.

Milky Way The name of our galaxy. Sometimes applied to the field of stars in our home spiral arm, which is correctly called the Orion arm.

mixed layer See *surface zone*.

mixed tide A complex tidal cycle, usually with two high tides and two low tides of unequal height per day.

mixing time The time necessary to mix a substance throughout the ocean, about 1,600 years.

molecule A group of atoms held together by chemical bonds. The smallest unit of a compound that retains the characteristics of the compound.

mollusks The category of animals that includes chitons, snails, clams, and octopuses.

monsoon A pattern of wind circulation that changes with the season. Also, the rainy season in areas with monsoon wind patterns.

motile Able to move about.

multicellular algae Algae with bodies consisting of more than one cell. Examples are kelp and ulva.

nanoplankton Very small members of the plankton community. Examples are coccolithophores and silicoflagellates.

Nansen, Fridtjof Pioneering Norwegian oceanographer and polar explorer.

natural system of classification A method of classifying an organism based on its ancestry or origin.

nautical chart A chart used for marine navigation.

neap tide The time of smallest variation between high and low tides, occurring when Earth, moon, and sun align at right angles. Neap tides alternate with spring tides, occurring at two-week intervals.

nekton Pelagic organisms that actively swim.

neritic Of the shore or coast. Refers to continental margins and the water covering them, or to nearshore organisms.

neritic sediment Sediment of the continental shelf, consisting mainly of terrigenous material.

niche Description of an organism's functional role in a habitat. Its "job."

Niskin bottle Water-sampling device suspended from a ship or platform.

NOAA National Oceanic and Atmospheric Administration, founded within the U.S. Department of Commerce in 1970 to facilitate commerical uses of the ocean.

node The line or point of no wave action in a standing pattern. See also *amphidromic point*.

nodule Solid mass of hydrogenous sediment, most commonly manganese or ferromanganese nodules and phosphorite nodules.

nonextractive resource Any use of the ocean in place, such as transportation of people and commodities by sea, recreation, or waste disposal.

nonrenewable resource Any resource that is present on Earth in fixed amounts and cannot be replenished.

nor'easter (northeaster) Any energetic extratropical cyclone that sweeps the eastern seaboard of North America in winter.

North Atlantic Deep Water Cold, dense water formed in the Arctic that flows onto the floor of the North Atlantic Ocean.

nutrient Any needed substance that an organism obtains from its environment except oxygen, carbon dioxide, and water.

ocean (1) The great body of saline water that covers 70.78% of the surface of Earth. (2) One of its primary subdivisions, bounded by continents, the equator, and other imaginary lines.

ocean basin Deep-ocean floor made of basaltic crust. Compare *continental margin*.

oceanic crust The outermost solid surface of Earth beneath ocean floor sediments, composed primarily of basalt.

oceanic ridge Young seabed at the active spreading center of an ocean, often unmasked by sediment, bulging above the abyssal plain. The boundary between diverging plates. Often called a mid-ocean ridge, though less than 60% of the length exists at mid-ocean.

oceanic zone The zone of open water away from shore, past the continental shelf.

oceanography The science of the ocean. See also *marine science*.

oceanus Latin form of *okeanos*, the Greek name for the "ocean river" past Gibraltar.

ooze Sediment of at least 30% biological origin.

orbit In ocean waves, the circular pattern of water particle movement at the air–sea interface. Orbital motion contrasts with the side-to-side or back-and-forth motion of pure transverse or longitudinal waves.

orbital wave A progressive wave in which particles of the medium move in closed circles.

Osteichthyes The class of fishes with bony skeletons.

outgassing The volcanic venting of volatile substances.

overfishing Harvesting so many fish that there is not enough breeding stock left to replenish the species.

ozone O_3, the triatomic form of oxygen. Ozone in the upper atmosphere protects living things from some of the harmful effects of the sun's ultraviolet radiation.

ozone layer A diffuse layer of ozone mixed with other gases surrounding the world at a height of about 20 to 40 kilometers (12 to 25 miles).

Pacific Ring of Fire The zone of seismic and volcanic activity that encircles the Pacific Ocean.

paleoceanography The study of the ocean's past.

paleomagnetism The "fossil," or remanent, magnetic field of a rock.

Pangaea Name given by Alfred Wegener to the original "protocontinent." The breakup of Pangaea gave rise to the Atlantic Ocean and to the continents we see today.

Panthalassa Name given by Alfred Wegener to the ocean surrounding Pangaea.

partially mixed estuary An estuary in which an influx of seawater occurs beneath a surface layer of fresh water flowing seaward. Mixing occurs along the junction.

passive margin Continental margin near an area of lithospheric plate divergence. Also called *Atlantic-type margin*.

pelagic Of the open ocean. Refers to the water above the deep-ocean basins, sediments of oceanic origin, or organisms of the open ocean.

pelagic sediment Sediments of the slope, rise, and deep-ocean floor that originate in the ocean.

pelagic zone The realm of open water. See also *benthic zone*.

pH scale A measure of the acidity or alkalinity of a solution. Numerically, the negative logarithm of the concentration of hydrogen ions in an aqueous solution. A pH of 7 is neutral; lower numbers indicate acidity, and higher numbers indicate alkalinity.

Phaeophyta Brown multicellular algae, including kelps.

photic zone The thin film of lighted water at the top of the world ocean. The photic zone rarely extends deeper than 200 meters (660 feet). Compare *euphotic zone*.

photosynthesis The process by which autotrophs bind light energy into the chemical bonds of food with the aid of chlorophyll and other substances. The process uses carbon dioxide and water as raw materials and yields glucose and oxygen.

physical factor An aspect of the physical environment that affects living organisms, such as light, salinity, or temperature.

physical resource Any resource that has resulted from the deposition, precipitation, or accumulation of a useful non-living substance in the ocean or seabed. Also called a *nonliving resource*.

phytoplankton Plantlike, usually single-celled members of the plankton community.

piston corer A seabed-sampling device capable of punching through up to 25 meters (80 feet) of sediment and returning an intact plug of material.

planet A smaller, usually nonluminous body orbiting a star.

plankter Informal name for a member of the plankton community.

plankton Drifting or weakly swimming organisms suspended in water. Their horizontal position is to a large extent dependent on the mass flow of water rather than on their own swimming efforts.

plankton net Conical net of fine nylon or Dacron fabric used to collect plankton.

plate One of about a dozen rigid segments of Earth's lithosphere that move independently. The plate consists of continental or oceanic crust and the cool, rigid upper mantle directly below the crust.

plate tectonics The theory that Earth's lithosphere is fractured into plates, which move relative to each other and are driven by convection currents in the mantle. Most volcanic and seismic activity occurs at plate margins.

plunging wave Breaking wave in which the upper section topples forward and away from the bottom, forming an air-filled tube.

polar cell The atmospheric circulation cell centered over each pole.

polar front Boundary between the polar cell and the Ferrel cell in each hemisphere.

polar molecule A molecule with an unbalanced charge. One end of the molecule has a slight negative charge, and the other end has a slight positive charge.

pollutant A substance that causes damage by interfering directly or indirectly with an organism's biochemical processes.

Polynesia A large group of Pacific islands lying east of Melanesia and Micronesia and extending from the Hawaiian Islands south to New Zealand and east to Easter Island.

polyp One of two body forms of Cnidaria. Polyps are cup-shaped and possess rings of tentacles. Coral animals are polyps.

poorly sorted sediment A sediment in which particles of many sizes are found.

population A group of individuals of the same species occupying the same area.

population density The number of individuals per unit area.

potable water Water suitable for drinking.

precipitation Liquid or solid water that falls from the air and reaches the surface as rain, hail, or snowfall.

primary consumer Initial consumer of primary producers or autotrophs; the second level in food webs.

primary producer An organism capable of using energy from light or energy-rich chemicals in the environment to produce energy-rich organic compounds. An autotroph.

primary productivity The synthesis of organic materials from inorganic substances by photosynthesis or chemosynthesis. Expressed in grams of carbon bound into carbohydrate per unit area per unit time ($gC/m^2/yr$).

principle of constant proportions The principle that the proportions of major conservative elements in seawater remain nearly constant though total salinity may change with location. Also called *Forchhammer's principle*.

progressive wave A wave of moving energy in which the wave form moves in one direction along the surface (or junction) of the transmission medium (or media).

proton A positively charged particle at the center of an atom.

protosun Our sun in the protostar stage—that is, as a tightly condensed knot of material that had not yet attained fusion temperature.

pteropod Small planktonic mollusk with a calcareous shell, which contributes to biogenous sediments.

pycnocline The middle zone of the ocean, in which density increases rapidly with depth. Temperature falls and salinity rises in this zone.

radioactive decay The disintegration of unstable forms of elements, which releases subatomic particles and heat.

radiolarian One of a group of usually planktonic amoeba-like animals with a siliceous shell, which contributes to biogenous sediments.

radiometric dating The process of determining the age of rocks by observing the ratio of unstable radioactive elements to stable decay products.

random distribution Distribution of organisms within a community whereby the position of one organism is in no way influenced by the positions of other organisms or by physical variations within that community. A very rare distribution pattern.

reef A hazard to navigation. A shoal, a shallow area, or a mass of fish or other marine life.

refraction Bending of light or sound waves as they move at an angle other than 90° between media of different optical or acoustical densities. See also *wave refraction*.

renewable resource Any resource that is naturally replaced on a seasonal basis by the growth of living organisms or by other natural processes.

residence time The average length of time a dissolved substance spends in the ocean.

restoring force The dominant force trying to return water to flatness after formation of a wave.

reverse estuary An estuary along a coast in which salinity increases from the ocean to the estuary's upper reaches because of evaporation of seawater and a lack of freshwater input.

Rhodophyta Red multicellular algae.

Richter scale A logarithmic measure of earthquake magnitude. A great earthquake measures above 8 on the Richter scale.

rip current A strong, narrow surface current that flows seaward through the surf zone and is caused by the escape of excess water that has piled up in a long-shore trough.

rogue wave A single wave crest much higher than usual, caused by constructive interference.

salinity A measure of the dissolved solids in seawater, usually expressed in grams per kilogram or parts per thousand by weight. Standard seawater has a salinity of 35‰ (parts per thousand) at 0°C (32°F).

salinometer An electronic device that determines salinity by measuring the electrical conductivity of a seawater sample.

salt gland Specialized tissue responsible for concentration and excretion of excess salt from blood and other body fluids.

salt wedge estuary An estuary in which rapid river flow and small tidal range cause an inclined wedge of seawater to form at the mouth.

sand Sediment particle between 0.062 and 2 millimeters in diameter.

sandbar A submerged or exposed line of sand accumulated by wave action.

sand spit An accumulation of sand and gravel deposited downcurrent from a headland. Sand spits often curl at their tips.

scattering The dispersion (or "bounce") of sound or light waves when they strike particles suspended in water or air. The amount of scatter depends on the number, size, and composition of the particles.

schooling Tendency of small fish of a single species, size, and age to mass in groups. The school moves as a unit, which confuses predators and reduces the effort spent searching for mates.

science A systematic process of asking questions about the natural world and testing the answers to those questions.

scientific method The orderly process by which theories explaining the operation of the natural world are verified or rejected.

scientific name The genus and species name of an organism.

sea Simultaneous wind waves of many wavelengths forming a chaotic ocean surface. Sea is common in an area of wind wave origin.

sea breeze Onshore movement of air as inland air heats and rises.

sea cave A cave near sea level in a sea cliff cut by processes of marine erosion.

sea cliff Cliff marking the landward limit of marine erosion on an erosional coast.

seafloor spreading The theory that new ocean crust forms at spreading centers, most of which are on the ocean floor, and pushes the continents aside. Power is thought to be provided by convection currents in Earth's upper mantle.

sea island Island whose central core was connected to the mainland when sea level was lower. Rising ocean separates these high points from land, and sedimentary processes surround them with beaches. Compare *barrier island*.

seamount Circular or elliptical projection from the seafloor, more than 1 kilometer (0.6 mile) in height, with a relatively steep slope of 20° to 25°.

Seasat First satellite dedicated to oceanic research, launched in 1978.

SEASTAR Satellite launched by NASA in 1977 carrying a color scanner designed, among other things, to measure the distribution of chlorophyll at the ocean surface.

sediment Particles of organic or inorganic matter that accumulate in a loose, unconsolidated form.

seiche Pendulum-like rocking of water in an enclosed area; a form of standing wave that can be caused by meteorological or seismic forces, or that may result from normal resonances excited by tides.

seismic sea wave Tsunami caused by displacement of earth along a fault. (Earthquakes and seismic sea waves are caused by the same phenomenon.)

seismic wave A low-frequency wave generated by the forces that cause earthquakes. Some kinds of seismic waves can pass through Earth.

seismograph An instrument that detects and records earth movement associated with earthquakes and other disturbances.

semidiurnal tide A tidal cycle of two high tides and two low tides each lunar day, with the high tides of nearly equal height.

sensible heat Heat whose gain or loss is detectable by a thermometer or other sensor.

sessile Attached. Nonmotile. Unable to move about.

sewage sludge Semisolid mixture of organic matter, microorganisms, toxic metals, and synthetic organic chemicals removed from wastewater at a sewage treatment plant.

shallow-water wave A wave in water shallower than $1/20$ its wavelength.

shelf break The abrupt increase in slope at the junction between continental shelf and continental slope.

shore The place where ocean meets land. On nautical charts, the limit of high tides.

siliceous ooze Ooze composed mostly of the hard remains of silica-containing organisms.

silt Sediment particle between 0.004 and 0.062 millimeter in diameter.

Sirenia The order of mammals that includes manatees, dugongs, and the extinct sea cows.

slack water A time of no tide-induced currents that occurs when the current changes direction.

sofar *So*und *f*ixing *and* *r*anging. An experimental U.S. Navy technique for locating survivors on life rafts, based on the fact that sound from explosive charges dropped into the layer of minimum sound velocity can be heard for great distances. See *sofar layer*.

sofar layer Layer of minimum sound velocity in which sound transmission is unusually efficient. Sounds leaving this depth tend to be refracted back into it. The sofar layer usually occurs at mid-latitude depths around 1,200 meters (3,900 feet).

solar nebula The diffuse cloud of dust and gas from which the solar system originated.

solar system The sun together with the planets and other bodies that revolve around it.

solar tide Tide caused by the gravitational and inertial interaction of the sun and Earth.

sonar *So*und *n*avigation *and* *r*anging.

sound A form of energy transmitted by rapid pressure changes in an elastic medium.

sounding Measurement of the depth of a body of water.

Southern Oscillation A reversal of airflow between normally low atmospheric pressure over the western Pacific and normally high pressure over the eastern Pacific. The cause of El Niño. See *El Niño*.

species diversity Number of different species in a given area.

spilling wave A breaking wave whose crest slides down the face of the wave.

spreading center The junction between diverging plates at which new ocean floor is being made. Also called *spreading zone*.

spring tide The time of greatest variation between high and low tides, occurring when Earth, moon, and sun form a straight line. Spring tides alternate with neap tides throughout the year, occurring at two-week intervals.

standing wave A wave in which water oscillates without causing progressive wave forward movement. There is no net transmission of energy in a standing wave.

star A massive sphere of incandescent gases powered by the conversion of hydrogen to helium and other heavier elements.

state An expression of the internal form of matter. Water exists in three states: solid, liquid, and gas. A solid has a fixed volume and fixed shape; a liquid has a fixed volume but no fixed shape; and a gas has neither fixed volume nor fixed shape.

stipe Multicellular algal equivalent of a vascular plant's stem.

storm Local or regional atmospheric disturbance characterized by strong winds often accompanied by precipitation.

storm surge An unusual rise in sea level as a result of the low atmospheric pressure and strong winds associated with a tropical cyclone. Onrushing seawater precedes landfall of the tropical cyclone and causes most of the damage to life and property.

stratigraphy The branch of geology that deals with the definition and description of natural divisions of rocks. Specifically, the analysis of relationships of rock strata.

subduction The downward movement into the asthenosphere of a lithospheric plate.

subduction zone An area at which a lithospheric plate is descending into the asthenosphere. The zone is characterized by linear folds (trenches) in the ocean floor and strong deep-focus earthquakes. Also called a *Wadati–Benioff zone*.

sublittoral zone The ocean floor near shore. The inner sublittoral extends from the littoral (intertidal) zone to the depth at which wind waves have no influence; the outer sublittoral extends to the edge of the continental shelf.

submarine canyon A deep, V-shaped valley running roughly perpendicular to the shoreline and cutting across the edge of the continental shelf and slope.

succession The changes in species composition that lead to a climax community.

surf The confused mass of agitated water rushing shoreward during and after the breaking of a wind wave.

surface current Horizontal flow of water at the ocean's surface.

surface zone The upper layer of ocean, in which temperature and salinity are relatively constant with depth. Depending on local conditions, the surface zone may reach to 1,000 meters (3,300 feet) or be absent entirely. Also called the *mixed layer*.

surf zone The region between the breaking waves and the shore.

sverdrup (sv) A unit of volume transport named in honor of oceanographer Harald U. Sverdrup: 1 million cubic meters of water flowing past a fixed point each second.

swash Water from waves washing onto a beach.

swell Mature wind waves of one wavelength that form orderly undulations in the ocean surface.

swim bladder A gas-filled organ that assists in maintaining neutral buoyancy in some bony fishes.

taxonomy In biology, the laws and principles that cover the classification and naming of organisms.

tektite A small, rounded, glassy component of cosmogenous sediments, usually less than 1.5 millimeters (¹⁄₁₆ inch) long. Thought to have formed from the impact of an asteroid or meteor on the crust of Earth or the moon.

Teleostei The osteichthyan order that contains the cod, tuna, halibut, perch, and other species of bony fishes.

temperature The response of a solid, liquid, or gas to the input or removal of heat energy. A measure of the atomic and molecular vibration in a substance, indicated in degrees.

terrane An isolated segment of seafloor, island arc, plateau, continental crust, or sediment transported by seafloor spreading to a position adjacent to a larger continental mass. Usually different in composition from the larger mass.

terrigenous sediment Sediment derived from the land and transported to the ocean by wind and flowing water.

territorial waters Waters extending 12 miles from shore and in which a nation has the right to jurisdiction.

thallus The body of an alga or other simple plant.

theory A general explanation of a characteristic of nature consistently supported by observation or experiment.

thermal equilibrium The condition in which the total heat coming into a system (such as a planet) is balanced by the total heat leaving the system.

thermal inertia Tendency of a substance to resist change in temperature with the gain or loss of heat energy.

thermocline The zone of the ocean in which temperature decreases rapidly with depth. See also *pycnocline*.

thermohaline circulation Water circulation produced by differences in temperature and/or salinity (and therefore density).

thermostatic property A property of water that acts to moderate changes in temperature.

tidal bore A high, often breaking wave generated by a tidal crest that advances rapidly up an estuary or river.

tidal current Mass flow of water induced by the raising or lowering of sea level owing to passage of tidal crests or troughs. See also *ebb current; flood current*.

tidal datum The reference level (0.0) from which tidal height is measured.

tidal range The difference in height between consecutive high and low tides.

tidal wave The crest of the wave causing tides. Another name for a tidal bore. Not a tsunami or seismic sea wave.

tide Periodic short-term change in the height of the ocean surface at a particular place, generated by long-wavelength progressive waves that are caused by the interaction of gravitational force and inertia. Movement of Earth beneath tide crests results in the rhythmic rising and falling of sea level.

tombolo Above-water bridge of sand connecting an offshore feature to the mainland.

top consumer An organism at the apex of a trophic pyramid, usually a carnivore.

TOPEX/Poseidon Joint French–U.S. satellite launched in 1992 to study the ocean.

tornado Localized, narrow, violent funnel of fast-spinning wind, usually generated when two air masses collide. Not to be confused with a cyclone. (The tornado's oceanic equivalent is a waterspout.)

trace element A minor constituent of seawater present in amounts less than 1 part per million.

trade winds Surface winds within the Hadley cells, centered at about 15° latitude, which approach from the northeast in the Northern Hemisphere and from the southeast in the Southern Hemisphere.

transform fault A plane along which rock masses slide horizontally past one another.

transform plate boundary Place where crustal plates shear laterally past one another. Crust is neither produced nor destroyed at this type of junction.

transverse current East-to-west or west-to-east current linking the eastern and western boundary currents. An example is the North Equatorial Current.

trench An arc-shaped depression in the deep-ocean floor with very steep sides and a flat sediment-filled bottom coinciding with a subduction zone. Most trenches occur in the Pacific.

trophic pyramid A model of feeding relationships between organisms. Primary producers form the base of the pyramid; consumers eating one another form the higher levels, with the top consumer at the apex.

tropical cyclone A weather system of low atmospheric pressure around which winds blow counterclockwise in the Northern Hemisphere and clockwise in the Southern Hemisphere. Originates in the tropics within a single air mass but may move into temperate waters if water temperature is high enough to sustain it. Small tropical cyclones are called *tropical depressions,* larger ones *tropical storms,* and great ones *hurricanes, typhoons,* or *willi-willies,* depending on location.

tsunami Long-wavelength shallow-water wave caused by rapid displacement of water. See also *seismic sea wave.*

turbidite A terrigenous sediment deposited by a turbidity current; typically, coarse-grained layers of nearshore origin interleaved with finer sediments.

turbidity current An underwater "avalanche" of abrasive sediments thought responsible for the deep sculpturing of submarine canyons and a means of transport for sediments accumulating on abyssal plains.

uniform distribution Distribution of organisms within a community characterized by equal space between individuals (the arrangement of trees in an orchard). The rarest natural distribution pattern.

United States Exclusive Economic Zone The region extending seaward from the coast of the United States for 200 nautical miles, within which the United States claims sovereign rights and jurisdiction over all marine resources.

United States Exploring Expedition The first U.S. oceanographic research voyage, launched in 1838.

upwelling Circulation pattern in which deep, cold, usually nutrient-laden water moves toward the surface. Upwelling can be caused by winds blowing parallel to shore or offshore.

valve In diatoms, each half of the protective silica-rich outer portion of the cell. The complete outer covering is called the *frustule.*

vertebrate A chordate with a segmented backbone.

voyaging Traveling (usually by sea) for a specific purpose.

water vapor The gaseous, invisible form of water.

wave Disturbance caused by the movement of energy through a medium.

wave crest Highest part of a progressive wave above average water level.

wave-cut platform The smooth, level terrace sometimes found on erosional coasts that marks the submerged limit of rapid marine erosion.

wave frequency The number of waves passing a fixed point per second.

wave height Vertical distance between a wave crest and the adjacent wave troughs.

wave period Time it takes for successive wave crests to pass a fixed point.

wave refraction Slowing and bending of progressive waves in shallow water.

wave shock Physical movement, often sudden, violent, and of great force, caused by the crash of a wave against an organism.

wave steepness Height-to-wavelength ratio of a wave. The theoretical maximum steepness of deep-water waves is 1:7.

wave trough The valley between wave crests below the average water level in a progressive wave.

wavelength The horizontal distance between two successive wave crests (or troughs) in a progressive wave.

weather The state of the atmosphere at a specific place and time.

Wegener, Alfred (1880–1930) German scientist who proposed the theory of continental drift in 1912.

well-mixed estuary An estuary in which slow river flow and tidal turbulence mix fresh and salt water in a regular pattern through most of its length.

well-sorted sediment A sediment in which particles are of uniform size.

westerlies Surface winds within the Ferrel cells, centered around 45° latitude, which approach from the southwest in the Northern Hemisphere and from the northwest in the Southern Hemisphere.

western boundary current Strong, warm, concentrated, fast-moving current at the western boundary of an ocean (off the east coast of a continent). Examples include the Gulf Stream and the Japan (Kuroshio) Current.

westward intensification The increase in speed of geostrophic currents as they pass along the western boundary of an ocean basin.

West Wind Drift The Antarctic Circumpolar Current, driven by powerful westerly winds north of Antarctica. The largest of all ocean currents, it continues permanently eastward without changing direction.

Wilson, John Tuzo (1908–1993) Canadian geophysicist who proposed the theory of plate tectonics in 1965.

wind The mass movement of air.

wind duration The length of time the wind blows over the ocean surface, a factor in wind wave development.

wind strength Average speed of the wind, a factor in wind wave development.

wind wave Gravity wave formed by transfer of wind energy into water. Wavelengths of from 60 to 150 meters (200 to 500 feet) are most common in the open ocean.

world ocean The great body of saline water that covers 70.78% of Earth's surface.

zone Division or province of the ocean with homogeneous characteristics.

zooplankton Animal members of the plankton community.

Credits

Chapter 1

Chapter opening photo © Pat Hermansen/The Image Bank/Getty Images **1.3a** Scripps Institution of Oceanography Images **1.3b** Tom Garrison **1.5** Courtesy of ESO **1.6** *The Universe Explained,* Colin A. Roman, ed. New York: H. Holt, © 1994 **1.7a** © William K. Hartmann **1.7b** Hubble Space Telescope/ STScI **1.9a** © William K. Hartmann **1.9b** © Don Dixon. All rights reserved **1.10** Ralph White/CORBIS **1.11** Department of Earth & Space Sciences/ UCLA **1.13a** NASA/Robert Sullivan **1.13b, c** Malin Space Science Systems/MGS/JPL/NASA

Chapter 2

Chapter opening photo Tom Garrison **2.1** © Taschen Verlag GMBH/de Espona Inforgraficia SL **2.5** © 1991 Herbert Kawainui Kane **2.7** James Ford Bell Library, University of Minnesota **2.8** The Granger Collection, New York **2.9** © National Maritime Museum, London **2.10** © 1991 Herbert Kawainui Kane **2.11** Reprinted by permission of the United States Naval Academy Museum **2.12, 2.13** Library of Congress **2.14** Reprinted by permission of the Royal Geographical Society, London, archives from the journal of Lt. Pelham Aldrich kept aboard H.M.S. Challenger, 1872–1875 **2.16** Archive, Scripps Institute of Oceanography **2.18** Deep Sea Drilling Program. Texas A&M University **2.19a** Hulton Deutsch Collection Limited/Liaison/Getty Images **2.19b** Alda Amundsen, Scott Polar Research Institute **2.19c** Courtesy, Dale Chayes, Lamont-Doherty Earth Observatory of Columbia University **2.20** Photo used by permission of JAMSTEC **2.21a, b** Tom Garrison **2.22** NASA

Chapter 3

Chapter opening photo Magnús Tumi Guomundsson, Science Institute, University of Iceland **3.1** © Hans-Peter Bunge, Princeton **3.2** Wilbur Garrett/NGS Image Collection **3.8** © Alfred Wegener Institute for Polar and Marine Research **3.11** From *The Earth's Dynamic Systems,* 9/e by Hamblin & Christiansen, Fig. 2.8. Reprinted by permission of Pearson Education, Inc., Upper Saddle River, NJ **3.13d** Courtesy, V. Courtillot **3.21** Dr. Peter Sloss, NOAA/National Geophysical Data Center **3.23a** © Fundamental Photographs **3.25** Dr. Peter Sloss, NOAA/National Geophysical Data Center, Boulder, Colorado

Chapter 4

Chapter opening photos **(a)** © Perry Thorsvik/National Geographic Society; **(b)** Gregory Matthew Allen **4.1** Archive, Scripps Institute of Oceanography **4.2b** Robert Profeta **4.3b** Andrew Goodwillie, Scripps Institute of Oceanography **4.4a** U.S. Department of the Navy **4.4c** Karen Marks, NOAA **4.11, 4.12** William Haxby, Lamont-Doherty Earth Observatory of Columbia University **4.14** Ray Sterner/ The Johns Hopkins University, Applied Physics Laboratory **4.15** Society for Sedimentary Geology **4.16** Reprinted by permission of U.S. Department of the Navy **4.18a** Courtesy, Alcoa, Inc. **4.18b** Courtesy, Lamont-Doherty Earth Observatory of Columbia University **4.20** Woods Hole Oceanographic Institution **4.22** Charles D. Hollister **4.23** Stanley Hart **4.26** Dr. Peter Sloss, NOAA/National Geophysical Data Center, Boulder, Colorado

Chapter 5

Chapter opening photos **(left)** © Richard Norris/Scripps Institute of Oceanography; **(right)** Don Davis **5.1, 5.2, 5.3** Charles D. Hollister **5.4** Charles Hollister, Heezen and Hollister **5.5a** Courtesy, USGS **5.5b** NASA **5.6** Courtesy, Department of Geology, University of Delaware **5.7a** Courtesy, Dr. Howard Spero **5.7b** © Wim van Egmond **5.7c** Jeremy Young, University of California **5.8a** © Wim van Egmond **5.8b** Greta Fryxell **5.9** Courtesy, Susumi Honjo, Woods Hole Oceanographic Institution **5.10a, b** Courtesy, Unocal **5.12a, 5.13a** Tom Garrison **5.14** Joint Oceanographic Institutes **5.15** Courtesy, Deep Sea Drilling Project. Texas A&M University **5.17** NASA/JPL/Malin Space Science Systems

Chapter 6

Chapter opening photo © Tom Stewart/CORBIS **6.8** From G. P. Kuiper, ed., *The Earth as a Planet,* © 1954 The University of Chicago Press. Reprinted by permission. **6.12c, d** © Bruce Hall **6.16b** Courtesy, EG7c Ocean Products **6.16c** Courtesy, WESMAR **6.19** Courtesy, Dr. William Cochlan, Hancock Institute for Marine Science, University of Southern California **6.20** Courtesy, David Breiter **6.22** Art by Lisa Starr

Chapter 7

Chapter opening photos **(left)** © PA News/CORBIS KIPA; **(right)** AP/Wide World Photos **7.1** © Stephen Gassman/Index Stock Imagery **7.2** From Levin Danielson, *Meteorology,* Fig. 3.4, p. 73 © 1998 McGraw-Hill Companies. Reprinted with permission. **7.10** NASA **7.11** Reprinted by permission of Alan D. Iselin **7.14** NASA and Tom Garrison **7.15** © Bettmann Archives/CORBIS **7.16** NASA **7.18b** NOAA/National Weather Service **7.19** © M. Laca/Weatherstock

Chapter 8

Chapter opening photo Tom Garrison **8.5** From *Laboratory Exercises in Oceanography,* 4/e by Pipkin, Gorslin, Casey, and Hammon. © 1987 by W. H. Freeman and Company. Reprinted by permission **8.7b** NASA/JPL **8.11** Reprinted by permission of O. Brown, R. Evans, and M. Carle, University of Miami Rosenstiel School of Marine and Atmospheric Science **Table 8.1** From M. Grant Gross, *Oceanography: A View of the Earth,* 5/e, p. 173, © 1990 Prentice-Hall. Reprinted with permission. **8.14, 8.16b, d** NASA **8.17** TOPEX/Poseidon Team, CNES/NASA **8.21b** Tom Garrison **8.21c** Courtesy, Philip Richardson, Woods Hole Oceanographic Institution

Chapter 9

Chapter opening photo NOAA **9.8** Tom Garrison **9.10** NASA **9.12** Art by Raychel Ciemma **9.13** NOAA **9.16a, b** © Dennis Junor, Creation Captured **9.17** Erich Lessing, Art Resource, NY: SO126532 40-15-01/23 Color transp., Crane, Walter (1845-1915). Neptune's Horses. Oil on Canvas. Neue Pinakothek, Munich, Germany. Reprinted by permission of Art Resource **9.18b** NASA **9.19a** Tom Garrison **9.19b** Courtesy, Thames Barrier Visitors Centre **9.22a–c** Courtesy, USGS **9.23** Bishop Museum, Hawaii **9.24** Photo by Kyodo Photo Service **9.26a** Tom Garrison **9.26b** AP/ Wide World Photos

Chapter 10

Chapter opening photo © Matt Hammersley/www.learn2climb.com **10.16** Courtesy of Scott Walking Adventures, Halifax, Nova Scotia **10.17** Used by permission of Cabrillo Marine Aquarium **10.18** Photo by Photothèque EDF

Chapter 11

Chapter opening photo Tom Garrison **11.2a, b** William Haxby, Lamont-Doherty Earth Observatory of Columbia University **11.3** © Ken W. Davis/Tom Stack & Associates **11.6** NASA **11.7** © Bob

Clemenz Photography **11.8** USGS/Hawaii Volcanoes National Park **11.9** © Alan Hoelzle **11.11a, b** Courtesy, Gerald G. Kuhn **11.12a, c** Courtesy, Dr. William Wallace **11.12b** Tom Garrison **11.12d** Courtesy, Dante de la Rosa **11.14b** NASA **11.16** Courtesy,© Jack Mertz **11.17** NASA **11.18a, b** Courtesy, Gerald G. Kuhn **11.18c** Used by permission of W. Demetrakas, O. C. Camera **11.19a** Courtesy, USGS **11.19b** USGS/EROS Data Center **11.20** Great Barrier Reef Marine Park Authority, Queensland, Australia **11.21** © Brian Parker/Tom Stack & Associates **11.24** Glen Wheless **11.25** © John S. Shelton **11.26a** The Fairchild Aerial Photography Collection at Whittier College, Flight C-1670, frame 4 **11.26b** The Fairchild Aerial Photography Collection at Whittier College, Flight C-14180, frame 3:61 **11.28** AP/ Wide World Photos **11.29** © John S. Shelton

Chapter 12

Chapter opening photo © Bruce Hall **12.2b** © Wim van Egmond **12.3** SeaWiFS Project NASA/Goddard Space Flight Center and ORBIMAGE **12.9** Art by L. K. Townsend **12.12** Tom Garrison **12.14a** NASA/USGS **12.14b** NASA **12.14c** © Don Davis. All rights reserved

Chapter 13

Chapter opening photo © Paul Ward **13.2b** Courtesy, Denny Kelly, Orange Coast College **13.3a, c** © Wim van Egmond **13.3b** From C. Shih and R. G. Kessel, *Living Images,*1982 Sudbury, MA: Jones & Bartlett Publishers. www .jbpub.com. Reprinted with permission **13.4a** Tom Garrison **13.4b** © Bill Curtsinger Photography **13.5a** Ian Probert **13.5b, 13.6** Seawifs Project NASA/Goddard Space Flight Center and ORBIMAGE **13.8** © Wim van Egmond **13.9** © Uwe Kils **13.10a–c** © Phillip Colla Photography **13.11** © Wim van Egmond **13.12** © David Paul Photography, Australia **13.13** © Chris Fallows/ www.apexpredators.com **13.14** © James D. Watt/Watt Wildlife Library **13.18** © David B. Fleetham/Visuals Unlimited **13.19** © SuperStock, Inc. **13.21b** Courtesy, Norman Cole **13.22a** Courtesy of the Friends of the Sea Lion **13.22b** © Bruce Hall **13.23** Courtesy of Friends of the Sea Otter, Laguna, California, Richard Bucich **13.24** © Tom & Therisa Stack/Tom Stack & Associates

Chapter 14

Chapter opening photo Tom Garrison **14.2b** © Bruce & Valerie Hall **14.4, 14.5** Tom Garrison **14.6** Courtesy, National Park Service **14.7, 14.8** Tom Garrison **14.9** Art by Darwen Hennings **14.10b** Reprinted from *Between Pacific Tides,* Fifth Edition, by Edward F. Ricketts, et al. Revised by David W. Phillips, with permission of the publishers, Stanford University Press, © 1985 by the Board of Trustees of the Leland Stanford Junior University **14.11a** Tom Garrison **14.11b** © Stan Elems/Visuals Unlimited **14.12** Courtesy, Pat Mason **14.15** Art by Darwen Hennings **14.16d** © Douglas Faulkner/ Science Source/Photo Researchers, Inc. **14.17, 14.18c** Woods Hole Oceanographic Institution, Charles Hollister **14.18a, b** Used By Permission of Woods Hole Oceanographic

Chapter 15

Chapter opening photo © Dan Parrett/ Alaska Stock **15.2a, b** Courtesy, Shell Oil Company **15.3** © Adam Schmedes/Loke Film **15.4** Courtesy, Cargill Salt, Newark, California **15.5** George Skene/Orlando Sentinel **15.6** *National Geographic,* January 2001. Reprinted with permission **15.7** Orbimage © Oribital Imaging Corporation **15.8b** © Shane Moore, Animals Animals/Earth Scenes **15.9** R. Watson, V. Christensen, and D. Pauly/Oceana. Reprinted with permission. **15.10** © Tom Campbell's Photographic **15.11** Tom Garrison **15.12** © Wedigo Ferchi/Bruce Coleman, Inc./PictureQuest **15.13** © Yann Arthus Bertrand/CORBIS **15.14** Courtesy, BC Salmon Farmers Association **15.15** Terrance Klassen Photography/ Top Stock Images **15.16** Tom Garrison **15.18** NASA/JPL **15.20** © 1989 Tony Dawson Photography **15.23** SeaWiFS/ NASA/ORBIMAGE **15.24** © Ken Sakamoto, Black Star Publishing **15.25** Tom Campbell's Photographic **15.26** SVA/TOM /NASA **15.30** Reprinted by permission of the Los Angeles Times Syndicate **15.31** © Hiroyuki Matsumoto/ Black Star Publishing/PictureQuest **15.32** SeaWiFS/NASA/ORBIMAGE **15.33** © 2002 Monterey Bay Aquarium Foundation. http://www.mbayaq.org/cr/ cr-seafood/download.asp. Reprinted with permission.

Appendix VI

Figures 1 and 2 Dan Dion

Index

Heat capacity, 101–102, *102*, 118
Heavy metals contamination, 301–302, *302*
Henry the Navigator (prince of Portugal), 22–23, *23*
Hermatypic corals, *275*, 275–276
Hermit crabs, 264, *264*
Hess, Harry, 45
Heterotrophs, 224, *225*, *226*
Hierarchy, classification based on, 232, *233*
High-energy coasts, 200
High seas, 326
High tides, 183, *184*
Hilo tsunami (Hawaii), 175, *175*
Himalaya Mountains, formation of, 50, *51*
Hipparchus, 20, 322
Holdfasts, 266, *267*
Hollow-Earth theory, 27
Holoplankton, 248
Holothurians, 280
Homing behavior, 255
Horse latitudes, *128*, 129
Hot spots, 52, *53*
Hubble, Edwin, 14
Hudson Canyon, 70, *70*
Humboldt Current, *144*, 146
Humpback whales, 257, *258*
Hurricane Agnes, 136
Hurricane Andrew, 136, 172
Hurricane Carla, *210*
Hurricane Dennis, *218*
Hurricane Florence, *134*
Hurricane Georges, *135*
Hurricane Kate, *136*
Hurricanes, 133–136, *134*, *135*, *136*
Hydrogen, 57, *100*, 100–101, *101*
Hydrogen bonds, 101, *101*, 103, *103*, 104
Hydrogenous sediments, *86*, 86–87, 90–91, *92*
Hydrologic cycle, *3*
Hydrostatic pressure, 230
Hydrothermal vents
    chemosynthesis at, 237
    communities at, *281*, 281–282
    description of, 329
    discovery of, *11*, 72, 74, *74*
    origin of, *75*
    water circulation through, 74–75
Hypatia, 20–21
Hypothesis, 5, *5*

Ice, *102*, 102–103, *103*, 307
Ice ages, 69, 95
Iceberg Detector, 61–62
Icebergs, 61–62, 88, *99*, 99–100
Iceland, *38*, 38–39, *72*, 79
Inertia, 182
Inman, Douglas, 217
Intelligence, of whales, 257
Interference, *168*, 168–169
Intermediate water mass, 153
International Whaling Commission, 294–295
Internet sites, 330–331
Interplanetary dust, *86*, 87
Intertidal communities, 269, *269*, 271–274, *272–273*, 282
Intertropical convergence zone (ITCZ), *128*, 129–130, *131*
Ionizing radiation, 307
Ions, 111, *112*

Iridium, 237
Iron, abundance in Earth, 57
Iron oxides, 91, *92*
Island arcs, 50, *51*, 78, *78–79*
Islands, global warming and, *308*
Isostatic equilibrium, *42*, 42–43, *43*, 79

Japan, *50*, *51*, 175, *176*, 178
Japan Marine Science and Technology Center (JAMSTEC), 33
Jason-1, 34–35, 63
Jellies, 248–249, *249*
Jellyfish, giant, 248
JOIDES Resolution, 31, 93, *94*
Joint Oceanographic Institutions (JOI), 34, 331
Juan de Fuca plate, *46*, 69

Kaiko, 33, *33*
Kelp beds, *224*, *267*, 267–268, *268*
Kermadec-Tonga Trench, *65*, *77*, 78, 79
Kingdoms, 232, *233*
Kobe earthquake of 1995, *51*, 58
Krakatoa eruption of 1883, 175
Krill, 248, *249*
Kuenen, Philip, 79
Kuroshio Current, 144, *144*
Kyoto Protocol, 307

Lagoons, 208, *208*, 277
Laie Point tsunami (Hawaii), *174*
Lamont-Doherty Earth Observatory, 33–34
Land. *See* Continents
Land breezes, 130
Landslides, tsunamis from, 178
La Niña, 130–131, 150
Laplace, Pierre-Simon, 182, 186
Larus californicus, *233*
Latent heat of fusion, *103*, 104
Latent heat of vaporization, *103*, 104, 118, 134–135
Latitude, 20, 124, *124*, 154, *154*, 321–323
Lava, 47, 49
Law of the sea, 326–327, *327*
Laws, scientific, 5, *5*
Lazcano, Antonio, 10
Lead pollution, 301–302
Leisure pursuits, 297, *297*
Length, units of, *316*, *317*
Leonardo da Vinci, 43
Leuresthes, 192, *192*
Library of Alexandria, 19–21
Life, origin of, 10–11, *11*, *12*, 14
Life in the ocean, 221–239. *See also* Benthic communities; Pelagic communities
    acid-base balance, 230
    changes in communities, 235
    chemosynthesis, 222–223, 237, 281, *281*
    classification systems, 232–233, *233*
    competition, 234–235
    deep-sea floor, 277–281, *280*
    dissolved gases, 229, 229–230
    dissolved nutrients and, 228–229
    energy and, 222–227, *223*
    evolution, *320*
    geological eras, *318–319*, 320
    hydrostatic pressure, 230
    hydrothermal vent communities, *281*, 281–282

limiting factors, 230
    marine communities, 233–235, *234*
    mass extinctions, 235–237, *236*, *237*
    photic zone, 108, *227*, 227–228, *228*, 231, *231*, 238
    photosynthesis, 222
    primary productivity, 223, *224*, *225*, 237, 262
    salinity and, 229
    temperature and, 228, *229*
    trophic relationships, 224–226, *225*, *226*
    zones, *231*, 231–232
Light
    colors, 84, 108, *109*, 118
    photic zone, 108, *227*, 227–228, *228*, 231, *231*, 238
    refraction, 107, *108*
Limestone (calcium carbonate), 89, 91, 287
Limiting factors, 230
Linnaeus, Carolus, 232, *232*
Lionfish, *253*
Lithification, 88
Lithosphere, *41*, 41–42, 57
Littoral zone, 232
Loihi, 52, *53*
Longitude lines, 20, 321–323
Longshore bars, 204, *204*
Longshore currents, 206, *206*, 216
Longshore drift, 206, *206*
Longshore transport, 205–206, *206*, *207*, 216
Longshore troughs, 204, *204*
Low-energy coasts, 200
Low tides, 183, *184*
Low tide terraces, 205
Lunar tides, 183–186, *184*, *185*

Mackerel, 254, *254*
Macrocystis, *267*, 268, *268*
Macroplankton, 248
Madhouse economics, 294
Magellan, Ferdinand, *24*, 24–25
Magma, 47, 49, 54
Magnesium resources, 288
Magnetic fields, 54, *55*
Magnetometers, 54
Maine coasts, 199, 200, 215, 219
Makarov, S. O., 30
Mako sharks, 251–252
Maldives, *308*
Mammals. *See* Marine mammals
Manatees, 261, *261*
Manganese nodules, 30, 86, 91, *92*, 96, 327
Mangrove coasts, 211, *212*, 270
Mangroves, 268, *269*, 269–270
Manicougan Crater (Quebec), 236, *236*
Mantle, 40–41, *41*, 45, *47*, *50*, 118
Mapmaking, 18–19, 20, *24*, *324*, 324–325, *325*
Maps, definition, 324
Mare Liberum (Grotius), 326
Marggraff, Lt. Frederick, 166–167
Mariana Trench, *77*, *77*, *79*, 95
Mariculture, 295–296, *296*
Marine biologists, 4, 329
Marine birds, 255–256, *256*, 304
Marine communities, 233–235, *234*. *See also* Benthic communities; Pelagic communities

Marine engineers, 4, 329
Marine geologists, 4, 329
Marine life. *See* Life in the ocean
Marine mammals
    Carnivora, 257–261, *260*
    Cetacea, 257, *258–259*, *260*
    characteristics of, 256–257
    pollution and, 304, *304*
    Sirenia, 261, *261*
Marine Mammals Protection Act of 1972, 295
Marine physicists, 329
Marine pollution. *See* Pollution
Marine reptiles, 255, *255*
Marine resources, 285–305
    aquaculture and mariculture, 295–296, *296*
    botanical, 295, *295*
    drugs from, 296–297
    exclusive economic zones, 326–327, *327*
    fish, crustaceans, and mollusks, *290*, 290–292, *291*, *292*, *293*
    fishery mismanagement, *292*, 292–293, *293*
    fresh water, 289, *289*
    government subsidies and, 294
    minimizing impact on, 310–311
    nonextractive, *297*, 297–298
    petroleum and natural gas, 286, *286*, 287
    pollution and, *298*, 298–305, *299*, *300*, *301*, *302*
    salts, *288*, 288–289
    sand and gravel, 286–288
    seafood intake recommendations, *311*
    types of, 285
    whaling, *294*, 294–295
Marine science, definition of, 4. *See also* Oceanography
Marine scientists, 4, *4*, *300*, 328–331
Marine technicians, 328
Mars, *13*, 13–14, 96, *96*
Mars drilling platform, 287
Masking pigments, 267
Mass, units of, *316*, *317*
Mass extinctions, 235–237, *236*, *237*
Mathematical Principles of Natural Philosophy (Newton), 181
Matthews, Drummond, 54
Maury, Matthew Fontaine, 27–28, *28*
Maximum sustainable yield, 292
Mean lower low water (MLLW), 188, 193
Mean low water (MLW), 188
Mean sea level, 188
Measurements, 315–317
Medical debris, 311
Medicines, from marine resources, 296–297
Mediterranean Deep Water, 153
Mediterranean Sea, 153, 209
Meiji Seamount, *53*
Melanocoetus, 255
Mercator, Gerhardus, 324
Mercator projections, *324*, 324–325, *325*
Mercury (planet), 8
Mercury pollution, 301–302
Meroplankton, 248
Mesopelagic zone, 232
Metabolic rate, 228
Metals, 91, *92*, 301–302, *302*